Perspectives in
PLANT BIODIVERSITY

The Editor

Renowned environmentalist and Vice-Chancellor of Periyar University, Salem **Prof. Dr. K. Muthuchelian** has been working with dedication since 1982 for the development of India by harnessing both traditional and frontier technologies. He earned his M.Sc. (Botany) from Madurai Kamaraj University with distinction and Ph.D. from the School of Biological Sciences, Madurai Kamaraj University, Madurai, India and then he did his Post-Doctoral research at the University of Ancona, Italy. Dr.K. Muthuchelian was a Faculty member of the School of Energy, Environment and Natural Resources, Madurai Kamaraj University, Madurai. He was conferred the prestigious *"Doctor of Science"* (D.Sc.) in recognition of his research accomplishment on biomass technology. He became the Director of the *'Centre for Biodiversity and Forest Studies'* in Madurai Kamaraj University in 2001. His laboratory has got International Recognition (Nitrogen Fixing Tree Associations, Hawaii, USA) for 'Erythrina Research'.

He served as a member in various prestigious professional bodies such as The New York Academy of Sciences - USA, Rural Development Forestry Network – UK, American Association for the Advancement of Science - USA and International Union for Conservation of Nature (IUCN) and Natural Resources, Switzerland (an affiliated body of UNO) for Southeast Asia. He has been an expert member of *'Man and Biosphere (MAB) programme'* of Ministry of Environment and Forests, Government of India, New Delhi, India. He has been nominated as *"Fellow of International Energy Foundation, Saskatchewan, Canada"* in recognition of his outstanding International contributions to the 'Biomass production and energy transfer technology'. He has been elected as *"Fellow of National Academy of Biological Sciences, India"* for his excellent contribution to the 'Environmental Sciences'. He also served as a member and Chairman of the Peer Team of NAAC (National Assessment and Accreditation Council, Bangalore) in many premier academic institutions in India.

Dr. Muthuchelian's scientific contribution has been recognized through numerous awards such as *'Tamil Nadu Best Scientist in Environmental Sciences (1999)'* by TNSCST, Government of Tamil Nadu, Chennai, India, *'Best Scientist in Environmental Sciences (2001-2002)'* by Nehru Yuvakendra, Government of India and Tamil Nadu Sports Authority, *'Best Scientist in Environmental Management (2005) (Karma Veerar Kamarajar Award)'* by Department of Environment, Government of Tamil Nadu, Chennai, India. He has been awarded the *'Best Teacher Gold Medal (First Prize)'* by the Madurai Kamaraj University, Madurai in 2008. He was awarded the prestigious *'76th Indian Science Congress Endowment – Eminent Scientist Award in Natural Sciences'* by Madurai Kamaraj University in 2010. He has been honored twice the *"Merit of Excellence Award"* in recognition of his outstanding contributions in the field of medicinal plants conservation and in plant science at international level. For his remarkable contribution to the society and dedicated involvement in the enlistment of the downtrodden, he has been awarded twice *'Best Vice-Chancellor Award'* by the Indian Red Cross Society, Tamil Nadu branch, Chennai.

Dr. K. Muthuchelian is the First Indian Scientist honored with the prestigious award of the University of Ancona, Italy for his outstanding contribution in 'Bioenergy production'. He was an expert invitee of Nitrogen Fixing Tree Association, Hawaii, USA. He has attended several International Conferences and Symposia in USA, Germany, Hungary, Bangladesh and Italy. He has published more than 150 research papers in refereed National and International Journals and presented many papers in National and International Symposia/Conferences. He has organized 32 National Scientific Meetings, Workshops, Seminars and Conferences. Recently, he has published a book in Tamil entitled *"Uyir Viriman"* (Biodiversity: Current Status and Management) widely appreciated in both the academic and other professionals. He has produced 26 Ph.D's and currently guiding 12 Ph.D. scholars.

While ensuring accelerated environmental development through applications of science and technology, Dr. K. Muthuchelian has been relentlessly focusing attention on ecological conservation and on sustainable development. The integrated approach has remained one of the distinguishing aspects of Dr. K. Muthuchelian's original and path breaking contributions towards the revolutionary growth for ecological security and environmental sustainability in India and indeed the world.

Perspectives in PLANT BIODIVERSITY

Editor

Dr. K. Muthuchelian
Ph.D., D.Sc., FNABS., FIEF (Canada)
Vice Chancellor
Periyar University
Periyar Palkalai Nagar
Salem – 636 011

2013
Daya Publishing House®
A Division of
Astral International Pvt. Ltd.
New Delhi – 110 002

Published by   :   **Daya Publishing House**®
A Division of
Astral International Pvt. Ltd.
-ISO 9001:2008 Certified Company
4760-61/23, Ansari Road, Darya Ganj
New Delhi-110 002
Ph. 011-43549197, 23278134
E-mail: info@astralint.com
Website: www.astralint.com

Laser Typesetting   :   **Classic Computer Services**
Delhi - 110 035

Printed at   :   **Salasar Imaging Systems**
Delhi - 110 035

PRINTED IN INDIA

Foreword

India has a rich and varied heritage of biodiversity covering ten biogeographical zones, the trans-Himalayan, the Himalayan, the Indian desert, the semi-arid zone(s), the Western Ghats, the Deccan Peninsula, the Gangetic Plain, North-East India, and the islands and coasts. India is rich at all levels of biodiversity and is one of the 17 mega diversity countries in the world. India's wide range of climatic and topographical features has resulted in a high level of ecosystem diversity encompassing forests, wetlands, grasslands, deserts, coastal and marine ecosystems, each with a unique assemblage of species.

Survey conducted so far in India have inventoried over 47,000 species of plants and over 89,000 species of animals over just 70 per cent of the country's total area. India's biogeographical location at the junction of the, Agrotrophical, Indo-Malayan and Paleo-Arctic realms has contributed to the biological richness of the country. The endemism of Indian biodiversity is high-about 33 per cent of the country's recorded flora is endemic to the country and is concentrated mainly in the North-East, Western Ghats, North- West Himalaya and the Andaman and Nicobar islands. About 62 per cent of the known amphibian species and 50 per cent of the lizards are endemic to India, the majority occurring in the Western Ghats.

The country is bestowed with immense agro-biodiversity and a rich diversity in landraces/traditional cultivars/farmer's varieties. A number of crop plants (384) are reported to be cultivated in India. A total of 49 indigenous major and minor crops have been reported in the "History of Agriculture in India" which include 5 cereals and minor millets, 4 pulses, 1 oilseed crop, 9 vegetables, 5 tuber crops, 11 fruits, 5 spices, 1 sugar yielding plant and 7 fiber crops. India is the centre of origin of 30,000-50,000 varieties of cultivated plants including rice, pigeon pea, mango, okra, bamboo etc.

The broad vision for biodiversity in Agenda 21st its conservation and sustainable use accompanied by equitable benefit sharing mechanism. This includes a focus on enhancing national biodiversity protection measures involving the development of national strategies; main streaming of biodiversity concerns; ensuring the fair and equitable sharing of the benefits accruing from biodiversity; country-wide studies on biodiversity; fostering traditional methods and indigenous knowledge; encouraging biotechnological innovations along with the suitable sharing of their benefits and promoting regional and international cooperation.

In a major advancement for the cause of biodiversity conservation in the country and in compliance with requirement of the Convention on Biological Diversity, the drafting of the country's National Biodiversity Strategy and Action Plan (NBSAP) with funding support from GEF, the Global Environmental Facility, is now underway. The strategy and action plan are very broad in scope and comprehensive in coverage and propose to prepare detailed action plans at sub-state, state, regional and national levels based on the framework Policy and Action Strategy on Biodiversity.

This book will bring out the first hand informations about diversity and bioresources both in land and water, and to manage the structured *in situ* and *ex situ* conservation measures.

I deeply indebted to thank all the contributors of this book for ready response within a short notice.

Dr. K. Muthuchelian

Preface

The term "Biodiversity" refers to the variety and variability among organisms and the ecological complexes in which they occur. Thus, biodiversity can be defined as 'the totality of genes, species and ecosystems of a region. India is one of the 17 mega diverse countries of the world. The genetic variation existing within a species is called genetic diversity. Species diversity refers to the variety of species within a region.

India is considered as one of the mega-biodiverse countries of the world. In India the forest cover is an actual area of 63.73 million ha. The Dense forest is 37.74 million ha., open forest is 25.51 million ha., mangrove forest is 0.487 million ha. and scrub forest is 5.19 million ha. The bio-geographical zone(s) are, The Western Ghats, The Deccan Peninsula, The Gangetic Plain, North - East India, The Islands and Coasts. The endemism of Indian biodiversity is high about 33 per cent of the country's recorded Flora and it is concentrated in North - East, Western Ghats, North - West Himalayas and Andaman Nicobar Islands.

This book addresses the economic, institutional and social challenges confronting scientists and policy makers in conserving biodiversity and ecosystem services that are critical for sustaining human well being and development. The contributors to this book are experts who have made significant contribution to biodiversity research. The volume encompasses a wide range of themes and issues such as Species biodiversity, Herb biodiversity, Tree biodiversity, Forest biodiversity, Orchids biodiversity, Fungal biodiversity, Crop diversity, ethnomedicinal informations, climate change etc. The book includes chapters with focus from various climatic zones of India covering diverse ecosystems like tropical evergreen forests, tropical deciduous forest, agroecosystems, arid, freshwater, estuarine, marine ecosystems.

In addition, we hope it will help natural resource managers, scientists and decision makers in overcoming their fear of models and help them in translating the model results into pro-active implementation to mitigate biodiversity loss.

This book also provides updated information about the current state of animal distribution across India covering a variety of habitats. The book will provide useful information to students, environmentalists, foresters as well to the general public at large.

My heartfelt thanks are due to all contributors who made this compilation possible as well as my research students tremendously helped me in shaping this book.

Dr. K. Muthuchelian

Contents

2013, Perspectives in Plant Biodiversity   Pages **1–5**
Editor: **Dr. K. Muthuchelian,** *Vice Chancellor, Periyar University, Salem*
Published by: **Daya Publishing House, NEW DELHI**

Chapter 1

Conservation of Plant Diversity in Kodai Hills (Southern Western Ghats) of South India

G. Selvaraj and K. Muthuchelian

HKRH College, Uthamapalayam (PO), Theni Dist.
Centre for Biodiversity and Forest Studies, Madurai Kamaraj University,
Madurai – 625 021

Introduction

The term "Biodiversity" refers to the variety and variability among organisms and the ecological complexes in which they occur. Thus, biodiversity can be defined as 'the totality of genes, species and ecosystems of a region. India is one of the 17 mega diverse countries of the world.

Genetic Diversity

The genetic variation existing within a species is called genetic diversity.

Species Diversity

It refers to the variety of species within a region.

Ecosystem Diversity

It refers the variations in the biological communities in which species live. These are a) Alpha diversity b) Beta diversity c) Gamma diversity. India in its geographical area include 1.8 per cent of forest area according to Forest Survey of India (2000). In India the forest cover an actual area of 63.73 million ha. The Dense forest is 37.74 million ha., open forest is 25.51 million ha., mangrove forest is 0.487 million ha. and scrub forest is 5.19 million ha. (MOEF–2002). The bio-geographical zone(s) are, The Western Ghats, The Deccan Peninsula, The Gangetic Plain, North–East India, The

Islands and Coasts (Rodgers : Panwar and Mathur, 2000). In India 47,000 species of plants and 80,000 species of Animals were reported (MOEF–1999).

The endemism of Indian biodiversity is high about 33 per cent of the country's recorded Flora and it is concentrated in North–East, Western Ghats, North–West Himalayas and Andaman Nicobar Islands.

About 62 per cent of the known Amphibian species and 50 per cent of the Lizards are endemic to India. They are occurring in Western Ghats (MOEF–1999). The MOEF is also the focal point for implementation of the convention on Biological Diversity. National Biodiversity Strategy and Action Plan (NBSAP) was acting with Global Environmental Facility (GEF). India has erected an Umbrella Legislation called the Biodiversity Act, 2002 (No. 18 of 2003) and also notified the Biological Diversity Rules, 2004.

Methods

Study Area

Kodai Hills is an Eastern offshoot of the Western Ghats. Kodai Hill station is situtated in the Palani Hills. The East West length of about 65 Kms and North South width of about 40 Kms. Kodai lake is one of the famous tourist attraction centres of South India. Berijam lake is one of the few lakes in India which is still non–polluting. Kodaikanal is connected with Palani (a famous temple town). It is also connected by road to Munnar (a famous hill resort in Kerala). The area of the Kodai taluk is 1050 Sq. Km. which has 10°7' to 10°25' N latitude and 77°16' to 77°44' E longitude. The maximum elevation is 2517 mts. (MSL). Temperature during summer is maximum at 19.8°C and minimum at 11.3°C. During the winter it shows maximum of 17.3°C and minimum of 8.3°C. The average annual rainfall is 1650 mm.

Dry Season (December–March)

Rainfall scares, Sun warm, Frosts.

Warm Season (April–June)

Showers occasional, Sun hot (season time).

South West Monsoon (June–September)

Moderate rainfall.

North East Monsoon (October–November)

Maximum rainfall.

Slope–Undulating terrain and steep slope towards south, south east and east.

Vegetation

Sholas and Grass lands are found in Upper Palani. Coffee, Fruit trees and Cash crops are in lower Palani Hills. Transport connected with all parts of South India by road. Population is 1,900 in 1901 and 52,931 in 2001. Socio Economic status–Agricultural labourers. Geomorphology–Plateau land forms and Fluvio denudational land forms.

Geology Charnockite

Tourist Spot–one of the hill stations of South India in 1987 the tourist visited (Domestic) is 7,30,237 and foreign tourist is 77,262. In 2007 the Domestic tourist is 18,60,301 and foreign tourist is 1,68,352. There are many sholas (Evergreen forest) are found in and around Kodaikanal. Among them Gundar Shola is 521.2 Sq.Km and the Picnic Shola is 7.4 Sq. Km.

Results and Discussion

The following changes had happened in Kodai hills because of the losses of biodiversity.

Landslips and Landslides

Landslide occurred in 1993 during heavy rainfall when Palani Ghat road was closed due to heavy rain.

Impact of Deforestation

Indiscriminate felling of valuable trees for monoculture plantation (Wattle, Eucalyptus).

Sl.No.	Land Use Type	1970 Area per cent	1977 Area per cent	2007 Area per cent
1.	Dense forest	48.61	35.99	33.21
2.	Open scrub	48.88	44.35	41.23
3.	Settlement	0.17	0.49	1.25

Tourism in Kodaikanal

Among the hill stations of India such as Shimla, Darjeeling, Srinagar and Ooty, Kodaikanal is till now preserved its natural beauty and sylvan surroundings.

Kodaikanal Lake

It is one of the beautiful and enchanting lakes.

Coaker's Walk

This place gives inspiration and relaxation to walkers.

Silver Cascade

It is an attractive falls which provides drinking water to Palani town.

Hazardous Effects of Water Pollution

Mushroom growth of boarding and lodges around Kodaikanal lake causing water pollution.

Biodiversity Under Stress at Kodai Hills

The grasslands are converted into Wattle Plant since 1950's. Even more disturbing is the loss of 25 per cent of forest cover during 1972–1980. Despite the loss, the Kodai hills are better preserved than the Nilgiris.

Habitat Loss

The population in 1909 was 4,043 and the population in 2001 was 52,931. Because of the increasing population many agricultural lands are converted into settlements.

Introduction of Alien Species

By the introduction of Alien species so far 14 species were extinct and 46 species are vulnerable.

Why to SAVE Biodiversity?

Atrocarpus hirsutus and *Cycas circinalis* are absolutely the last sentinels of a long bygone age. The Sholas and grasslands are watershed for Amaravathi and Vaigai basins. The floor of the Sholas packed with decaying leaves retain 33 per cent of rain water through sponge and release this water gradually during the year.

How to Save Biodiversity?

The endangered species should be covered with fence. Kodai hills should be declared as National Park. Rare and endangered species seeds must be collected and stored.

The virtual disappearance of certain native ornamental orchids around Kodaikanal in recent decades is an indication of the denudation that had been taking place. (Fr. E. Gombert of Shembaganur in 1902).

Conservation of Biodiversity

1. By maintaining seed bank.
2. Avoid indiscriminate forest felling and forest fire.
3. Collection of germination of rare and endangered species.
4. Wildlife existence is necessary for environmental balance.
5. It is important to conserve existing biodiversity in terms of increasing and protecting natural population and density.

Agriculture and Eco-Development

It is to replace annual crops by perenninal fruits. Soil conservation work must be executed. Grazing live stock in dense forest area is completely stopped. Village common land should not be diverted for other purposes. Natural forests and grasslands are protected. Conversion of grasslands into monoculture is banned (Eucalyptus). Commercial forestry is totally banned. Afforestation programme is emphasised. There should be no expansion of the plantation in the hill areas. Road works which open up good forest area should be firmly discouraged. To organise special programmes to rehabilitate lakes, streams and marshes. Greenway programme is encouraged.

References

Blasco and Ignacimuthu, 1975. Biological changes at Kodaikanal, 1949–1974. *Trop. Ecol.*, 16: 147–162.

Gupta, R.K., 1960a. Vegetation of Kodaikanal in South India: Systematic list of trees, shrubs and herbs. *J. Bombay Nat. Hist. Soc.*, 57: 45–65.

Gupta, 1962b. Studies on some shola forests on the Palani hills near Kodaikanal. *Indian Forester*, 88: 848–853.

Gupta, 1962c. Vegetation Kodaikanal in South India. A supplementary list of trees, shrubs and herbs. *J. Bombay Nat. Hist. Soc.*, 59: 185–199.

Matthew, K.M., 1959, The flora of Kodaikanal. *Bull. Bot. Surv.*, 4: 95–104.

Mathew, K.M., 1965b. The exotic flora of Kodaikanal. *J. Bombay Nat. Hist. Soc.*, 62: 56–75.

Mathew, 1969. The exotic flora of Kodaikanal, Palani hills. *Rec. Bot. Surv., India*, 20: 1–241.

Mathew, K.M. and Mathew, K.J., 1993. The flora of the Palani Hills: A conservational and situation report. In: *Proc. Symp. on Rare Endangered and Endermic Plants of the Western Ghats*, Thiruvananthapuram, p. 159–166.

Pallithanam, J., 1957. Observations on the flora of Kodaikanal. *J. Bombay Nat. Hist. Soc.*, 54: 835–844.

2013, Perspectives in Plant Biodiversity   Pages 6–10
Editor: Dr. K. Muthuchelian, Vice Chancellor, Periyar University, Salem
Published by: Daya Publishing House, NEW DELHI

Chapter 2

A Floristic Study of Six Angiospermous Parasites of Tirunelveli District, Tamil Nadu

M.S.J. Charles[1] and A. Saravana Ganthi[2]

[1]St. Joseph's Institute of Management,
St. Joseph's College (Autonomous),
Tiruchirapalli – 620 002, Tamil Nadu
[2]Department of Botany, Rani Anna Govt. College for Women,
Tirunelveli – 627 009, Tamil Nadu

ABSTRACT

Angiospermous parasites, a distinct plant group, about < 1 *per cent* of all flowering plants, possess a complex nutritional mode and association with other plants. This paper deals with six species of such angiospermous parasites surveyed over the plains of Tirunelveli District, Tamil Nadu. Further, their morphology, taxonomical characters in technical terms, distribution and medicinal importance from literature are explained here. Anatomical studies have been carried out to investigate the type of interaction and the intensity of host-parasite relationship to understand the dynamics of host-parasite interaction.

Keywords: Angiospermous parasites, Host-parasite interaction, Tirunelveli District, Medicinal importance.

Introduction

Insectivorous and parasitic plants have been the subject of great curiosity for man since ages. These groups of plants have a typical complex nutritional mode different from other green plants that are photosynthetic, motionless and oblivious to other organisms. The vast majority of angiosperms produce all their food through photosynthesis, whereas a significant number of plants having a heterotrophic mode obtain their food from other organism(s) like the mycotrophs and haustorial parasites (Furman and Trappe, 1991). Parasites form modified roots called haustoria making morphological and physiological connection to another plant facilitating the movement of nutrients between the host (donor) and the parasite (receiver) (Kuijt *et al.,* 1978)

Hemiparasites and Holoparasites are the two basic parasite types, the former are chlorophyllous and photosynthetic yet dependent on the hosts for water and nutrients through the haustorial connections. The hemiparasites are divided again into facultative as well as obligate parasites depending upon their degree of dependence on the host. Facultative hemiparasites are photosynthetic may not require the host to complete their life cycles but when present at the roots of hosts they extract water and nutrients through haustoria, whereas the obligate hemiparasites must attach themselves to host plants to complete their life cycles. The obligate hemiparasites are both primitive and advanced. The primitive ones are photosynthetic xylem feeders on the host (stem parasites of Loranthaceae). The advanced obligate hemiparasites obtain carbon via phloem connections besides other nutrients. Holoparasites on the other hand are totally achlorophyllous, non-photosynthetic and wholly depend on the host xylem and phloem for water and nutrients inhabiting mostly on the host roots.

The haustorial parasites are < 1 per cent of all the flowering plants with 3900 species in 275 genera, among 18 families found in every habitat of the world. A good parasite does not kill its host but some parasites do pathogenitically affect the host physiologically and in fecundity. These parasites possess host ranges, host preference and host specificity. Host preference refers to host that are parasitized in nature. When these parasites occur in other hosts under artificial conditions indicates a broader host range. Some parasites do have preference in terms of specific hosts even among the species of the same genus.

Materials and Methods

Table 2.1: Six angiospermous parasites under this study.

Sl.No.	Name	Family	Habit
1.	*Cassytha filiformis*	Lauraceae	Parasitic twiner on shrub
2.	*Dendrophthe falcata*	Loranthaceae	Woody parasite on tree
3.	*Santalum album*	Santalaceae	Partial root parasite on tree
4.	*Striga asiatica*	Scrophulariaceae	Herbaceous root parasite
5.	*Taxillus tomentosus*	Loranthaceae	Woody parasite on tree
6.	*Viscum articulatum*	Viscaceae	Leafless parasite on tree

The above parasites were surveyed and collected from the foot hills of Courtallam, Pavoorchatram, Surandai and Palayamkottai, the plains of Tirunelveli District, the southern end of the Western Ghats, South India. The specimens were collected during January–February after the monsoon rains. These were identified and confirmed using Gamble and Fitscher: Flora of the Presidency of Madras (1965), Matthew, K. M. The Flora of the Tamil Nadu Carnatic (1988), Matthew, K. M. The Flora of the Palni Hills (1998).

Morphological and Taxonomical Characters in Technical Terms

Cassytha filiformis L. (Tamil: Antharakkodi) Leafless **twiner**. **Spikes** terminal or axillary; peduncle upto 4.5 cm. **Flowers** sessile, in spikes, 3-merous, bisexual, 3 mm across. **Drupe** globose; seed 1, globose. **Distribution**: Throughout India.

Dendrophthe falcata (L). Etting. (**Syn**: Loranthus falcatus Kurz). Parasitic **shrub**; branches terete. **Leaves** opposite, lanceolate-elliptic-ovate, 6-13 x 2.5-9 cm; petiole up to 1.5 cm. **Flowers** in axillary dense racemes, 3-8 cm long, green/red, curved. **Berry** oblong to 1 cm long with tubular calyx. **Distribution**: Peninsular India.

Santalum album L. (Tamil: Santhanam) **Tree** to 10 m. **Leaves** opposite below, alternate above, elliptic-ovate to lanceolate, 4-7 x 2.5-4 cm, subcoriaceous; petiole to 1.5 cm. **Inflorescence** paniculate cymes, 3-chotomous to 2.5 cm, terminal and axillary. **Flowers** bisexual, 5-merous, brownish purple, 6 mm across. **Drupe** globose. **Distribution**: Peninsular India.

Striga asiatica (L). Kuntze (**Syn**: Striga lutea Lour). Parasitic scabrid **herb**, 15-30 cm high. **Leaves** 2-3 x 0.2-0.4 cm, linear. **Flowers** white, yellow or pink, in 6-9 cm long spikes. **Capsule** ovoid, 5-6 mm across. **Distribution**: Peninsular India.

Taxillus tomentosus (Roth). Tiegh (**Syn**: Loranthus tomentosus B. Heyne ex Roth). Parasitic **shrub**; branches drooping. **Leaves** obovate-oblanceolate, rounded, glabrous above, rusty-tomentose below, 3-5 x 1.5-3.5 cm. **Flowers** rusty villous in short peduncled axillary cymes. **Berry** ovoid, 6mm across. **Distribution**: Peninsular Sri Lanka.

Viscum articulatum Burm F. (Tamil: Ottu) Leafless, compact, massive **clumps** to 1 m across. Lower **internodes** terete, upper ones decussately flattened, narrow below, wide above, 3-7 x 0.5-9 cm, yellowish green. **Flower** triads in axillary fascicles; ♂ flowers fewer than ♀ flowers. **Berry** ovoid-globose, glassy-white. **Distribution**: Throughout India.

Host–Parasite Interaction

The anatomical studies of host-parasite regions indicate the degree of intensity of parasitism varies from plant to plant as well as the host preference and specificity. The parasites interact with host plants by means of developing a penetrating and absorbing organ called haustorium, internal and external suckers connected to the vascular systems of the hosts. Table 2.2 will indicate the host preference of the included species.

Medicinal Importance

The medical importance of the included parasitic plants collected from the available literature is as shown Table 2.3.

Table 2.2

Sl. No.	Parasite	Host Plant
1.	*Cassytha filiformis*	*Morinda tinctoria*
2.	*Dendrophthe falcata*	*Azadirachta indica , Angisus latifolia, Tectona grandis*
3.	*Santalum album*	*Angisus latifolia*
4.	*Striga asiatica*	*Cyanodon dactylon*
5.	*Taxillus tomentosus*	*Zizyphus zygoperus*, on trunks of *Sapindus* sp..
6.	*Viscum articulatum*	*Zizyphus zygoperus, Dalbergia latifolia*

Table 2.2

Sl. No.	Parasite	Medicinal Use
1.	*Cassytha filiformis*	Urethritis, chronic dysentery, eye and skin infections
2.	*Dendrophthe falcata*	Menstrual troubles and asthma
3.	*Santalum album*	Oil is widely used in perfumes, wood and oil are diuretic, diaphoretic, refrigerant and expectorant
4.	*Striga asiatica*	Diabetes
5.	*Taxillus tomentosus*	Not known
6.	*Viscum articulatum*	Febrifuge and aphrodisiac

Conclusion

The study of the interaction between the parasites and their host plants brings out an intrinsic and positive interdependence exists which is beneficial to partners, the parasite and the host, vis-à-vis to most life forms, help in the sustainability and maintaining the eco-friendliness in nature. The man made destruction of the environment under the guise of development has endangered the long held interrelationships and interdependence between plants and animals and with environment. All our efforts to safeguard the nature should be oriented to ensure the prime state of interrelationships and interdependence existing among the various life forms.

References

Ambasta, S.P., 1992. *The Useful Plants of India*. National Institute of Science Communication, New Delhi.

Furman, T.E. and Trappe, J.M., 1991. Phylogeny and ecology of mycotrophic achlorophyllous angiosperms. *Quart. Rev. Biol.*, 46: 219–225.

Gamble, 1965. *Flora of the Presidency of Madras*.

Kuijt, J., Viser, J.H. and Weber, H.C., 1978. Morphological observations on leaf haustoria and related organs of the South African genus *Hyobanche. Can. J. Bot.*, 56: 2981–2986.

Matthew, K.M., 1988. *The Flora of the Tamil Nadu Carnatic.* The Rapinat Herbarium, St. Joseph's College, Tiruchirapalli.

Matthew, K.M., 1998. *The Flora of the Palni Hills.* The Rapinat Herbarium, St. Joseph's College, Tiruchirapalli.

2013, Perspectives in Plant Biodiversity

Editor: **Dr. K. Muthuchelian,** *Vice Chancellor, Periyar University, Salem*

Published by: **Daya Publishing House, NEW DELHI**

Chapter 3

Common and Cultivated Oil Yielding Plants in Western Ghats of Southern Part of Tamil Nadu

K. Rajendran[1] and M. Jothibasu[2]

[1]Post Graduate and Research Department of Botany, Thiagarajar College, Madurai – 625 009, Tamil Nadu
[2]Department of Botany, Distance Education Centre, Alagappa University, Karaikudi, Tamil Nadu

Introduction

Wild and cultivated oil yielding plants are economically and commercially valuable source of food, medicine and fuel because of its wide availability and it also improve the income of village and tribal people. An attempt was made in Southern districts of Tamil Nadu and it was found that there are 20 species belongs to 20 genera spreading over 18 families of angiosperms. Out of which 4 plant species are used as edible oil plants. All of 20 species are used as medicine and 1 species predominantly cultivated for fuel. Among these, *Pongamia pinnata, Vitex negundo* and *Cympopogan citratus* are commonly found in common and *Anacardium occitentale, Azadirachta indica, Jatropha curcas, Citrus limon* and *Cocos nucifera* are cultivated in large scale plantations. Hence to conserve and improve the oil yield, an appropriate agro-biotechnology using organic and biofertilizer must be developed and can we practiced in avenue and degraded lands available in the village.

Materials and Methods

Description of the Study Area

The present study was carried out in Thalaiyanai hills of Tirunelveli forest Division in Southern part of Tamil Nadu. The Thalaiyanai hills is located between 8 52'–9 27' N latitude and 77 10'–77 25' E longitude at an elevation about 850m above mean sea level. The temperature at this site ranges 21–34°C and the average precipitation averages 600–1200 mm.

Methodology

Survey was conducted in remote villages situated at the foot hills of the Thalaiyanai hills from March 2010 to June 2010. During the study information regarding both cultivated and wild oil yielding plants and their local name are recorded with the help of the local tribal people and aged farmers. The plants were botanically identified with the help of the Flora of Tamil Nadu Carnatic (Matthew, 1983–1983) and An Excursion flora of Central Tamil Nadu (Matthew, 1991).

Results and Discussion

The present data were obtained through direct field visits and interviews with the local tribal people and farmers. In the present study 20 species of 20 genera belonging 18 families of important oil yielding plants includes both wild and cultivated species were identified and the list of the recorded plants with their family and local name are given in the Table 3.1. Out of which 4 plant species are used as edible oil plants. All of 20 species are used as medicine and 1 species predominantly cultivated for fuel. Among these, *Pongamia pinnata, Vitex negundo* and *Cympopogan citratus* are commonly found in common and *Anacardium occitentale, Azadirachta indica, Jatropha curcas, Citrus limon* and *Cocos nucifera* are cultivated in large scale plantations.

In view of greater importance of oil yielding plants, many of the progressive farmers are switching for cultivation of these species as they found it to be more profitable than traditional crops. Apart from the above species, there is much scope to cultivate many other oil yielding plants in agro-climatic regions. To bring more oil yielding plants in cultivation and for their increased productivity, further research is needed as regards systematic and scientific cultivation methods, which includes organic farming, irrigation, harvesting, marketing and preservation. To enhance the economic condition of rural people and progressive farmers, it is essential to impart necessary training to them in mass cultivation practices. By understanding the important oil yielding plants being utilized by people traditionally for various products available in the market and the raw material require for it, we can evolve a system to grow them in farm lands, wastelands, lakes and riverbanks which will reduce the pressure on the natural forests and thereby conserve the biodiversity and improve the socio-economic of the rural people.

Acknowledgements

The authors are cordially grateful to the tribal and other local people inhabiting in Thalaiyanai hills for their kind and valuable support during our field visits.

Table 3.1: Oil yielding plants of Thalaiyanai hills.

Sl. No.	Botanical Name	Family	Local Name
	Cultivated oil yielding plants		
1.	*Anacardium occitentale* L.	Anacrdiaceae	Mundhiri
2.	*Arachis hypogaea* L.	Fabaceae	Nilakkadalai
3.	*Azadirachta indica* Andr. Juss.	Meliaceae	Vembu
4.	*Citrus limon* (L). Burm. f.	Rutaceae	Yelumitchai
5.	*Cocos nucifera* L.	Arecaceae	Thennai
6.	*Helianthus annus* L.	Asteraceae	Sooriyagaandhi
7.	*Jatropha curcas* L.	Euphorbiaceae	Kaattaamanakku
8.	*Sesamum indicum*		Yell
9.	*Simarouba gluca* Roxb	Simaroubaceae	
	Wild oil yielding plants		
10.	*Calophyllum inophyllum*		Punnai
11.	*Cymbopogan citratus* (DC). Stapf.	Poaceae	
12.	*Eucalyptus tereticornis* Sm.	Myrtaceae	Thailamaram
13.	*Madhuca longifolia* (Koen). Macbr.	Sapotaceae	Yiluppai
14.	*Michelia champaca* L.	Magnoliaceae	Shenbagam
15.	*Ocimum basilicum* L.	Lamiaceae	Thiruneettruppachilai
16.	*Pongamia pinnata* (L). Pierre	Fabaceae	Pungam
17.	*Ricinus communis* L.	Euphorbiaceae	Aamanakku
18.	*Santalum album* L.	Santalaceae	Sandhanam
19.	*Schieicheria oleosa*		
20.	*Vitex negundo* L.	Verbenaceae	Notchi

References

Matthew, K.M., 1983–1986. *Flora of Tamil Nadu Carnatic*. Rapinant Herbarium, St. Joseph College, Tiruchirapalli, Tamil Nadu.

Matthew, K.M., 1991. *An Excursion Flora of Central Tamil Nadu, India*. Oxford and IBH Publishing Co., New Delhi.

2013, Perspectives in Plant Biodiversity   *Pages* **14–21**
Editor: **Dr. K. Muthuchelian,** *Vice Chancellor, Periyar University, Salem*
Published by: **Daya Publishing House, NEW DELHI**

Chapter 4

Effect of High Power Transmission Line on Herbaceous Species Composition and Forest Biodiversity

S. Somasekaran[1] and K. Muthuchelian[2]

[1]*Department of Physics, Rajapalayam Rajus' College, Rajapalayam*
[2]*School of Energy, Environment and Natural Resources,*
Madurai Kamaraj University, Madurai – 625 021

ABSTRACT

In the present study the status and distribution of plant communities and its composition due to the passage of high voltage power transmission lines through the forest lands of Ayyanar koil hills of Western Ghats were studied. From our study it was observed that the number of herbaceous species was greater in the site II natural forest area compared to the site I area which is under the high power transmission lines. Taxonomically, the numbers of families were more in site II (natural forest) compared to site I (under the high power transmission line). Herb diversity indices were also lesser in site I (under the high power transmission lines) compared to the site II (natural forest).

Introduction

Almost all types of forests, ranging from scrub forest to the tropical evergreen rain forest, coastal mangroves to alpine scrub occur in India. A comprehensive attempt has been made to classify the forests of India by *Champion and Seth* (1968). They have recognized 16 major forest types, which in turn comprise 221 minor types. Tropical moist and tropical dry deciduous forests constitute the bulk of total forest cover in India, 37 per cent and 28.6 per cent respectively; none of the remaining 14 types reaches 10 per cent. Tropical moist forests, popularly known as the rain forests, have species rich habitats and support the maximum biodiversity (Wilson, 1989).

The growing population, human greed and lack of knowledge about ecological balance of nature have destroyed the environment in such a manner that there is already a vast degradation of soil, water and air biodiversity and even light/temperature. Development activities like construction of roads, bridges, railway lines and erection of high voltage power transmission lines etc., also destroy a number of plant habitats. In India most of the high voltage power transmission lines pass through the agricultural and forest lands. The flow of electricity along a conductor/wire establishing an electromagnetic field around it influences the living organisms like plants, animals and human beings by the way of electromagnetic induction (Markus, 1999).

The present investigation is to study the status and distribution of plant communities and its composition due to the passage of high voltage power transmission lines through the forest lands of Ayyanar koil hills of Western Ghats, Tamil Nadu, South India. The high voltage power transmission line paves the way for the exotic plant invasion and also reduces the species richness, which inhibits the regeneration of native species.

Study Area

The Western Ghats, a chain of mountains in the western peninsular India extending from Tapti river valley in Gujarat to Kanyakumari in Tamil Nadu, is about 1600 km long in North–South direction. The study area concentrates in and around the Virudhunagar district forest areas located in Tamil Nadu, South India. Ayyanar koil hills, in reserve forests of Giant squirrel wildlife sanctuary, is located 85 km south of Madurai between 9° 42', N 77° 37', E is at 1234–1355 M MSL and 9° 42', N 77° 38', E is at 1255–1355 M MSL elevation respectively in site I and site II in Virudhunagar district, Tamil Nadu. The mean average rainfall recorded in the study site is 900 mm. The average maximum and minimum temperature recorded in the study areas were 39°C and 25°C in summer and 30°C and 21°C in winter respectively. This study area is situated at Eastern slopes of Western Ghats, which is a rain shadow area of the South West monsoon. All the streams are seasonal and carry water for limited periods. Climate is generally hot. In our forest study area a high tension 110 KV power transmission line is passing through from Periyar hydal power station to Kayathar power grid station.

Materials and Methods

Phytosociological studies were done under the (Periyar–Kayathar feeder) high voltage power transmission lines of 110 KV (between Kaluthaikadavu tower and Neeravai tower) inside the forest area. This was taken as site I and the area, which was about 500 m away from the high voltage power transmission line inside the natural forest area, served site II. Density, frequency, basal area, and Importance

Value Index (IVI) were estimated in each study area using 20 randomly placed quardrats (1m × 1m) for herbs (Kershaw, 1973; Misra, 1968).

The Species Diversity Index was calculated using a formula given by Margalef (1968). The index of dominance of community was calculated by Simpson Index (Simpson, 1949). The index of the Species Richness (d) was calculated using Menhinick (1964). The Evenness Index of the community (e) was calculated using Pielou, (1982). The Physiochemical properties of soil were determined by methods of analyzes of soils, plants, waters and fertilizers (Tandon, 1994) (Table 4.1).

Table 4.1: Soil characteristics under 110 KV high power transmission lines (Site I) and natural forest (Site II) study sites at Ayyanar hills of Southern Western Ghats, South India.

Variables	Site I (Under 110 KV power transmission lines)	Site II (Natural Forest)
Water Holding Capacity	29.37	35.43
EC (m.mhos/cm)	0.202	0.296
pH	7.35	7.15
Organic Carbon (per cent)	3.69	3.2
Nitrogen (per cent)	0.218	0.238
Phosphorus (per cent)	0.180	0.15
Potassium (per cent)	0.285	0.290

Results and Discussion

A total of 43 plant species belonging to 27 genera and 18 families were recorded from our study areas (Table 4.2). The number of herbaceous species were greater in the site II natural forest area (27 species) compared to the area which is under the 110 KV high voltage power transmission line *i.e.*, site I (16 species) (Table 4.3). Exotic plant species such as *Lantana camara* and *Chromolaena odorata* were dominant in site I and open habitat of moist deciduous forests at Ayyanar koil hills of Western Ghats, Tamil Nadu. Taxonomically, the number of families were more in the site II compared to site I. Herb diversity indices were lesser in the site I compared to the site II while dominance index showed a reverse trend.

This variation in herb diversity and dominance indices may be due to the effect of an EMF from the high voltage power transmission line (110 KV) and human disturbance by means of cutting trees below the transmission lines by the electricity department people in order to avoid short circuit current. The enormity of magnitude of biodiversity, heterogeneity in its spatial and temporal distribution, limited expertise in its assessment and management constitute one of the most pressing challenges in assessing and managing the biodiversity in the tropics (Gadgil, 1996; Utkarsh *et al.*, 2001). The type of disturbance may influence variation in these conditions, and thus their effect on plants. Forest clear cutting can affect the distribution of plants, some species may increase and others may decrease and still others may experience a change from forest edge to forest interior. Due to over exploitation of valuable plant species due to the random biotic disturbances, the present forest sites encompass a total of 43 species only.

Table 4.2. Importance Value Index (IVI) of herbaceous species enumerated in the two 0.1 ha area of tropical moist deciduous forest in Ayyanar koil hills of Southern Western Ghats, Tamil Nadu, India. Natural forest (Site II).

Botanical Name	Relative Frequency	Relative Density	Relative Dominance	Importance Value Index (IVI)
Abutilon indicum (L) Sweet	2.198	0.600	0.599	3.397
Acacia leucophloea (Saplings)	2.198	0.240	0.240	2.678
Blepharis maderaspatensis (L).B.Heyne.	4.393	1.080	1.002	6.478
Canavalia virosa Wight and Arn	1.099	0.840	0.479	2.778
Cerapegia juncea Roxb	2.198	0.240	0.240	2.678
Chromolaena odorata (L). Kind and Rob.	4.371	20.39	20.31	45.071
Curculigo orchioides Gaertn	2.198	1.200	1.198	4.596
Desmodium triflorum (L). DC.	3.297	0.720	0.719	4.736
Diospyros melanoxylon Roxb.	2.198	0.360	0.359	2.917
Dolichos falcata L.	2.198	0.600	0.599	3.39
Helicteres isora L.	5.495	1.080	1.078	7.653
Hybanthus enneaspermus (L). F.V.Mull	2.198	0.960	0.959	4.117
Ichnocarpus frutescens (L). R.Br	5.495	2.160	2.157	9.812
Imperata cylindrical (L). Reausch	3.297	9.484	9.467	22.248
Ipomoea staphylina Roem and Shult	1.099	0.240	0.240	1.579
Lantana camara L.	6.593	12.605	12.582	31.825
Melochia corchorifolia L.	1.099	0.960	0.959	3.018
Merremia tridentat L.	2.198	0.240	0.240	2.678
Mimosa pudica L.	3.297	1.681	1.681	6.656
Ocimum gratissimum L.	1.099	0.120	0.120	1.339
Oxalis latifolia Kunth.	10.989	8.643	8.625	28.26
Panicum sps	7.692	10.684	10.665	29.041
Pavonia odorata Willd.	3.297	0.840	0.839	4.976
Phyllanthus amarus Shum and Thon	3.297	0.600	0.299	4.496
Ruellia tuberos L.	8.791	3.000	2.996	14.787
Urginea indica (Roxb) O Kunth.	1.099	0.120	0.120	1.339
Vernonia cinerea (L). Nees.	1.099	0.360	0.359	1.818
Others	5.445	20.38	20.31	46.135

Anthropogenic disturbance has played an important role not only in creating unhealthy vegetation by also in altering the structure, species composition and natural functions of ecosystems (Cambell and Liegel, 1996; Whitmore and Burslem, 1996;

Table 4.3: Importance Value Index (IVI) of herbaceous species enumerated in the two 0.1 ha area of tropical moist deciduous forest in Ayyanar koil hills of Southern Western Ghats, Tamil Nadu. India. Under 110 KV power transmission lines (Site I)

Botanical Name	Relative Frequency	Relative Density	Relative Dominance	Importance Value Index (IVI)
Arbus precatorius L.	8. 000	2.95	1.83	12.78
Abutilon indicum (L). Sweet	2. 000	2.21	2.04	6.25
Acalypha fruticosa Forssk	2. 000	0.98	0.272	3.252
Achyranthus bidentata L.	2. 000	0.655	0.182	2.837
Barleria pronitis L	2. 000	0.98	0.272	3.252
Blepharis maderaspatensis (L). B.Heyne.	10. 000	4.26	1.18	15.54
Canthium reheedei DC.	6. 000	1.31	2.26	9.57
Helicteres isora L.	8. 000	3.28	3.62	14.9
Holoptelea integrifolia Pl.	4. 000	1.64	1.02	6.66
Imperata cylindrical (L). Reausch	2. 000	2.62	2.147	6.767
Jasminum grandiflorum L.	2. 000	0.327	0.089	2.416
Chromolaena odorata (L). King and Rob.	2. 000	60.59	61.71	122.3
Phyllanthus amarus Shum and Thon.	4. 000	2.95	0.205	7.155
Lantana camara L.	14. 000	17.38	11.04	42.42
Vernonia cinerea (L). Nees.	2. 000	1.31	0.089	3.399
Zizyphus xylopyrus (Retz) Willd.	2. 000	0.327	0.205	2.532
Others	10.000	12.22	13.85	36.07

Table 4.4: Consolidated details of quantitative plant diversity analyses in two 0.1 ha tropical moist deciduous forest ecosystems at Ayyanar koil hills of Southern Western Ghats in Tamil Nadu, India.

Parameter	Site I (Under 110 KV power transmission lines)	Site II (Natural Forest)
Number of Species	16	27
Number of Genera	16	27
Number of Families	11	18
Diversity Indices		
Shannon Weiner's index of diversity Herbs (cm²/m²)	1.7422	2.1056
Simpson index of dominance Herbs (cm²/m²)	3.9061	4.7196
Species richness Herbs (cm²/m²)	0.565	0.712
Evenness index Herbs (cm²/m²)	1.768	2.019
Basal Area Herbs (cm²/m²)	0.234	0.295
Density (Stems ha¹) Herbs	9,35,00	10,44,00

Table 4.5: Family wise contribution to genera and species in the tropical moist deciduous forest ecosystem at Ayyanar koil hills of Southern Western Ghats Tamil Nadu, India.

Family Name	Genera	Species
Malvaceae	Abutilon	indicum (L). Sweet
	Pavonia	odorata Willd
	Abrus	precatorius L.
Mimosaceae	Acacia	leucocephloa Willd
	Mimosa	pudica L.
Acanthaceae	Blepharis	maderaspatensis (L). B.Heyne
	Ruellia	tuberos L.
	Barleria	pronitis L.
	Blepharis	maderaspatensis (L). B.Heyne
Fabaceae	Canavalia	virosa Wight and Arn
	Desmodium	triflorum (L). D.
	Dolichos	falcata L.
Asclepiadaceae	Ceropegia	juncea Roxb.
Asteraceae	Chromolaena	odorata (L). Kind and Robinson.
	Vernonia	cinerea (L). Nees
Hypoxidaceae	Curculigo	orchioides Gaertn
Ebenaceae	Diospyros	melanoxylon Roxb.
Sterculiaceae	Helicteres	isora L.
	Melochia	corchorifolia L.
Violaceae	Hybanthus	enneaspermus (L). F.V.Mull
Apocynaceae	Ichnocarpus	frutescens (L). R.Br
Poaceae	Imperata Panicum	cylindrical (L). Reausch
Convolvulaceae	Ipomoea	staphylina Roem and Shult
	Merremia	tridentat L.
Verbenaceae	Lantana	camara L.
Lamiaceae	Ocimum	gratissimum L.
Oxalidaceae	Oxalis	latifolia Kunth
	Imperata	cylindrical (L). Reausch
Euphorbiaceae	Phyllanthus	amarus Shum and Thon
	Acalypha	fruticosa Frrosk
Liliaceae	Urginea	indica (Roxb) O Kunth
Amaranthaceae	Achyranthus	bidentata L.
Rubiaceae	Canthium	reheedei D.
Ulmaceae	Holoptelea	integrifolia Wight and Arn.
Oleaceae	Jasminum	grandiflorum L.
Rhamnaceae	Zizyphus	xylopyrus (Retz) Willd

Swamy *et al.,* 2000; Radha Veach *et al.,* 2003). In the present study, lower species richness in terms of number and diversity indices of herbs were observed under the site I area (high power transmission line 110 KV) compared to (natural forest) site II area (Tables 4.4 and 4.5). Margutti *et al.* (1996) and Reddy (1998) reported similarly lowest numbers of species in disturbed sites. Several studies have agreed that the gaps created by the natural and human disturbance provide good shelter for the establishment and growth of natives as well as serving the most likely point of invasion by exotics (Whitmore and Burslem 1996; Denslow *et al.,* 1980).

Conclusion

From our study we have concluded that Biodiversity is essential for human survival and economic well being and for the ecosystem function and stability. The biodiversity inventories of Western Ghats of Ayyanar koil forest, particularly those of the poor known moist deciduous forests of Virudhunagar district of Tamil Nadu are needed for conservation management. The present study areas under the 110 KV high voltage power transmission lines (site I) and natural forest (site II) are moist deciduous forests and due to the passage of 110 KV high voltage power transmission lines and the human disturbance by the electricity board department people, the present forest sites encompass a total of 43 species only. The density and diversity status of these forests are not comparable to many other tropical forests situated under similar macroclimatic conditions. So further studies are needed to safe guard the plants from EMF and the real pattern of regeneration and dynamic changes due to the effect of EMF from high power transmission lines. Hence, for better ecological status conservation of the forest species is suggested.

Acknowledgements

One of the author thanks to the University Grants Commission, New Delhi, for having facilitated to do research under Faculty Development Programme.

References

Champion, H.G. and Seth, S.K., 1968. *A Revised Survey of Forest Types of India*. Delhi.

Campell, S. and Liegel, L., 1996. Disturbance and forest health in Oregon and Washington. *USDA for Serv. Gen. Tec. Rep.*, PNW, p. 381–405.

Denslow, J.S., 1980. Patterns of plant species diversity during succession under different disturbance regimes. *Oecologia,* 46:18–21.

Gadgil, M., 1996. Documenting diversity: An experiment. *Curr. Sci.*, 70: 36–44.

Kershaw, K.A., 1973. *Quantitative and Dynamic Plant Ecology,* 2nd Edn. Edward Arnold, London, pp. 308.

Margalef, R., 1968. *Perspective in Ecological Theory*. University of Chicago Press, Chicago, pp. 111.

Margutti, L., Ghermandi, L. and Rapoport, E.H., 1996. Seed bank and vegetation in a patagonian roadside. *Intern. J. Ecol. Environ. Sci.*, 22: 159–175.

Markhus, Z., 1999. Electric and magnetic fields in the environment. In: *Encyclopedia of Energy Technology and the Environment.* John Wiley & Sons, New York, 2: 1089–1110.

Menhinick, E.F., 1964. A comparison of some species diversity indices applied to samples of field insects. *Ecol.*, 45: 859–861.

Misra, R., 1968. *Ecology Workbook.* Oxford and IBH Publication, New Delhi, pp. 244.

Pielou, G.T., 1982. *Biological Diversification in the Tropics.* Columbia University Press, New York.

Radha, V., Lee, D. and Tomphilippi, 2003. *Human Distubance and Forest Diversity in the Tansa Valley, India.* Kluwer Academic Publishers, Printed in the Netherlands, p. 1051–1072.

Reddy, K.B., 1998. Effect of long-term disturbance on herbaceous plant community. *Intern. J. Ecol. Environ. Sci.*, 24: 131–139.

Simpson, E.H., 1949. Measurement of diversity. *Nature,* 163: 688.

Swamy, P.S., Sundarapandian, S.M., Chandrasekar, P. and Chandrasekaran, S., 2000. Plant species diversity and tree population. Structure of a humid tropical forest in Tamil Nadu. *Indian Biodiver. Conserv.*, 9: 1643–1669.

Tandon, B.N., 1994. Food practices and nutritional status among tribals of India: An appraisal. In: *Tribal Health in India*, (Ed.) S.K. Basu. Manak Publications, New Delhi, pp. 106–115.

Utkarsh, G., Pramod, P., Kunte, K., Achar, K.P., Bhatta, G.K., Pandi, P. and Sivaramakrishan, K.G., 2001. Decentralized biodiversity assessment in the Western ghats. In: *Tropical Ecosystem: Structure, Diversity and Human Welfare, Proceeding of the International Conference on Tropical Ecosystem*, (Eds.) Ganeshaiah, R. Umashankar and K.S. Bhawa. Oxford and IBH, New Delhi, p. 535–538.

Whitemore, T.C. and Burslem, D.F.R., 1996. Major disturbance in tropical rainforest. In: *Dynamics of Tropical Communities,* (Eds.) D.M. New Berg, H.H.T. Prins and N.D. Brown. Blackwell Science, U.K., p. 549–565.

Wilson, J., 1989. Working plant for the Coimbatore south division Proc. Chief conservator of forest, Madras. Proc. Mis. No. 186–317.

2013, Perspectives in Plant Biodiversity   Pages 22–26
Editor: Dr. K. Muthuchelian, Vice Chancellor, Periyar University, Salem
Published by: Daya Publishing House, NEW DELHI

Chapter 5

Studies on the Distribution of Certain Medicinal Plants in Semi Arid Fallow Lands of Kovilpatti

Murugesan Ramamoorthy*

G.V.N.College, Kovilpatti, Thoothukudi District, T.N.

ABSTRACT

Dry land agriculture of Kovilpatti was depending upon seasonal rains. The agriculturists often abandon their marginal dry lands as fallow fields due to unpredictable marginal yield and drought. The wild vegetations emerging in the fallow lands are utilized for grazing by nomadic goat herders, medicinal plant collectors and fire wood collectors. The study area (90 ha–black soil) is an abandoned fallow land since 2002, situated in the fallow fringes of Kovilpatti (9°11'23"N 77°51'35"E). About 10 permanent quadrates (4 meter wide and 25 meters long) were marked random places in the study site. Herbal collectors and goat herders were freely allowed in the qudrat areas, to study the anthropological impacts on density of plant species. The highest number plant species was noted during November, December and January in the study site. The data collected in the months of December during 2004, 2005, 2006, 2007 and 2008 were analyzed and taken for discussion. During initial study period (2004) some 34 species of plants were identified along with 14 medicinal plants. During the year 2006, only 12 species were noted in the study site. In December 2007 *Prosopis juliflora* shrubs and the grass *Ophiuros exaltatus*

* Corresponding Author: E-mail: mramurty@gmail.com

found along with other few species in the study site. The aggressive growth of *Prosopis juliflora* and *Ophiuros exaltatus* changed the study site scenario by their dominance in December 2008. The herbal collectors uproot and harvest the medicinal plants, and most of the annual and perennial vegetations are utilized by the browsing goats. The common medicinal plants slowly and totally vanished from the study site owing to the human activities.

Keywords: *Medicinal plants, Rangelands, Fallow land, Prosopis juliflora, Ophiuros exaltatus.*

Introduction

The dry land agriculturists often abandon their lands in Kovilpatti (9°11'23"N 77°51'35"E) area, as uncultivated lands due to unpredictable marginal income and drought. The fallow lands become heterogeneous pasture fields and give access to livelihoods of nomadic goat herders, medicinal plant collectors and fire wood collectors. Also public demands for herbal medicine are very high. The diverse source of off form income make the people depend on these sites. Definitely the fallow land pastoral system is multifunctional with grazing land value, carbon sequestration, soil fertility, biodiversity and rural economy. The clear knowledge about biodiversity of plant communities is necessary for proper land management (Burrows 1998). This paper presents and scrutinizes the changes in biodiversity of medicinal plants in the study site.

Materials and Methods

The study area (90 ha–black soil) is an abandoned fallow land since 2002, situated in the fallow suburbs of Kovilpatti (9°11'23"N 77°51'35"E). The area taken for study is surveyed from December 2002. About 10 numbers of permanent qudrats (4 meters wide and 25 meters long) were marked randomly in study site. The Vegetation Density and frequency per cent of medicinal plants along with the other plant species is calculated (Philips, 1959). Anthropological activities are not restricted throughout the study period. The data collected on the density and frequency per cent of medicinal plants, in the months of December during 2004, 2005, 2006, 2007 and 2008 were analyzed and taken for discussion.

Results

The highest number plant species was noted during December of every year in the study site and the results are shown in Table 5.1. During initial study period (December, 2004) some 34 species of plants were identified along with 14 medicinal plants utilized by herbal healers and siddha medicine practitioners. The goats of nomadic herders nibbled and reduced most of the plant species including medicinal plants.

The medicinal plants diversity fallen during 2005 when compared to 2004. During the year 2006, only 12 species of medicinal plants were noted in the study site. Among them only 5 plants are targeted for bulk harvesting and the medicinal plant collectors uproot the whole plant for marketing from September to February. *(Viz–*

Table 5.1: Medicinal plant Frequency per cent and Density in the study site.

Medicinal Plants Name of the species	Period of Data Density									
	Dec 2004		Dec 2005		Dec 2006		Dec 2007		Dec 2008	
	Frequency per cent	Density	Frequency per cent	Density	Frequency per cent	Density	Frequency per cent	Density	Frequency per cent	Density
Abutilon indicum	100	24	100	29	100	15.3	100	17.2	100	19.2
Aerva lantana	100	16.5	80	18.9	80	11.2	100	6.3	80	7.8
Boerhavia diffusa	70	6.4	40	4.3	30	1.2	0	0	0	0
Cassia auriculiformis	30	0.2	20	0.1	20	0.1	0	0	0	0
Enicostemma littorale	100	8.2	60	6.2	0	0	0	0	0	0
Gloriosa superba	100	7.3	30	4.2	10	0.2	0	0	0	0
Indigofera tinctoria	100	26.5	100	11.6	20	2.3	0	0	0	0
Leucas aspera	80	6.9	0	0	0	0	0	0	0	0
Momordica tuberosa	100	6.5	100	8.9	100	4.2	0	0	0	0
Ocimum sanctum	80	8.2	90	5.3	20	2.4	0	0	0	0
Phyllanthus nirurii	40	0.3	10	1.0	10	1.0	10	1.0	0	0
Physalis minima	30	0.4	10	0.2	0	0	0	0	0	0
Solanum xanthocarpum	30	0.3	0	0	0	0	0	0	0	0
Ziziphus species	10	0.1	10	0.1	0	0	0	0	0	0

Boerhavia diffusa, Solanum xanthocarpum, Phyllanthus maderaspatensis, Gloriosa superba and *Ocimum sanctum).* The species like *Momordica tuberosa, Ziziphus species* and *Physalis minima* are used for harvesting the fruits alone.

In December 2007 *Prosopis juliflora* shrubs and the grass *Ophiuros exaltatus* found along with other few species in the study site. During the study period of December 2008, the aggressive growth of *Prosopis juliflora* and *Ophiuros exaltatus* changed the study site scenario by their dominance.

Discussion

The emerging vegetations in the fallow lands are utilized for grazing by nomadic goat herders, medicinal plant collectors and fire wood collection. Sheep and goat grazing definitely causes impacts on the species diversity (Chandraprakash Kala *et al.,* 2002). Many studies have focused on the impact of environmental factors on plant species distribution and composition (Campagne *et al.,* 2006). Most of the medicinal plants and palatable forage species disappeared from the study site within the six years of study period.

The irregular rain and drought prevailed in Kovilpatti also one of the many reason for the disappearance of the medicinal plants. The fallow land pastoral system shows a changeable biodiversity of plants may also due to mismanagement of anthropological activities.

Due to over exploitation, many plant species slowly disappeared year after year. Also the native species of plants could not surmount the dominance of *Prosopis juliflora* and it changed the study site scenario.

The plant resources of the fallow lands help to alleviate poverty of the dependent poor people and it is one of the important criterion for consideration. The live stocks are being recognized as part of integrated solutions for sustainable natural resource management in the broader development context (World Bank 2007). The agricultural and environment measures of European Union were unable to restore biodiversity and even to slow down its decline (Wilson *et al.,* 2007). The lack of sustained utilization may lead to vulnerable loss of medicinal plant biodiversity in the dry land. The unpredictable rain, environmental factors, nomadic goat grazing, dominance of *Prosopis juliflora* and anthropological activity together caused decline in the biodiversity plants in the study site.

References

Burrows, W.S., 1998. *Woodland Monitoring 'Traps' Transect Recording and Processing System,* (Eds.) P.V. Bacj, E.R. Anderson, W.H. Burrows, M.J. Jkennedy and J.O. Carter. Department of Primary Industries, Queensland.

Campagne, P., Roche, P. and Tatoni, T., 2006. Factors explaning shrub species distribution in hedgerows of a mountain landscape. *Agricultural Ecosystems and Environment,* 116: 244–250.

Kala, Chandraprakash, Singh, Sanja Kumar and Rawat, Gopalsingh, 2002. Effect of sheep and goat grazing on the species diversity in the alpine meadows of western Himalaya. *The Environmentalist,* 22: 183–188.

Phillips, E.A., 1959. *Methods of Vegetation Study.* Holt, Reinhart and Winston, New York.

Wilson, A., Vickery, J. and Pendlebury, C., 2007. Agri-environment schemes as a tool for reversing declining populations of grassland waders: Mixed benefits from environmentally sensitive area in England. *Biological Conservation,* 136(1):128–135.

World Bank, 2007. Agriculture for development. *World Development Report 2008.* The World Bank, Washington D.C., US., 365 pages.

2013, Perspectives in Plant Biodiversity   *Pages* **27–33**
Editor: **Dr. K. Muthuchelian,** *Vice Chancellor, Periyar University, Salem*
Published by: **Daya Publishing House, NEW DELHI**

Chapter 6

Some Endemic and Threatened Plants of the Agasthiyamalai Biosphere Reserve, Tamil Nadu

T.J.S. Rajakumar, R. Selvakumari, S. Murugesan and N. Chellaperumal
Centre for Botanical Research,
St. John's College, Palayamkottai – 627 002, Tamil Nadu

Introduction

The concept of endemism is quite old. A. P. De Candole (1855) and Engler (1882) have given a preliminary idea of endemism and its types. Later different workers have done research on the concept of endemism. But the concept of threatened plants is comparatively new and people got interested in this field when International Union for Conservation of Nature and Natural Resources (IUCN) published their Red Data Book in 1966.

In 1988 Myers introduced the term "Hot spots" for the geographical regions particularly rich in endemic, rare, and threatened species found in relatively small areas. The International Union for Conservation of Nature and Natural Resources (IUCN) has identified 234 centres of plant diversity sites all over the world. The twenty five terrestrial biodiversity of the world contain a total number of 1, 33, 149 species *i.e.* 44 per cent of all vascular plant species. In 2000 Myers *et al.* suggested that out of the 25 hot spots 9 were recognized as leading hotspots while 8 as hottest hotspots based on an analysis of some factors *i.e.* number of endemic plant and animal species, endemic species area ratios and the habitat loss.

India with an area of about 3.287 million km². and coast line of over 7500 km. is the second largest country in Asia and seventh in the world that represents 2.46 per cent of the total world land mass. This great geographical expense of the country, with its extra-ordinary diversity of climate, soil and topography is coterminous with almost all types of ecosystems found anywhere in the world. India is recognized as one of the 12 mega diversity centres in the world. The forest area in India is 6.33 lakhs km². *i.e.* 19.27 per cent of the total land mass in India. There are about 4700 plants in India including 17500 flowering plants about 320 families representing 6.8 per cent in global population. Among different biogeographical zones in India, the Eastern Himalayan and Western Ghats are botanically rich areas of world significance due to high rainfall, moist and cold climate coupled with factors like altitude, latitude and longitude.

Globally about 44 per cent of vascular plants and 35 per cent of vertebrates are considered endemic species in 25 hot spots. These endemics are confined to an aggregate expanse of 28 million km² or 14 per cent of the earth land surface. These areas remain under tremendous human pressure. The flora of the Indian subcontinent is extremely rich in diversity and endemism. Out of the 17500 species of flowering plants, there are about 5725 species endemic to this subcontinent *i.e.* 25–30 per cent of total vascular flora *i.e.* 2500 species of plants are under different categories of endangerment.

In 1940 Chatterjee pioneered the work on endemism in the Indian context. According to him the British India had 133 endemic genera of Angiosperms. He also pointed out that about 61. 5 per cent of the Indian flora is endemic. In 1980 Nayar pointed out that there are 141 endemic genera in India. In 1964 Nayar considered that the Western Ghats hills tops resemble islands so far as the distribution of endemic species is concerned and contains more than 1600 endemic species. The Western Ghats is also a rich germplasm centre for a number of wild relatives of cultivated plants. The Southern Western Ghats are a conglomerate of hill ranges *i.e.* Travancore hills of Kerala and Nilgiri, Anamalais, Palani and Tirunelveli hills *i.e.* Agasthiyamalai Biosphere reserve.

Agasthiyamalai Biosphere Reserve (ABR) is located at the Southern extreme portion of province 5B of the biogeographical region of the Western Ghats. This Biosphere reserve consists of different zones in Tamil Nadu and Kerala and has a total area of 3373. 36 km². The Tamil Nadu portion which lies in the Eastern part of the Western Ghats has an area of 1672. 36 km². It is recognised as one of the three mega diversity centres in India and as one of the 25 global hotspots of Biodiversity. This region is one of the richest centres of endemic species. ABR are represents diverse ecosystems with almost all types of vegetation found in the Western Ghats. It is moderately free from human intervention due to its challenging topography, inclement weather and fairly good protection. These factors coupled with the high rain fall, high relative humidity, suitable temperature and high altitudinal variation ranging from 30 m to 1868 m msl resulted in the evolution of several ecological niches which are occupied by flora and fauna with narrow endemism.

ABR has many key habitats that are the result of microclimatic condition. It provides the spatial niches for many unique, rare, endangered and threatened species. The area represents diverse ecosystems with almost all types of vegetation found in the Western Ghats. The natural barriers, varied altitudes, habitat, climate and rainfall resulted in the development of a rich and diverse population. More than 30 new taxa have been discovered from here in recent years. Out of the 179 families in Tamil Nadu as many as 157 families are found in this region. In 1970 Blasco has shown that South Indian hill tops are abundant in endemic species and he has also pointed out that the Nilgiris is an important centre of speciation, next to the Tirunelveli–Travancore hills *i.e.* Agasthiyamalai Biosphere Reserve area.

During the exploration of flora of Agasthiyamalai Biosphere Reserve many endemic and threatened species have been identified. Some are

Acranthera grandiflora Beddome–Rubiaceae

Status
IUCN (Revised): Endemic and rare.

Distribution
It is confined to Tirunelveli and Travancore hills in the districts of Tirunelveli in Tamil Nadu and Thiruvananthapuram in Kerala. During the present explorations only 4 populations, each with less than 3 or 4 individuals in an area less than 5 km^2 were observed. It is necessary to study its ecology and reproductive bilogy of this species.

Acrotrema arnottianum Wight–Dilleniaceae

Status
IUCN (Revised): Rare.

Distribution
Endemic to Southern Western Ghats of Kerala and Tamil Nadu. During the present exploration only were 3 populations, each with less than 5 individuals in an area of less than 5 Km^2 were observed in Valzhaiyar at Mundanthurai forest. On no occasion was it seen in fruits.

Diotacanthus grandis (Beddome) Benth.–Acanthaceae

Status
IUCN (Revised): Endemic.

Distribution
It is a narrow endemic and is confined to Western Ghats. of Kanyakumari and Tirunelveli districts in Tamil Nadu and Thiruvananthapuram district in kerala. During the present explorations 5 populations each with 7 or 8 individuals were observed near Kannikatti, Mundanthurai forest range. These populations monitoring and causal factors identified for future conservation measures.

Eugenia singampattiana Beddome–Myrtaceae

Status
IUCN (Revised): Endemic.

Distribution
It is a narrow endemic and is confined to the Tirunelveli hills in Tami Nadu. During the present explorations 2 plants were collected in the vicinity of Thambaraparani reservoir. It is found that the number of fruits per tree is very less, its reproductive biology needs to be studied.

Euphorbia santhapaui A. N. Henry–Euphorbiaceae

Status
IUCN (Revised): Critically endangered.

Distribution
It is a narrow endemic collected at Poongulam in Tirunelveli hills. Single population of 7 trees was observed. This population is to be periodically monitored and research on all aspects are to be carried out.

Eriocaulon ensiforme C. E. C. Fischer–Eriocaulaceae

Status
IUCN (Revised): Critically endangered.

Distribution
A critically endangered species of the Southern Western Ghats. In the present study it is found to be extremely rare, collected on the edges of Poongulam, the place of origin of the Thambaraparani river.

Homalium jainii Henry and Swamin.–Flacourtiaceae

Status
IUCN (Revised): Critically endangered.

Distribution
Along streamsides at an elevation of 500 to 900m. This species seems to be an endemic in that it has been collected so far in Muthukuzhivial in Kanyakumari district and in the vicinity of the present study area, on the way to Vanatheertham and Nagapothigai. During the present study it was collected on the way to Vazhaiyar from Kuthalapari other than the type localities.

Memecylon subramanii A. N. Henry–Melastomataceae

Status
IUCN (Revised): Endemic and rare

Distribution
It is distributed in Tirunelveli hills in Tamil Nadu. During the present explorations only 3 populations, each with 2 individuals were observed. Plants usually occur in deep forest in shade adjoining rivulets. As the species is niche–specific the populations need periodical monitoring.

Piper barberi Gamble–Piperaceae

Status
   IUCN (Revised): Critically endangered

Distribution
   The original type collection was made by Barber from Kannikatti in May 1901. Subsequently it could only be collected from the type locality after a lapse of over six decades. Later Henry located it at Balamore in Kanyakumari districts in 1976. Now it was collected near Vazhaiyar in Mundanthurai forest range in 2007.

Popowia beddomeana Hook. f. and Thomson–Annonaceae

Status
   IUCN (Revised): Endangered.

Distribution
   Endemic to the higher altitudes (1000–1300) of the evergreen forests in the Southern Western Ghats. In the present exploration it was collected on the way to Aduppukal in Mundanthurai forest range.

Schefflera bourdillonii (Gamble)–Araliaceae

Status
   IUCN (Revised): Endemic

Distribution
   A narrow endemic of the Tirunelveli hills in Tamil Nadu. During the exploration single population of 8 trees was observed. This population is to be periodically monitored and research on all aspects are to be carried out.

Syzygium bourdillonii (Gamble) Rathakr. and Nair–Myrtaceae

Status
   IUCN (Revised): Critically endangered

Distribution
   Evergreen forests at an altitude of 500 to 1000 m. It has been recorded only from Thiruvanandapuram and Quilon districts in Kerala. Now it was collected in Kanikatti, Mundanthurai forest in Tirunelveli districts.

Vernonia gossypiana Gamble–Asteraceae

Status
   IUCN (Revised): Critically endangered

Distribution
   Evergreen forests at 1000 to 1500 m. It is an endemic to Southern end of the Western Ghats in Thiruvananthapuram district in Kerala. Now it was collected in open rocky areas of upper Kodaiyar and Agasthiyamalai peak in Tirunelveli districts.

References

Ahmedullah, M. and Nayar, M.P., 1987. *Endemic Plants of Indian Region, Vol. 1 Peninsular India*, Kolkata.

Beddome, R.H., 1877. The forests and flora of the Tirunelveli District. *Indian Forester,* 3: 19–25.

Blasco, F., 1970. Aspects of the flora of and ecology of Savanas of the South Indian hills. *J. Bombay Nat. Hist. Soc.*, 67: 522–534.

Bourdillon, T.F., 1908. *The Forest Trees of Travancore*, Trivandrum.

Caldwell, R., 1881. *A History of Tinnevelly*, New Delhi.

Chatterjee, D., 1940. Studies on the endemic flora of India and Burma. *J. Roy. Asiat. Soc. Bengal,* 5: 19–67.

Decandole, A.P., 1855. *Geographie Botanique Raisonnee Vols. 1 and 2.* Geneva.

Engler, A., 1882. *Versuh ciner Enturick Lungsgeschichte der pflangenwett Leipzig.*

Fyson, P.F., 1932. *The Flora of the South Indian Hill Stations*, 2 Vols. Chennai.

Gopalan, R., 1997. Plant diversity in Agasthiyamalai Hills, Southern Western Ghats. In: *Plant Diversity Hotspots in India: An Overview*, (Eds.) P.K. Hajra and V. Mudgal, Kolkata.

Govindarajalu, E. and Swamy, B.G.L., 1958. Enumeration of plants collected in Mundanthurai and its neighbourhood (Tirunelveli district). *J. Madras Univ.*, 28B: 161–177.

Henry, A.N. and Subramayam, K., 1981. Studies on the flora of Agasthiyamalai and surrounding regions in Tirunelveli district, Tamil Nadu. *Bull. Bot. Surv., India*, 23: 42–45.

Henry, A.N., Vivekananthan, K. and Nair, N.C., 1979. Rare and threatened flowering plants of South India. *J. Bombay. Nat. Hist. Soc.*, 75: 684–697.

Hooker, J.D., 1872–1897. *The Flora of British India*, Vols.1–7. London.

Mabberley, D.J., 1997. *The Plant Book*. London.

Myer, N., 1988. Threatened biotas "Hot spots" in tropical forests. *The Environmentalist,* 8: 187–208.

Myer, N., Mittermeier, R.A., Mittermeter, C.G., da Fonseca, G.A.B. and Kent, J., 2000. Biodiversity hotspots for conservation priorities. *Nature,* 403: 853–858.

Nair, N.C. and Daniel, P., 1986. The floristic diversity of the Western Ghats and its conservation: A review. *Proc. Indian Acad. Sci. (Suppl)*, p. 127–163.

Nayar, M.P., 1996. *'Hot Spots' of Endemic Plants of India, Nepal and Bhutan,* Thiruvanthapuram.

Nayar, M.P., 1980. Endemism and patterns of distribution of genera (Angiosperms) in India. *J. Econ. Tax. Bot.*, 1: 99–110.

Ramaswamy, M.S., 1914. A botanical tour in Tinnevelly Hills. *Rec. Bot. Surv. India*, 6: 105–171.

Rangachariya, K., 1919. A note on the flora of Tirunelveli district. *Madras Agricultural Department Yearbook,* p. 95–109.

Rao, R.R., 1994. *Biodiversity in India: Floristic Aspects.* Dehradun.

Rao, Rama, 1914. *Flowering Plants of Travancore,* Trivandrum.

Sebastine, K.M. and Henry, A.N., 1960. Studies on the flora of Singampatti R.F.in Tirunelveli. *Bull. Bot. Surv. India,* 2: 22–24.

Sebastine, K.M. and Saroja, T.L., 1962. A further contribution to the flora of Kannikatty forest, Tirunelveli District. In: *Indian Sci. Congr. Assoc.* 49th session, Part III, Abstract 321.

Shankaranarayanan, K.A., 1960. The vegetation of Tirunelveli district. *J. Indian Bot. Soc.,* 39: 474–479.

Vajravelu, E., Joseph, J. and Rathakrishnan, N.C., 1987. Flora of Kalakkad hills, Tirunelveli District, Tamil Nadu. *J. Econ. Tax. Bot.,* 10: 249–305.

Vajravelu, E. and Daniel, P., 1983. Enumeration of threatened plants of Peninsular India. In: *Materials for a Catalogue of Threatened Plants of India,* (Eds.) S.K. Jain and A.R.K. Sastry, p. 8–43.

Wight, R., 1835–1836. Observations on the flora of Courtallum. *Madras J. Litt. and Sci.* 2: 380–391: 3: 94–96: 4. 57–66.

World Conservation Monitoring Centre, 1992. *Global Biodiversity Status of Earth's Living Resources.* Chapman and Hall, London.

2013, Perspectives in Plant Biodiversity   *Pages* **34–39**
Editor: **Dr. K. Muthuchelian,** *Vice Chancellor, Periyar University, Salem*
Published by: **Daya Publishing House, NEW DELHI**

Chapter 7

Diversity and Opportunities with Indian Acclimatized Jatropha Species

P. Ratha Krishnan* and S.P. Ahlawat
National Research Centre for Agroforestry,
Jhansi – 3, U.P

ABSTRACT

Suitability of vegetable oils for the production of biodiesel is gaining national and international importance. Tree-borne oilseeds are the best and potential alternative to mitigate the current and future energy crisis and also to transform the vast stretches of wasteland into green oil fields. The potential sources identified so far include *Jatropha curcas, Pongamia pinnata, Madhuca latifolia, Azadirachta indica, Calophyllum inophyllum, Simarouba glauca, etc.* Among these, *J. curcas* emerges as the most promising tree-borne oilseed on the basis of its adaptability to a wide range of edapho-climatic conditions coupled with the suitability of *Jatropha* oil as a source of biodiesel. This article highlights the diversies of Indian acclimatized Jatropha species and the opportunities available with them for yield improvement of *J. curcas*.

Introduction

Abundant availability of natural resources and the technical know how largely determines the economic well being of any country. When our total area under degraded and wastelands stands at 120.41 M ha (Maji *et al.,* 2010), we requires proper planning

---

* Corresponding Author: E-mail: ratha_forestry@yahoo.co.in

to address the resource management issues. Biomass is a source of fundamental renewable energy since civilization. Since energy independence is our first and highest priority (Abdul Kalam, 2005) we are in way of exploring the possible alternate energy sources including bio-energy. Among the renewable energy sources, liquid bio-energy from vegetable oils is one of the best options to reduce GHG emissions. Among the plant category, the properties of physic nut (*Jatropha curcas*) have won over the interest of various agencies and Planning Commission of India. The genus Jatropha belongs to Euphorbiaceae family was reported nearly 170 known species distributed in the tropical and subtropical Africa and America. This genus possesses poisonous substance in the sap/seed. Indumentums simple hairs, sometimes glandular and leaves alternate, often digitately lobed. Fruits capsular to tardily dehiscent and sub–drupaceous.

Materials and Methods

The tool used for diversity assessment is by an extensive survey on literature and the Indian soils from Jammu to Kanyakumari which makes the possibility of collection of Indian acclimatized Jatropha species. Among the Jatrophas the physic nut (*Jatropha curcas*), most primitive form is potential to be cultivated for bio-diesel and medicinal properties. The diversity of Jatropha and their potential made options for yield improvement over *J. curcas*. After analysis on important morphological traits available with each Jatropha species (Ratha Krishnan and Paramathma, 2009), attempts were made towards yield improvement by inter and intra specific hybridization, inter and intra specific grafting and quality *J. curcas* seedling production with short nursery period were attempted and the results of that experiments were given below.

Results and Discussion

Jatropha are herbs, shrubs or trees, monoecious (rarely dioecious), exudates watery to white. This genus possesses poisonous substance in the sap/seed. Indumentum simple hairs and sometimes glandular hairs, leaves alternate, often digitately lobed. Flowers are terminal cymes with a single pistillate flower at the end of the primary axis. Sepals-5, free, imbricate; petals-5, mainly free; staminate disc annular or 5 free glands, stamens 6–10, in two whorls; pistillate foliaceous annular, 5 –lobed; fruits capsular to tardily dehiscent and sub-drupaceous. Even though 12 Jatropha species were notified by several Indian floras, research has been confined with 9 species while, the physic nut (*J. curcas*), the most primitive form has potential to be cultivate for bio-diesel and medicinal properties. Following are the important traits available with each species while Table 7.1 contains the detail information on important morphological features and desirable exploitable traits of different Jatropha species. a) *J. curcas*–High seed yield and oil content; b) *J. gossypifolia*–Drought tolerant, profuse growth and year round fruiting; c) *J. glandulifera*–Profuse fruiting and drought-tolerant; d) *J. multifida*–Bigger fruit size and diseases resistance; e) *J. tanjorensis*–Robust and drought hardy; f) *J. podagrica*–Bigger fruit size and Fusarial wilt resistance; g) *J. integerrima*–Semi-hard wood stem and disease-resistant, h) *J. pandurifolia*–year round flowering; i) *J. villosa*–Evergreen and rhizomatous plant; j) *J. nana*–Hard and woody root system; k) *J. heynei*-Tuberous root stock, l) *J. maheswari*–Drought hardy and rhizomatous plant.

Table 7.1: Important morphological features and desirable traits of different Jatropha species.

Sl.No.	Species	Native Place	Distribution in India	Important Morphological Features	Oil (per cent) Content	Propagation Methods	Desirable Traits
1.	Jatropha curcas	Tropical America	Got introduced in all the states of India	Large shrub, highly branching, cordate–palmately lobed leaves, greenish-yellow flowers, distinct coflorescence, tardily dehiscent fruits with black, ecarunculate seeds	30–42	Seed, cutting, grafting, air layering and tissue culture	High seed yield and oil content
2.	Jatropha gossypifolia	Brazil	Common in disturbed soils of all states	Fertile large shrub, profuse branching, cordate leaves, glandular plant parts, dark crimson–purple flowers, violently dehiscent capsules with small brown carunculate seeds	28–30	Seed and cuttings	Drought-tolerant and profuse, year round fruiting
3.	Jatropha glandulifera	India	Deccan and Carnatic Black cotton soils	Fertile smaller plant, spread and dichotomously branched, narrow leaves with serrated margin, have smooth papery park, profuse fruiting, but dehisce before maturity	20–27	Seed and cuttings	Profuse fruiting and drought-tolerant
4.	Jatropha multifida	South America	Ornamental nurseries	Fertile shrub, uniform branching, leaves divided into 5–11 lobes, long petiole and pedunculate, flat-topped cyme, coral-red flowers and fruits are non-dehiscent capsules	32–40	Seed and by cuttings during spring	Bigger fruit size and resistant to diseases
5.	Jatropha tanjorensis	India	Tanjore, Trichy, and Ramnad district of Tamil Nadu	Sterile shrub, profuse branching, cordate–palmately lobed leaves, margins distinctly serrate, greenish-yellow flowers with crimson-red tinge, no fruit-set	Sterile	Cuttings	Robust and drought hardy
6.	Jatropha podagrica	Panama	Ornamental nurseries in Southern and Central India	Fertile, caudiciform shrubs, cordate leaves with peltate base, flat-topped corymbose cyme, bright scarlet flowers, fruits dehiscent capsule with big brown ecarunculate seeds	Up to 54	Seed and by division of branches	Bigger fruit, Fusarial wilt resistant
7.	Jatropha integerrima	West Indies	Ornamental nurseries of south India	Fertile shrub, sparsely branched, ovate fiddle-shaped leaves, crimson-red flowers, dehiscent capsules, seeds small carunculate and brown with spots	No report	Cuttings	Semi-hard wood stem and disease-resistant

Contd...

Table 7.1–Contd...

Sl.No.	Species	Native Place	Distribution in India	Important Morphological Features	Oil (per cent) Content	Propagation Methods	Desirable Traits
8.	Jatropha pandurifolia	Cuba	Ornamental nurseries (but rare)	Dioecious shrub with slender, graceful branches. Leaves alternate, shallowly cordate at the base, inflorescence terminal cyme, calyx purplish red in color. Petals twisted in the bud, white hairs inside at the base. Flowering through out the year, fruits are capsule and purplish green	No report	Seed and cuttings	Flowering through the year
9.	Jatropha villosa	India	Kongan region, Nilgiri, Kanya-kumari, and Ramnad districts of Tamil Nadu	Fertile Undershrub, shoots rusty-villow, profuse branching, drought-tolerant, evergreen, rhizomatous plant	No report	No report	Evergreen and rhizomatous plant
10.	Jatropha nana	India	Poona and Mumbai. Endemic to the Deccan,	Shrub with woody root as thick as finger, stem round, smooth. Leaves 3 lobed/entire with the largest middle lobe. Flowers pedicellate, and few flowered terminal paniculate cymes. Capsule fruit.	No report	Seed	Woody root system
11.	Jatropha heynei	India	Indian Peninsula	Shrub, branching from a tuberous rootstock (weigh about 1 Kg). Leaves deeply 3–fid, lobes oblanceolate. Flowers unisexual and small. Both flowers and fruit capsules are green in color.	No report	Seed	Tuberous root stock
12.	Jatropha maheswari	India	Naturally occurs in southern districts of Tamil Nadu	Fertile evergreen, drought-hardy and rhizo-matous plant, leaves long, elliptical and resembles mango leaves	No report	No report	Drought-hardy and rhizomatous plant

A decade back itself few native Jatropha species were utilized in castor improvement programme and inter-specific hybridization has been attempted between Jatrophas with limited success (Dehgan, 1984; Sujatha and Prabakaran, 2003). Occurrence of natural hybrids was reported by Pax (1910) in some South American Jatropha species; *J. cinerea–J. canescens* complex in Mexico (Dehgan and Webster, 1979) and a hybrid complex of *J. integerrima–J. hastata* in Cuba (Dehgan, 1984). Artificial hybridization attempted by Parthiban *et al.* (2009), between *Jatropha curcas* and eight other *Jatropha* species, resulted *J. curcas* X *J. integerrima* as successful hybrid with more seed set, while the other crosses failed to produce seeds due to the existence of crossability barriers either in pre-zygotic state or in post-zygotic state.

Studies conducted at NRCAF, Jhansi, the intra-specific reciprocal crossing between selected 10 parents of *J. curcas* has resulted for significant increase in fruit yield and female: male flower ratio per inflorescence (1:8) of the F_1 hybrids (Kumar *et al.*, 2009). Inter-specific cleft grafting between *J. curcas* (scion) and *J. gossypifolia* (root stock) produced early union in 18–21 days with > 80 per cent success both in *ex* and *in-situ* conditions. 1.5 yr grafts planted in the nursery revealed the good grafting compatibility between species, early fruiting and intermediate performance of grafts over mother plants (Ratha Krishnan *et al.*, 2007). The combined application of Azospirillum + Phosphobacteria + VAM was identified as a best nursery treatment for getting good growth, productivity and quality of *Jatropha curcas* seedlings due to their synergistic effect. This treatment assured the production of quality seedlings in 72 days while reducing nursery maintenance period and cost.

Conclusion

A decade back itself, literatures were available for the usage of few Jatropha species on castor improvement programme in India (Prabakaran and Sujatha, 1999) while Non-toxic Jatropha from Mexico was obtained and the *in vitro* propagation method was standardized for its multiplication and the protocol was published by Sujatha *et al.* (2005). But the improved variety/end products were not reached the ground. Even though Indian search/research on Jatropha were more than a Century old the co-coordinated research towards yield improvement is lacking which is need of the hour.

References

Abdul Kalam, A.P.J., 2005. *Independence Day Address to the Nation* on 15th August, 2005.

Dehgan B., 1984. Phylogenetic significance of interspecific hybridization in Jatropha (Euphorbiaceae). *Syst. Bot.,* 9: 467–478.

Dehgan, B. and Webster, G.L., 1979. *Morphology and Intrageneric Relationships of the Genus Jatropha (Euphorbiaceae).* University of California Publications in Botany.

Kumar, R.V., Ahlawat, S.P., Gupta, V.K. and Palsania, D.R., 2009. Genetics and breeding of Jatropha species. *NRCAF, Jhansi–Annual Report* 2008–09, p. 130.

Maji, A.K., Obi Reddy., G.P. and Sarkar, Dipak, 2010. *Degraded and Wastelands of India: Status and Spatial Distribution.* Indian Council of Agricultural Research, New Delhi, p. 158.

Parthiban, K.T., Senthil Kumar, R., Thiyagarajan, P., Subbulakshmi, V., Vennila, S. and Govinda Rao, M., 2009. Hybrid progenies in *Jatropha*: A new development. *Curr. Sci.*, 96: 815–823.

Pax, F., 1910. Euphorbiaceae–Jatropheae. In: *Das Pflanzenreich IV*, (Ed.) A. Engler. Leipzig: Verlag von Wilhelm Engelmann, 147(42).

Prabakaran, A.J. and Sujatha, M., 1999. Jatropha tanjorensis Ellis and Saroja: A natural interspecific hybrid occurring in Tamil Nadu, India. *Genet. Resour. Crop Evol.*, 46: 213–218.

Ratha Krishnan, P. and Paramathma, M., 2009. Potentials and Jatropha species wealth of India. *Current Science*, 97(7): 1000–1004.

Ratha Krishnan, P., Kumar, R.V., Handa, A.K. and Gupta, V.K., 2007. Grafting of *Jatropha curcas on Jatropha gossypifolia* to induct drought hardiness and adaptability. *Agroforestry Newsletter*, 19(4): 2–3.

Sujatha, M. and Prabakaran, A.J. 2003. New ornamental Jatropha hybrids through interspecific hybridization. *Genet. Resour. Crop Evol.*, 50: 75–82.

Sujatha, M., Makkar, H.P.S. and Becker, K., 2005. Shoot bud proliferation from axillary nodes and leaf sections of non-toxic *Jatropha curcas* L. *Plant Growth Regulation*, 47: 83–90.

2013, Perspectives in Plant Biodiversity   *Pages* **40–46**

Editor: **Dr. K. Muthuchelian,** *Vice Chancellor, Periyar University, Salem*

Published by: **Daya Publishing House, NEW DELHI**

Chapter 8

Bioreclamation of Tsunami Affected Agricultural Land by *Heliotropium curassavicum* Linn.

K.C. Ravindran[1], A. Indrajith[1], K. Sanjiviraja[1], D. Ayyappan[1] and V. Balakrishnan[2]

[1]Department of Botany, Annamalai University, Annamalainagar – 608 002, Tamil Nadu
[2]Department of Biotechnology, K.S.R. College of Technology, Tiruchengode – 637 215, Tamil Nadu

ABSTRACT

A Tsunami on 26th December, 2004 by an earthquake of magnitude M 9.0 occurred, along the plate boundary marked by subduction zone between the Indian plate and the Burmese microplate near Sumatra Island of Indonesia with the epicenter located on the shallow depths of seabed. In the Tsunami aftermath, agricultural land has degraded through several processes including salinization of soil and water, de-surfacing of landscape due to the deposition of sand to clay sediments and destroying the dike of paddy fields and destroying irrigation/drainage infrastructures and roads. Nearly 8460 hectares of agricultural land were affected in Tamil Nadu. Leaching and chemical amendments are primarily used for reclamation. However the cost of reclamation by this technique is higher and an alternative source of reclamation is phytoremediation. In recent years, phytoremediation a plant based amelioration strategy has emerged as a low-cost and environmentally acceptable technique. Salinity is not inimical to all plants. A wide range of plants species especially halophytes, survive salt concentration equal or greater than that of seawater. Vascular halophytes accumulate high levels of sodium and other salt in their vacuole and compatible solutes in the cytoplasm. The objective of the present study is to utilizing *Heliotropium curassavicum* a halophyte to assess the feasibility of salt bioaccumulation for restoration of Tsunami affected saline agricultural lands as an alternative method to other leaching techniques. From the results it is observed that *Heliotropium curassavicum* significantly reduced the soil salinity by absorbing sodium ions. The EC of soil had a 40 per cent saturation in a single growth cycle. The reduction in sodium absorption ratio also confirms the reduction of salinity.

Introduction

Apart from Tsunami affected agricultural soil, salt affected soils occur with in the boundaries of atleast 75 countries (Szabolcs, 1994). These soils also occupy more than 20 per cent of the global irrigated area (Ghassemi *et al.,* 1995); in some countries, they occur on more than half of the irrigated land (Cheraghi, 2004). Salinity can inhibit plant growth due to various factors, including ion toxicity, impairment of mineral nutrition and changes in the water relations. The extent to which each of these factors can affect growth depends on plant genotype and on environmental conditions (Munns, 2002). The major saline ions, Na^+ and Cl^- can suppress net nutrient uptake due to competitive ionic interactions or affect membrane integrity. For example, high levels of Na^+ often induce K^+ deficiencies (Tester and Davenport, 2003). Salinity may cause two kinds of stress in plant tissues; osmotic and ionic. The latter is often associated with high Na^+/K^+ and Na^+/Ca^{2+} ratios and accumulation of Na^+ and Cl^- in tissues, which in harmful to general metabolism of cells (Blumwald *et al.,* 2000). Under salt stress, maintenance of cytosolic K^+ and homeostasis of intracellular ion concentration is even more crucial (Zhu, 2003).

Reclamation of saline soils uses many different methods as physical amelioration (deep ploughing, subsoiling, sanding, profile inversion), chemical amelioration (amending of soil with various reagents: gypsum, calcium, iron sulphate), electro-reclamation (treatment with electric current). The most effective, hydrotechnical amelioration methods are based on the removal of exchange and soluble sodium and changing the ionic composition of soils by added chemicals with parallel leaching of sodium salt out of the soil profile (Chhabra, 1994).

Phytoremediation is the use of plants to partially or substantially remediate selected contaminants in contaminated soil, sludge, sediment, groundwater, surface water and wastewater. It utilizes a variety of plant biological processes and the physical characteristics of plants to aid in site remediation (Pivetz, 2001). The efficiency of different plant species used in phytoremediation of sodic and saline sodic soils has been found to be highly variable. In general, the species with greater production of biomass together within the ability to withstand ambient soil salinity and sodicity as well as periodic inundation have been found to be efficient in soil amelioration (Qadir *et al.,* 2002). Some vascular halophytes accumulate high levels of sodium and other slats in their above ground tissue while others do not (Gorham *et al.,* 1987). Boyko (1966) was the first person to suggest that halophytic plants could be used to desalinate soil and water. The hypothesis set forth by Boyko does not distinguish between sodium and other salts. Bio-reclamation of saline-sodic soil by Amshot grass in Northern Egypt. Helalia *et al.* (1990) reported that Amshot grass (*Echinochloa stagninium*) when compared to ponding and gypsum treatment reduced the exchangeable sodium per cent of the surface layer of the soil.

Materials and Methods

Experimental Site for Reclamation Studies

The experimental site (Tsunami affected Agricultural Land) was located at T.S. Pettai village nearer to Pichavaram Mangrove Forest (11°24′ N and 79°44′E) on the northeast coast of Tamil Nadu, India.

Table 8.1: Chemical characteristics of saline surface soil.

Sl.No.	Physico-chemical Properties	Values
1.	EC	4.9 dS m^{-1}
2.	PH	8.8 meq/lit
3.	SAR	15.62
4.	Na	63.1 meq/lit
5.	Cl	55.04 meq/lit
6.	Ca	10.3 meq/lit
7.	Mg	12.0 meq/lit
8.	K	8.5 meq/lit

Selection of Species and Preparation of Land

The saline field was ploughed with tractor drawn disc plough followed by a through harrowing to break the clods. The field was properly leveled and 5 ′ 10 M size plots were earmarked with raised bunds and 30 cm spacing between 2 seedlings. The seedlings *Heliotropium curassavicum* were collected and planted at experimental field. Proper care was taken to maintain the plants. Plant samples were harvested for experimental purpose at an intervals of 30, 60, 90 and 120 days.

Heliotropium curassavicum Linn. (Boraginaceae)

A glaucous fleshy herb with leaves linear or linear-spathulate or lanceolate. Stigma conical, apex bifid and nut lets glabrous, smooth on the back and the margins are corky.

Measurement of EC and pH of the Soil Sample

The soil samples from 0-15 cm depth were collected from the experimental field for soil analysis. Soil samples were also taken at 0 days and 30 days intervals upto 120 days from the individual plots. These soil samples were dried and powdered gently with wooden wallet and passed through 2 mm sieve. The sieved soil samples were then taken up for analysis. The analytical methods are given in Table 8.2.

Measurement of EC and pH of the Plant Sample

Five gram of fresh plant samples were ground in mortar and pestle by using water (1:3) and then filtered through cheese cloth. This crude extract was used to determine the EC in Elico EC meter and pH was measured in Elico pH meter.

Table 8.2

Sl.No.	Property	Estimation Procedure	Reference
1.	pH	1:2:5 soil-water paste using Elico-pH meter	Jackson (1973)
2.	Electrical conductivity (dS m⁻¹)	Saturation extract of the soil with water using Elico conductivity bridge	Jackson (1973)
3.	SAR	$SAR = \dfrac{Na^+}{\sqrt{\dfrac{Ca^{2+}+Mg^{2+}}{2}}}$	Richards (1954)

Results

Electrical Conductivity (EC) in Saline Soil and Plants

Saline soils are formed from the accumulation of salts. There are different ways in the which salts accumulate in soil. The salt content of a soil can be estimated by measuring the electrical conductivity (EC) of the soil. In *Heliotropium* cultivated soil, EC reduced from 4.7 to 3.6 dS m⁻¹ and increased in plant samples from 2.7 to 6.00 dS m⁻¹.

pH in Soil and Plant

In saline soil after 120 days period of cultivation, soil pH declined from 8.4 to 7.5 per cent (10.8 per cent reduction) and increased in *Heliotropium* from 6.5 to 7.9 (17.8 per cent).

Sodium Absorption Ratio (SAR)

SAR is usually a good indicator of the structural ability of soil. In general sodium absorption ratio (SAR) is employed to understand the equilibrium relation between soluble and exchangeable cations. In general, SAR content gradually declined in natural saline soil. SAR content was reduced from 16.0 to 7.5 meq/lit (53 per cent reduction) and it indicate the reduction of salinity in the soil.

Table 8.3: Effect of salinity on soil EC and pH, plant EC and pH and SAR on *Heliotropium curassvium* growing in salt affected soil.

Days	Soil EC (dS m⁻¹)	Plant EC (dS m⁻¹)	Soil pH	Plant pH	SAR
0	4.7	2.7	8.4	6.5	16
30	4.4	2.8	8.2	6.8	15
60	3.9	4.4	7.8	7.4	10
90	3.7	5.3	7.6	7.7	8
120	3.6	6.00	7.5	7.9	7.5

Values are shown are mean ± SD for five replicate experiments.

Discussion

In general, EC was reduced in the soil samples and corresponding increase in the plant samples was observed in saline soil grown halophytes after 120 days cultivation. Similar to EC, pH was also reduced in the soil samples and corresponding increase in the plant samples. Sodium absorption ratio is usually a good indicator of the structural ability of the soil. In present study, it was noticed that SAR content was reduced to 53 per cent in *Heliotropium curassavicum* cultivated soil.

Several studies involving phytoremediation for improving the sodic soil have been carried out in various parts of the world (Ghaly, 2002). Zahran *et al.* (1982) observed that *Juncus rigiolus* decreased the soil EC from 33 to 22 dS m^{-1}. Singh *et al.* (1989) conducted a long term field study on an alkaline soil in order to improve such soils by growing *Prosopis julifera* and *Leptochloa fusca* and they concluded that the soil EC decreased from 2.20 to 0.42 dS m^{-1} and pH from 10.6 to 9.5.

The establishment of plantations of *Eucalyptus* and *Acacia* have resulted decline in soil pH, EC and increase in soil organic carbon on soils affected by high levels of sodicity (Gill and Abrol, 1993). From bio-reclamation studies of saline-sodic soil by Amshot grass (*Echinochloa stagninium*), Helalia *et al.* (1990) reported that when compared to ponding and gypsum on reducing the alkalinity and salinity of highly-sodic soil in Northern Egypt, showed higher reduction of the exchangeable sodium per cent (ESP). Reduction in exchangeable sodium was accompanied by a decrease in the sodium absorption ratio within the upper 45 cm (18 inches) of soil, reducing SAR 42-45 per cent. In addition, Amshot grass significantly reduced the salinity of the soil, when compared to either ponding or gypsum and produced higher yield than clover (*Melilotus officinalis*) cultivated in such soils.

Holmes (2001) has conducted an extensive laboratory and field investigations of the ecology of plants in extreme environments in an effort to select plants that are suitable for phytoremediation in saline sites. She has successfully used native halophyte plants to reclaim salt contaminated soils in Ohio, Oklahoma and Texas. She reported that the content of sodium in the soil was decreased by 65 per cent two years after planting with salt accumulating plants. In India, Chaudhri *et al.* (1964) investigated the ability of *Suaeda fruticosa* to accumulate sodium and other salts and reported that the leaves of this plant were found to contain 9.06 per cent salt on a fresh weight basis.

Conclusion

Heliotropium curassavicum could therefore be used successfully to reclaim highly salinized areas and rendering them more suitable for crop production after a few repeated cultivation.

Acknowledgements

Financial assistance received from the University Grants Commission, New Delhi is gratefully acknowledged.

References

Blumwald, E., Aharon, G.S. and Apse, M.P., 2000. Sodium transport in plant cells. *Biochem. Biophys. Acta,* London, 1465(1): 140–151.

Boyko, H., 1966. Basic ecological principles of plant growing by irrigation with highly saline or seawater. In: *Salinity and Aridity,* (Ed.) H. Boyko. Dr. W. Junk Publishers, The Hauge.

Chaudhri, I.I., Shah, B.H., Naqvi, N. and Mallick, I.A., 1964. Investigation on the role of *Suaeda fruticosa* Forsk in the reclamation of saline and alkaline soils in West Pakistan Plains. *Plant Soil,* 21: 1–7.

Cheraghi, S.A.M., 2004. Institutional and scientific profiles of organizations working on saline agriculture in Iran. In: *Prospects of Saline Agriculture in the Arabian Peninsula,* (Eds.) F.K. Taha, S. Ismail and A. Jaradat. *Proceedings of the International Seminar on Prospects of Saline Agriculture in the GCC Countries,* March 18–20th, 2001, Dubai, United Arab Emirates, pp. 399–412.

Chhabra, R., 1994. *Soil Salinity and Water Quality.* Oxford and IBH Publication Co., New Delhi.

Ghaly, F.M., 2002. Role of natural vegetation in improving salt affected soil in northern Egypt. *Soil Tillage Res.,* 64: 173–178.

Ghassemi, F., Jakeman, A.J. and Nix, H.A., 1995. *Salinisation of Land and Water Resources: Human Causes, Extent, Management and Case Studies.* CABI Publishing, Wallingford, United Kingdom.

Gill, H.S. and Abrol, I.P., 1993. Afforestation and amelioration of salt-affected soils in India. In: *The Productive Use of Saline Land,* (Eds.) N. Davidson and R. Galloway, *Proceedings of a Workshop* held in Perth, Western Australia, ACIAR Proceedings No. 42: 23–27.

Gorham, J., Hardy, C., Jones, R.G. Wyn, Joppa, L.R. and Law, C.N., 1987. Chromosomal location of the K/Na discriminating characters in the D Genome of wheat. *Theor. Appl. Genet.,* 74: 584–588.

Helalia, A.M., El-Amir, S., Abouzeid, S.T. and Nagholoul, K.F., 1990. Bioremediation of saline-sodic soil by amshot grass in northern Egypt. *Tillage and Soil Research,* Elsevier Science Publ., B.V. Amsterdam.

Holmes, P.M., 2001. Mycorrhizal colonization of halophytes in central European salt marshes, Referenced by E.P. Glenn in Scientific Amer, pp. 112–114.

Jackson, M.L., 1973. *Soil Chemical Analysis.* Prentice Hall of India Pvt. Ltd., New Delhi.

Munns, R., 2002. Comparative physiology of salt and water stress. *Plant Cell Environment, Logan,* 25(5): 659–662.

Pivetz, B.E., 2001. *Phytoremediation of Contaminated Soil and Groundwater at Hazardous Waste Sites,* EPA/540/5–01/500.

Qadir, M., Qureshi, R.H. and Ahmad, N., 2002. Amelioration of calcareous saline–sodic soils through phytoremediation and chemical strategies. *Soil Use Manag.,* 18: 381–385.

Richards, L.A. (Ed.), 1954. *Diagnosis and Improvement of Saline and Alkali Soils.* USDA Hbk, 60: 160.

Singh, K., Chauhan, H.S., Rajput, D.K. and Singh, D.V., 1989. Report of a 60 month study on litter production, changes in soil chemical properties and productivity under poplar (*P. deltoidse*) and Eucalyptus (*E. hybrid*) interplanted with aromatic grasses. *Agroforestry Syst.,* 9: 37–45.

Szabolcs, I., 1994. Soils and salinization. In: *Handbook of Plant and Crop Stress,* (Ed.) M. Pessara Kil. Marcel Dekker, New York, pp. 3–11.

Tester, M. and Davenport, R., 2003. Na$^+$ resistance and Na$^+$ transport in higher plants. *Ann. Bot.,* Ottawa, 91: 503–527.

Zahran, M.A. and Wahid, A.A.A., 1982. Halophytes and human welfare. In: *Tasks of Vegetation Science, Vol. 2: Contributions of the Ecology of Halophytes,* (Eds.) D.M. Sen and K.S. Rajpurohit. D.W. Junk Publishers, The Hauge.

Zhu, J., 2003. Regulation of ion homeostasis under salt stress. *Curr. Opin. Plant Biol.,* Dorderecht, 6: 441–445.

2013, Perspectives in Plant Biodiversity
Pages 47–58
Editor: **Dr. K. Muthuchelian,** *Vice Chancellor, Periyar University, Salem*
Published by: **Daya Publishing House, NEW DELHI**

Chapter 9

Traditional Use of Medicinal Plants and its Status Among the Tribes in Mananthavady of Wayanad District, Kerala

T.B. Shyma* and A.G. Devi Prasad

*Department of Studies in Environmental Science,
University of Mysore, Manasagangotry,
Mysore – 570 006, Karnataka*

ABSTRACT

Documenting the indigenous knowledge through botanical studies is important for the conservation and utilization of biological resources. The information on the traditional uses of plants and its status is lacking in Mananthavady of Wayanad district. The paper documents the traditional knowledge of medicinal plants that are in use by the indigenous tribes residing in pockets at Mananthavady forests. About 295 medicinal plant species belonging to 93 families have been recorded. The five major tribes namely Kurichia, Kuruma, Kattunaika, Adiyan and Paniya residing at Thirunelly, Kartikkulam, Thalappuzha, Kuruva, Kunhome and Mangalassery forest areas are using these plants for the treatment of various health problems. Among these 5 species are endangered, 22 species are vulnerable, 11 species are rare and 220 species are abundant.

Keywords: *Traditional uses, Tribes, Medicinal plants.*

---

* Corresponding Author: E-mail: envimys2009@yahoo.com

Introduction

India is rich in ethnic diversity and traditional knowledge (TK) that has resulted in a considerable body of ethno botanical research, of which one study has revealed a deep understanding of medicinal plants supported by high consensus. The term ethno botany was coined in 1895 by the North American botanist John Harsh Berger to describe the studies of plants used by primitive and aboriginal people (Balick and Cox, 1996) living in rural and forest areas. Rural people of the study area have strong relations with the surrounding environment.

We can distinguish two different goals or approaches to this study: the contribution to the knowledge of a part of human cultural heritage and search for new drugs or useful plant derived products. On the other hand the most recent term ethno pharmacology has undergone only slight evolution in meaning since its original definition as a multidisciplinary area of research concerned with the observation description and experimental investigation of indigenous drugs and their biological activities (Rivier and Bruhn, 1979). Due to the side effects of modern allopathic drugs in the present days, people are attracted towards herbal medicines and their consumption.

Documenting the indigenous knowledge through ethno botanical studies is important for the conservation and utilization of biological resources. Ethnobotanical survey has been found to be one of the reliable approaches in drug discovery (Fabricant and Farnsworth, 2001). Traditionally this treasure of knowledge has been passed on orally from generation to generation without any written document (Perumal Samy and Igacimuthu, 2000) and is still retained by various indigenous groups around the world. Today, it is estimated about 64 per cent of the total global population remains dependent on traditional medicines. (Farnsworth, 1994; Sindiga, 1994). Traditional medicine is becoming popular in the world today. The global market for traditional therapies estimated to be at $60 billion a year and is steadily growing. (WHO, 2002).

The traditional medical practices are an important part of the primary health care system in the developing world (Ghosh, 2003). According to World Health Organization (WHO). about 65-80 per cent of the world's population in developing countries, due to the poverty and lack of access to modern medicine, depended essentially on plants for their primary healthcare (Calixto, 2005).

However the knowledge of medicinal plant is rapidly dwindling due to the influence of western life styles, reduction in number of traditional healers and the lack of interest of younger generations to carry on the tradition (Bussman *et al.,* 2006; Muthu *et al.,* 2006).

Kerala the Southernmost state of India is known for its biodiversity reserve, high cultural heterogeneity and high rate of literacy. There are five major ethnic groups namely Kurichia, Kuruma, Kattunaika, Adiyan and Paniya groups scattered along the western Ghats forest (Silja *et al.,* 2008). These tribes use different medicinal plants for their primary healthcare and other purposes on the basis of their traditional knowledge. Present study was devised to document the traditional knowledge of medicinal plants used by the tribes in Mananthavady forests for sustainable utilization and development.

Materials and Methods

The study area Mananthavady taluk is located in Wayanad district of Kerala (Figure 9.1). It lies between North latitude 11° 45´ and 11° 58´ and East longitude 75° 50´ and 76° 5´. It lies at an altitude of 700-2100 M above sea level. The mean maximum and minimum temperature were 29° C and 18° C. Periodic field trips of ethno botanical exploration was undertaken between 2007-2009 in Mananthavady forests of Wayanad district. During the surveys personal interviews were conducted with the herbal medicine practitioners, traditional healers, elderly tribal people and village dwellers.

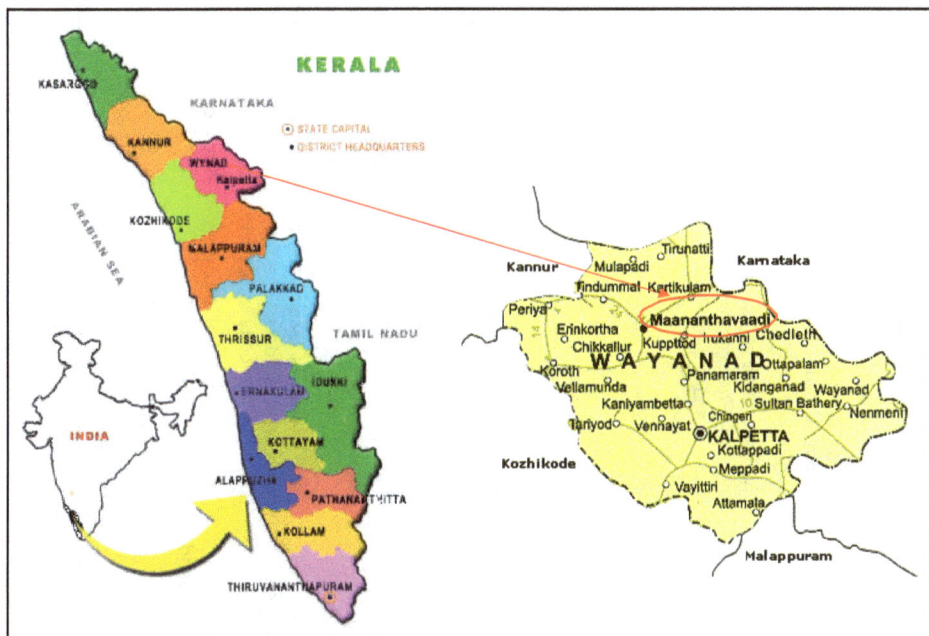

Figure 9.1: Map showing the study area Mananthavady Taluk.

The plant specimens were collected and identified using Local flora (Sasidharan, 2004), available field keys and with the help of taxonomists at M. S. Swami Nathan Research Foundation, Kalpetta and Calicut University, Kerala. Data on the plant species, local names, and parts used, disease treated, and mode of preparation and administration of medicine were recorded. The collected data were processed. A list of Threatened medicinal plants and its present status in the region with their vernacular names, parts used and therapeutical applications is shown in Table 9.1.

Results and Discussion

The present study documented 295 medicinal plant species belonging to 93 families used by the Kurichia, Kuruma, Adiyan and Paniya tribes residing at different forests of Mananthavady taluk. Among these 39 species are used for the treatment of skin problems, followed by 33 species for abdominal problems, 20 species for arthritis, 19 species for fever, cough and cold, 12 species for head ache, diabetes and piles,

Table 9.1: List of threatened medicinal plants of Mananthavady.

Botanical Name	Local Name	Family	Parts Used	Therapeutic Use	Tribes Associated	Status
Acorus calamus L.	Vayambu	Araceae	Whole plant	The plant juice is administered orally to treat abdominal pain and diarrhoea. Tuber paste is taken orally to control blood circulation.	All	Endangered
Actinodaphne malabarica Balakr.	Vatt Kambilyvirinji	Lauraceae	Bark	The bark paste of this plant with the bark paste of *Pterospermum suberifolium*,) mixed together and applied topically to bruised part of the body.	Kurichia	Rare
Amorphophallus paeoniifolius (Dennst) Nicols *var. paeoniifolius*	Kattuchena	Araceae	Corm.	Edible. Medicine for piles.	All	Vulnerable
Aristolochia tagala Cham.	Garudakkody	Aristolochia-ceae	Whole plant	The plant paste taken internally to cure abdominal pain	All	Rare
Artocarpus hirsutus Lam.	Anjili	Moraceae	Leaf	Burn the leaves of *Artocarpus hirsutus*. The ash is taken internally to treat abdominal problems.	Paniya Kurichia Adiyan Kattunaika	Vulnerable
Baliospermum montanum (Willd.) Muell.-Arg.	Nagadenthi	Euphorbia-ceae	Root Seed Leaf	Root paste is applied externally on the painfull swelling of piles.	Kurichia Kuruma Kattunaika	Vulnerable
Calophyllum apetalum Willd.	Cherupunna	Clusiaceae	Seed Flower	Seed powder is taken internally to cure menstrual disorders. The flower paste is applied on the body to get releaf from itching.	Kuruma	Vulnerable
Celastrus paniculatus Willd.	Palulavam	Celastera-ceae	Bark Leaf , Seed	Brain tonic. Seed oil promotes intelligence.	Kurichia	Vulnerable
Chonemorpha fragrans (Moon) Alston	Perumkurumba	Apocyna-ceae	Root	The rubbed root paste is applied on the skin cures skin diseases. The rubbed root paste is taken internally to purify blood.	Kurichia	Vulnerable

Contd...

Table 9.1–Contd...

Botanical Name	Local Name	Family	Parts Used	Therapeutic Use	Tribes Associated	Status
Cinnamomum macro-carpum Hook.f.	Karuva	Lauraceae	Bark Root Leaf	The oil extracted from the root, bark and leaf is used to prepare massaging oil for rheumatism. The powdered bark is taken internally with honey to treat cough.	All	Vulnerable
Cinnamomum sulphuratum Nees	Kaattu karuva	Lauraceae	Bark Leaf	The leaf and bark paste is taken internally to treat cough and head ache	All	Vulnerable
Cinnamomum zeylanicum Blume	Karuva	Lauraceae	Bark	The bark powder mixed with honey is taken internally to treat cough and dysentery	All	Vulnerable
Coscinium fenestratum (Gaertn.) Colebr.	Maramanjal	Menispermia-ceae	Underground part Stem	By applying the grinded underground part cures skin diseases and reduces leucoderma. The grinded stem is used to treat bleeding piles.	Kurichia, Kuruma	Critically endagered
Curcuma zedoaria (Christm.) Rosc.	Kaattu manjal	Zingibera-ceae	Tuber	Grinded tuber is a good medicine for psoriasis	All	Vulnerable
Cycas circinalis L.	Eanth	Cycadaceae	Bark Seed	Eating the seed powder with ghee increases sperm production	Kurichia	Vulnerable
Dipterocarpus bourdillonii Brandis	Karanjili Kalpayin	Dipterocar-paceae	Gum	The smoke coming from the burned resin. cures swelling due to rheumatism on the legs	Kurichia	Endangered
Dysoxylum mala-baricum Bedd.ex Hiern	Akil, Vellakil	Meliaceae	Bark	Anticancerous. Medicine for rheumatism.	Kuruma	Vulnerable
Aegle marmelos (L.) Correa.	Koovalam	Rutaceae	Root, Bark, Leaf, Fruit	Medicine for dysentery, fever, vomiting, body pain, piles and jaundice. Promote digestion	All	Rare
Embelia ribes Burm.f.	Vizhal	Myrsinaceae	Leaf	Grinded leaf is applied on the skin to cure skin problems	Kurichia	Rare
Embelia tsjeriam-cottam (Roem. and Schult.) DC.	Kattu vizhal	Myrsinaceae	Root, Leaf	Grinded root juice is a component of medicine for diarrhoea and pneumonia	Kurichia	Vulnerable

Contd...

Table 9.1–*Contd...*

Botanical Name	Local Name	Family	Parts Used	Therapeutic Use	Tribes Associated	Status
Garcinia gummi gutta (L.) Robs.*var gummi-gutta*	Kudampuli	Clusiaceae	Seed Leaf Fruit	The seed is fermented and the solution mixed with salt and garlic and apply 2 drops inside the throat. And the remaining smears outside the throat. It is used to cure tonsilites, ulcer-bleeding piles, dysentery, and diarrhoea.	Kurichya	Rare
Gloriosa superba L.	Menthonni	Amarillida-ceae	Tuberous root	Bleeding piles, White discharge, skin diseases	Kurichia	Rare
Goniothalamus wynaadensis (Bedd.) Bedd.	Anappanal	Annonaceae	Bark	Medicinal	Kurichia	Rare
Hedychium coronarium Koening	Kalyana-souganthikam	Zingibera-ceae	Rhizome	The rhizome paste is applied on the bruised part	Kuruma Kurichia Adiyan	Rare
Holigarna beddomei Hook.f.	Malancheru	Anacardia-ceae	Bark	Dried bark is used to cure dysentery	Kurichia	Vulnerable
Hopea glabra Wight and Arn.	Irumbakam	Dipterocar-paceae	Leaf	Antiviral	Kuruma	Vulnerable
Humboltia brunonis Wall.	Malayasokam	Caesalpinia-ceae	Leaf Bark	Medicinal	Kurichia Kattunaika	Vulnerable
Hydnocarpus macro-carpa (Bedd.)	Malamarotty	Flacourtia-ceae	Seed	The oil extracted from the seeds of this plant is used to treat skin diseases	Kurichia	Endangered
Madhuka nerifolia	Aattu Elippa	Sapotaceae	Honey	Medicinal	Kurichiakuruma Kattunaika	Vulnerable

Contd...

Table 9.1–*Contd...*

Botanical Name	Local Name	Family	Parts Used	Therapeutic Use	Tribes Associated	Status
Nothapodytes nimmo-niana (Graham) Mabb.	Ulukkuvetty	Icacinaceae	Leaf	Cut the leaves of Ulukkuvetty in to small pieces and fry it in a pan and take one egg and fry it in another pan and both mix together and fry again and inhale the smoke coming from it with eyes opened. It cures head ache after delivery. The leaves of ulukkuvetty and pachamanjal grind together and apply on the affected part and massage properly. It reduces swells	Kurichya	Vulnerable
Pseudarthria viscida (L.) Wight and Arn.	Moovila	Fabaceae	Root	Root juice is taken in to treat fever, diarrhea, asthma, worms and piles	Kurichia, Kattunaica, Kuruma	Vulnerable
Raphidophora pertusa (Roxb.) Schott.	Anachakkara	Araceae	Flower Leaf	Mumps and tonsilitis	Kurichia Kattunaika	Rare
Rauvolfia serpentina (L.) Benth. ex Kurz	Sarpagandhi	Apocyna-ceae	Leaves	A leaf of sarpagandhi and nagam is a good medicine for snake poisons	Kurichya Kuruma	Endangered
Saraca asoka (Roxb.) deWilde.	Ashokam	Caesalpinia-ceae	Bark	Used in the treatment of menorrhagia, uterine infection, haemorrhagic dysentery, internal bleeding	All	Vulnerable
Smilax zeylanica L.	Kariyilanchi	Smilacaceae	Root	The root juice is used to cure rheumatism, skin troubles and blood dysentery	Kurichia Kuruma Kattunaika	Vulnerable
Solenocarpus indicus Wight and Arn. tree	Kattanbazham	Anacardia-ceae	Fruit Bark	Medicinal (did not get more information related to medicine preparation from the tribes)	All	Rare
Symplocos cochinchi-nensis (Loureiro) S. Moore	Paachotti	Symploca-ceae	Leaf	The leaf paste is applied on the forehead to cure head ache. A mixture of coconut oil and the juice of pachotti leaves boiled and prepare the massaging oil applied on the affected part to cure Leucorrhea and Psoriasis	Kurichia	Vulnerable

Contd...

Table 9.1–*Contd...*

Botanical Name	Local Name	Family	Parts Used	Therapeutic Use	Tribes Associated	Status
Symplocos racemosa Roxb.	Valiya pachotty	Symploca-ceae	Bark	Leucorrhea, Liver diseases	All	Vulnerable
Tinospora sinensis (Lour.) Merr.	Kaattamruthu	Menisper-maceae	Stem	The stem juice is taken internally to treat piles and ulcer	Kurichia	Rare

9 species for circulatory disorders and uterine problems, 8 species for snake bite and asthma, 7 species for worm infection and poison, 6 species for urinary problems and ulser, 4 species for veterinary problems, anticancerous, jaundice, wound, bruise and sprain and 3 species for toothache (Figure 9.2). The most prominent disease treated by the tribes in the study area were skin problems like psoriasis, itching and boils. A previous study on Mullukuruma tribes in Noolpuzha village panchayath of Wayanad district by Silja *et al.* (2008) documented 14 plants for the treatment of skin diseases. In the present study plant species reported for the treatment of skin diseases in Mananthavady taluk shows a frequent usage of different plant parts like the leaves and roots of *Bauhinia tomentosa*, leaves of *Cassia tora*, roots, leaves and seeds of *Clitorea ternata*, whole plant of *Cynodon dactylon*,roots of *Gloriosa superba*, leaves of *Indigofera tinctoria*, leaves of *Argyreia nervosa*, leaves of *Asclepias curassavica*, flower paste of *Calophyllum apetalum*, whole plant of *Chassalia curviflora*, bark of *Cinnamomum camphora*, tuber of *Curcuma zedoaria*, Leaves of *Embelia ribes*, whole plant of *Emilia sonchifolia*, whole plant of *Euphorbia thymifolia*, whole plant of *Hedyotis brachypoda*, seeds of *Hydnocarpus macrocarpa*, tuber of *Ipomoea marginata*, Leaves of *Leucas species* etc.

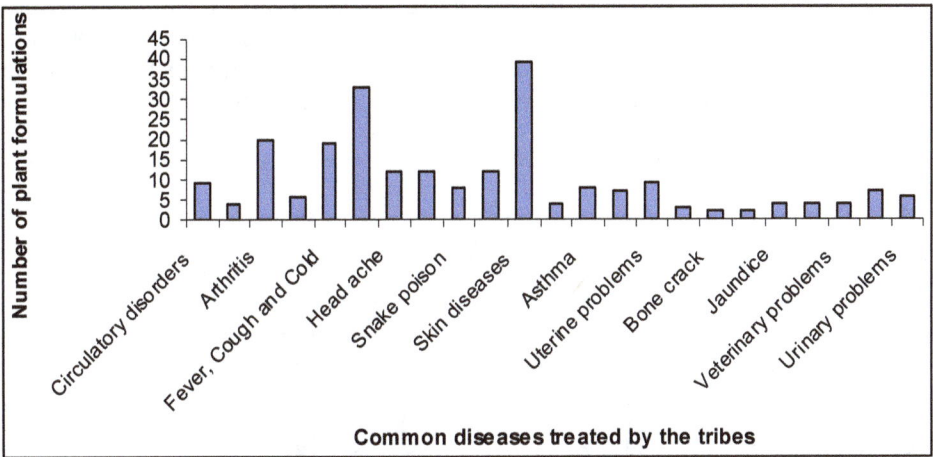

Figure 9.2: The diseases treated with the number of plant formulations by the traditional healers Mananthavady taluk.

The Kurichia tribes are using the leaf juice of *Atalantia racemosa* Wight *var.racemosa* internally to treat acidity. The present study reports that, Kurichia and Adiya tribes use leaf, root and latex of *Tabernaemontanum divaricata* for the treatment of intestinal worms. The most common medicinal use of *Tabernaemontanum divaricata* involves its antimicrobial action against infectious diseases such as syphilis, leprosy and gonorrhoea as well as its antiparasitic action against worms, dysentery, diarrhoea and malaria (Van *et al.,* 1984). The Kurichia use the sun dried bark powder of *Oroxylum indium* to cure gastric ulcer. The Kurichia tribes use the fresh leaves of *Lantana camera* to make the massaging oil for the treatment of psoriasis. The leaf oil of this plant is used in the treatment of itches of skin (Anon, 2003). The leaves and bark of *Pittosporum neelgherrense* are used as antidotes for snake bite by the Kurichia tribes.

Similar type of treatment for snake bite has been practiced by Malayar tribes (Saradamma *et al.,* 1990).

Even though all plant parts are valuable, the preparation of a formulation requires specific parts of plants. The tribes are using various plant parts such as leaves (110 species), barks (58 species), whole plants (56 species), roots (35 species), fruits (37 species), seeds (26 species) and underground parts (25 species) of different species. The paper provides knowledge related to edible parts of 35 species of plants, amongst which a few are of medicinal value. This includes *Diospyros malabarica* (Fruit), *Elaeagnus conferta* (Fruit), *Elaeocarpus tectorius* (Fruit), *Moringa oleifera* (Leaf), *Spondias pinnata* (Fruit), *Syzygium cumini* (Fruit), *Syzygium laetum* (Fruit), *Syzygium tamilnadensis* (Fruit) etc.

The data obtained from our results showed that about 149 plants have external use and 81 plants are internally applied. We observed that most of the tribal medicines in this region are being prepared from the members of Fabaceae (17 species) followed by Euphorbiaceae, Asteraceae and Apocynaceae (12species), Rutaceae and Caesalpiniaceae (11species) (Figure 9.3).

Our studies also revealed that about 2 per cent of the existing species are endangered, 9 per cent are vulnerable, 4 per cent are rare and 85 per cent are abundant (Figure 9.4). The exploitation of these species for medicinal purposes has picked up without much attention for their conservation. The present documentation about the

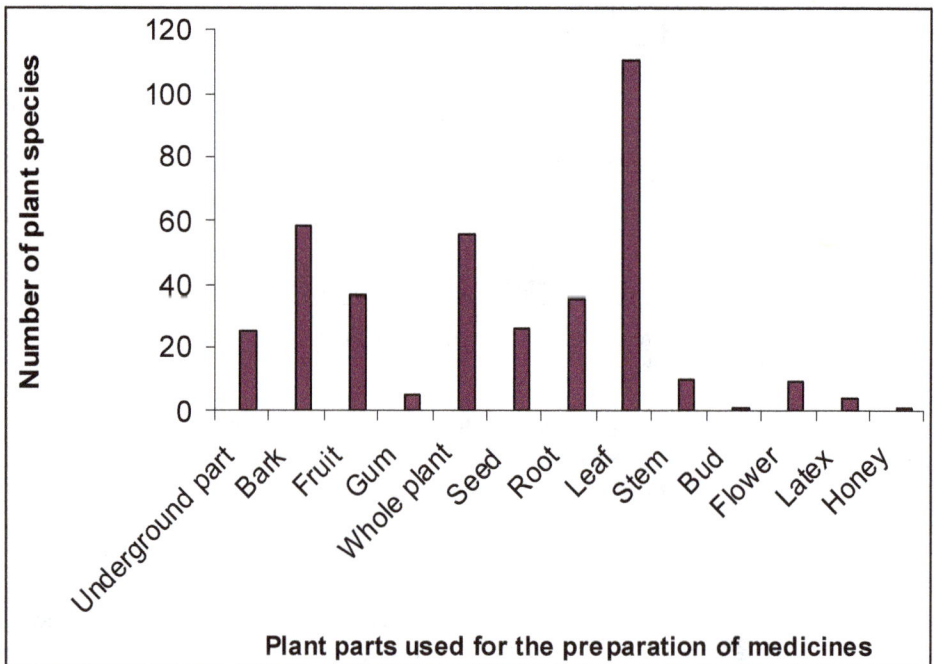

Figure 9.3: Plant parts used from different plant species for the preparation of medicines, Mananthavady taluk

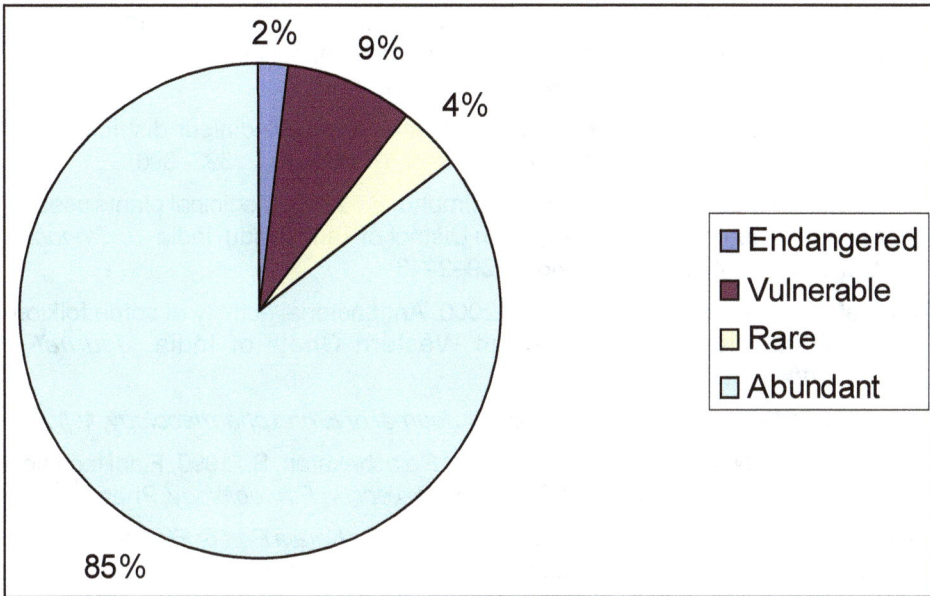

Figure 9.4: The number of plants which are coming under Endangered, Vulnerable, Rare and Abundant category. Its percentage.

traditional knowledge of medicinal plants and their status in Mananthavady of Wayanad district, therefore, serves as a valuable information tool for future sustainable use and conservation.

Acknowledgements

We are very grateful to all those tribal medical practitioners of Mananthavady taluk who were willing to share their folk knowledge of plants. Our gratitude is also addressed to forest division Mananthavady, taxonomists of research centers, and personalities who helped to complete our work.

References

Anon, 2003. *The Wealth of India: Raw Materials*. CSIR, New Delhi, 6: 31.

Balick, M.J. and Cox, P.A., 1996. *Plants, People and Culture: The Science of Ethnobotany*. Scientific American Library, New York.

Bussmann, R.W., Gilbreath, G.G., Solio, J., Lutura, M., Lutuluo, R., Kunguru, K., Wood, N. and Mathenge, S.G., 2006, Plants use of the Maasai of Sekenane Valley, Maasai Maru, Kenya. *J. Ethnobio. Ethnomed.*, 1186/1746–4269–2–22.

Calixto, J.B., 2005. Twenty five years of research on medicinal plants in Latin America: A personal review. *J. Ethnopharmacol.*, 100: 131–134.

Fabricant, D.S. and Farnsworth, N.L., 2001. The value of plants used in traditional medicine for drug discovery. *Environmental Health Perspectives (Supplement)*, 109: 69–75.

Farnsworth, N., 1994. Ethno-pharmacology and drug development. In: *Bioactive Compounds from the Plants* (Cifa Foundation Symposium No. 185), (Eds.) D.J. Chadwick and J. Marsh. Wiley, Chichecter, p. 42–51.

Ghosh, A., 2003. Herbal folk remedies of Bankura and Medinipur districts, West Bengal (India). *Indian Journal of Traditional Knowledge,* 2: 393–396.

Muthu, C., Ayyanar, M., Raja, N. and Igacimuhu, S., 2006. Medicinal plants used by traditional healers in Kancheepuram District of Tamil Nadu, India. *J. Ethnoboil. Ethnomed.,* 2: 43doi:1186/1746–4269–2–43.

Perumal Sámi, R. and Igacimuthu, S., 2000. Antibacterial activity of some folklore medicinal plants used tribals in Western Ghats of India. *Journal of Ethnopharmacology,* 69: 63–71.

Rivier, L. and Bruhn, J.G., 1979. Editorial. *Journal of Ethno pharmacology,* 1: 1.

Saradamma, L., Nair, C.P.R., Bhat, A.V. and Rajashekaran, S., 1990. Final technical report. *All India Co-ordinated Research Project on Ethnobiology,* Phase 1: 128.

Sasidharan, N., 2004. *Biodiversity Documentation for Kerala Part 6: Flowering Plants.* Kerala Forest Research Institute, Kerala, India, pp. 702.

Silja, V.P., Samitha, V.K. and Mohanan, K.V., 2008. Ethnomedicinal plant knowledge of the Mullukuruma tribe of Wayanad District, Kerala. *Indian Journal of Traditional Knowledge,* 7(4): 604–612.

Van Beek, T.A., Verpoorte, R., Svendsen, A.B., Leeuwenberg, A. J. and Bisset, N.Gr., 1984. *Tabernaemontana* L. (Apocynaceae): A review of its taxonomy.

World Health Organization, 2002. *WHO Traditional Medicine Strategy 2002–2005.* http:www.who.int/(2002).

2013, Perspectives in Plant Biodiversity   *Pages* **59–66**
Editor: **Dr. K. Muthuchelian,** *Vice Chancellor, Periyar University, Salem*
Published by: **Daya Publishing House, NEW DELHI**

Chapter 10

Medicinal Plant Diversity in Foot Hills of Karandamalai of Southern Eastern Ghats of Tamil Nadu

N. Yasoth Kumar[1]* and K. Rajendran[2]
[1]*Department of Botany, Yadava College, Madurai – 625 014*
[2]*Department of Botany, Thiagarajar College, Madurai – 625 009*

ABSTRACT

A survey was carried to collect information from traditional healers on the use of medicinal plants in Karandamalai, Southern Eastern Ghats of Tamil Nadu. A village inhabited by Valaiyan communities. The indigenous knowledge of local traditional healers and the native plants used for medicinal purposes were collected through questionnaire and personal interviews during field trips. Medicinal plants used by the Valaiyan were mostly used to cure skin diseases, poison bites, stomachache and nervous disorders. The main objective of the study was to bring the wild plants species from traditional healers. This study showed that many people in the area of Karandamalai still continue to depend on medicinal plants at least for the treatment of primary healthcare. The traditional healers are dwindling in number and there is a danger of traditional knowledge disappearing soon since the younger generation is not interested to carry on this tradition.

* Corresponding Author: E-mail: yasoth2002@yahoo.co.in

Introduction

Plants have been used in traditional medicine for several thousand years. The knowledge of medicinal plants has been accumulated in the course of many centuries based on different medicinal systems such as Ayurveda, Unani and Siddha. In India, it is reported that traditional healers use 2500 plant species and 100 species of plants serve as regular sources of medicine. During the last few decades there has been an increasing interest in the study of medicinal plants and their traditional use in different parts of the world. Documenting the indigenous knowledge through ethno botanical studies is important for the conservation and utilization of biological resources.

According to the World Health Organization (WHO), as many as 80 per cent of the world's people depend on traditional medicine for their primary healthcare needs. There are considerable economic benefits in the development of indigenous medicines and in the use of medicinal plants for the treatment of various diseases. Most of these people form the poorest link in the trade of medicinal plants. In the developed countries, 25 per cent of the medical drugs are based on plants and their derivatives. Traditional medical knowledge of medicinal plants and their use by indigenous cultures are not only useful for conservation of cultural traditions and biodiversity but also for community healthcare and drug development in the present and future.

Ethnobotany is not new to India because of its rich ethnic diversity. There are over 400 different tribal and other ethnic groups in India. The tribals constitute about 7.5 per cent of India's population.

During the last few decades there has been an increasing interest in the study of medicinal plants and their traditional use in different parts of India and there are many reports on the use of plants in traditional healing by either tribal people or indigenous communities of India. The objective of this study was to interact with local traditional healers and document their knowledge on medicinal plants, their usage and the types of diseases treated etc. The present-day traditional healers are very old. Due to lack of interest among the younger generation as well as their tendency to migrate to cities for lucrative jobs, wealth of knowledge in this the area is declining. So far no systematic ethno botanical survey has been made in this area and this is the first report on the medicinal plants used by the local traditional healers. A perusal of the literature reveals that, some of the ethno medicinal works has been done in the forests of nearest districts in Tamil Nadu in the last two decades. During the course of exploration of ethno medicinal plants of the district, the information have been gathered from the healers of rural villages found near forest areas where the people depend mostly on forests for their need and have sound knowledge of herbal remedies.

Materials and Methods

The Study Area and Medicinal Plant Survey

Tamil Nadu is the 11th largest state in India with a geographical area of 130058 km^2 and lies between 11°00' to 12°00' North latitudes and 77°28' to 78°50' East longitudes. The total forest cover Tamil Nadu is 21482 km^2 (16.52 per cent). This includes 12,499 km^2 of dense forests (9.61 per cent) and 8,963 km^2 of open forests (6.91 per cent). Of the total forest area of Tamil Nadu, 3305 km^2 are under protected

Table 10.1: Medicinal plants used by the Valaiyan tribes.

Therapeutic Indication and Associated Plants	Local Name	Family	Parts Used and Ethnomedicinal Preparation
Asthma			
Solanum trilobatum L.	Thoodhuvalai	Solanaceae	Juice of leaves is taken orally for seven days.
Adhatoda zeylanica Medicus.	Adathodai	Acanthaceae	Leaf paste is taken orally.
Cold			
Adhatoda zeylanica Medicus.	Adathodai	Acanthaceae	Leaf powder is mixed with water and taken orally in the morning.
Plectranthus coleoides Benth.	Omavalli chedi	Lamiaceae	Juice of leaves is taken internally.
Solanum trilobatum L.	Thoodhuvalai	Solanaceae	Juice of leaves is taken orally for seven days early morning until cure.
Cough			
Adhatoda zeylanica Medicus.	Adathodai	Acanthaceae	Leaf powder is mixed with water and taken orally in the morning.
Terminalia chebula Retz.	Kadukkai maram	Combretaceae	Powdered fruit is mixed with the water or cow's or goat's milk and taken internally.
Vitex negundo L.	Notchi	Verbenaceae	Fresh leaves are boiled with water and the vapour is inhaled twice a day.
Diabetes			
Andrographis lineata Wallich ex Nees.	Siriyanangai	Acanthaceae	Leaf powder is mixed with cow's or goat's milk and taken orally.
Gymnema sylvestre (Retz). R. Br. ex Roem. and Schult.	Sirukurinjan	Asclepiadaceae	Powdered leaves are mixed with cow's milk and boiled rice, kept over night and taken internally twice a day.
Dysentery			
Acalypha fruticosa Forsskal.	Chinni chedi	Euphorbiaceae	Decoction of leaves taken orally.
Fever			
Hemidesmus indicus H.f.	Nannari	Asclepiadaceae	Decoction of whole plant is taken internally.
Vitex negundo L.	Notchi	Verbenaceae	Fresh leaves are boiled with water and the vapour is haled twice a day.

Contd...

Table 10.1–Contd...

Therapeutic Indication and Associated Plants	Local Name	Family	Parts Used and Ethnomedicinal Preparation
Headache			
Ceropegia candelabrum L.	Perun kodi	Asclepiadaceae	Paste of leaves is applied on forehead.
Pergularia daemia (Fors). Chiov.	Veli parutthi	Asclepiadaceae	Fresh leaves are boiled with water and the vapour is inhaled.
Jaundice			
Centella asiatica (L). Urban.	Vallarai	Umbelliferae	Juice of leaf is mixed with equal amount of goat's milk and taken orally for seven days.
Nervous disorders			
Blepharis maderaspatensis (L). Roth.	Vettukaaya pachilai	Acanthaceae	Leaf paste is mixed with the powdered black gram, crushed onion and white yolk of one egg and the mixture is applied topically over the fractured bones.
Gymnema sylvestre (Retz). R. Br. ex Roem. and Schult.	Sirukurinjan	Asclepiadaceae	Paste of leaves is applied externally.
Phlebophyllus kunthianum Nees.	Kurinji chedi	Acanthaceae	Fresh leaves and bark are heated with gingelly oil and applied externally on affected part of the body.
Pimples			
Acalypha paniculata Miq.	Paruva thazhai	Euphorbiaceae	Leaf paste is applied over pimples regularly once a day until cure.
Poison bites			
Andrographis lineata Wallich ex Nees.	Siriyanangai	Acanthaceae	Paste of leaves is applied externally on bitten site of scorpion and snake.
Andrographis paniculata (Burm.f). Wall. ex Nees.	Periyanangai or Nilavembu	Acanthaceae	Paste of leaves is applied externally on bitten site of scorpion sting and snakebites.
Tylophora indica (Burm. f). Merr.	Nangilai	Asclepiadaceae	Paste of leaf and root is mixed with equal amount of root paste of *Rauvolfia serpentina* and applied externally on the spot of snakebite. Leaf juice alone is also taken internally to cure snakebite.

Contd...

Table 10.1–Contd...

Therapeutic Indication and Associated Plants	Local Name	Family	Parts Used and Ethnomedicinal Preparation
Skin diseases			
Acalypha fruticosa Forsskal.	Chinni chedi	Euphorbiaceae	Leaf and root paste is applied topically on the affected places.
Anisochilus carnosus (L.f). Wallich.	Saetthupun thazhai	Lamiaceae	Paste of leaves is applied over the affected places.
Stomachache			
Acalypha paniculata Miq.	Paruva thazhai	Euphorbiaceae	Juice of leaves is taken orally.
Solanum nigrum L.	Mana thakkali	Solanaceae	Fresh leaves are cooked with onion bulbs and cumin seeds and taken along with food.
Terminalia chebula Retz.	Kadukkai maram	Combretaceae	Powdered fruit is mixed with water or cow's or goat's milk and taken internally.
Throat infection			
Acorus calamus L.	Vasambu	Araceae	Dried rhizome is rubbed on stone with water and one or two drops of watery paste are given orally to the children for clarity of speech. Increased dosage will affect speech.
Piper nigrum L.	Milagu	Piperaceae	The dried seeds are taken orally.
Toothache			
Solanum erianthum D.Don	Malai sundai	Solanaceae	The ripened or unripened fruits are boiled with water and the vapour is inhaled once or twice a week through mouth.
Toddalia asiatica (L). Lam.	Kindu mullu	Rutaceae	Powder of root and stem bark is used as tooth powder.
Wounds			
Acacia caesia (L). Willd.	Nanjupattai	Mimosaceae	Bark is ground with water and applied topically over the affected part.
Acacia leucophloea (Roxb). Willd.	Sarayapattai maram	Mimosaceae	Paste of fresh bark is applied topically on cuttings until cure.
Anisomeles malabarica (L). R. Br. Ex. Sims.	Paei miratti	Lamiaceae	Paste of stem is mixed with coconut oil and applied over the affected places.
Blepharis maderaspatensis (L). Roth.	Vettukaaya pachilai	Acanthaceae	Paste of leaves is mixed with lime juice and applied on cuts.
Clausena dentata (Willd). Roem.	Anai thazhai	Rutaceae	Paste of leaves is applied over the affected parts.
Cryptolepis buchananii Roem and Schul.	Paalkodi/Karunkodi	Asclepiadaceae	Stem latex (5–10 drops) is applied on the affected places.

area (15 per cent) which includes, 8 Wildlife sanctuaries, 12 Bird sanctuaries, 5 National parks, 3 Biosphere reserves and one Tiger reserve. 'Valaya' or 'Valaiyans' are one of the oldest groups inhabiting Southern Tamil Nadu. Valaiyans are skilled hunters and forest product gatherers. Their name is believed to be derived from the word 'valai' (or net), since this implement is constantly employment by them in the capturing of jungle game.

Karandamalai is situated 43 Km from the District of Madurai. It is adjoined by Ariyalur hills in the west and towards the northwest and north east it is surrounded by the Sirumalai and Perumalai hills. It lies between 10° 15 to 10° 21 north latitude and 78° 9 to 78° 15 east longitude. The altitude from foot hills to the highest Jandamedu range from 180 to 916 M. Forest types range from tropical thorn forests to mixed deciduous forests and moist deciduous forest.

Local traditional healers having practical knowledge of plants in medicine were interviewed. They were requested to collect specimens of the plants they knew or to show the plant species on site. These informants were traditional healers themselves or their families and had knowledge of the medicinal use of the plants. Fuel wood from the surroundings was the main energy source for cooking and eating. The wealth of medicinal plant knowledge among the people of this district is based on hundreds of years of beliefs and observations. This knowledge has been transmitted orally from generation to generation; however it seems that it is vanishing from the modern society since younger people are not interested to carry on this tradition.

Interview with Traditional Healers

Adopting the survey methods, ethno medicinal data were collected through general conversations with the informants. The questionnaires were used to obtain information on medicinal plants with their local names, parts used, mode of preparation and administration. A total of 10 informants, comprising 8 males and 2 females were identified between the ages of 48 and 74. They were selected based on their knowledge of medicinal plants either for self-medication or for treating others. Informants were asked to come to field and show the plants with local name; the species mentioned by the informants were taxonomically identified.

Conclusion

The survey indicated that, the study area has plenty of medicinal plants to treat a wide spectrum of human ailments. Earlier studies on traditional medicinal plants also revealed that the economically backward local and tribal people of Tamil Nadu prefer folk medicine due to low cost and sometimes it is a part of their social life and culture. It is evident from the interviews conducted in different villages; knowledge of medicinal plants is limited to traditional healers, herbalists and elderly persons who are living in rural areas. This study also points out that certain species of medicinal plants are being exploited by the local residents who are unaware of the importance of medicinal plants in the ecosystem.

This study concluded that even though the accessibility of Western medicine for simple and complicated diseases is available, many people in the studied area is still continue to depend on medicinal plants, at least for the treatment of some simple diseases such as, cold, cough, fever, headache, poison bites, skin diseases and tooth infections. Well-knowledged healers have good interactions with patients and this would improve the quality of healthcare delivery. The present-day traditional healers are very old. Due to lack of interest among the younger generation as well as their tendency to migrate to cities for lucrative jobs, there is a possibility of losing this wealth of knowledge in the near future. It thus becomes necessary to acquire and preserve this traditional system of medicine by proper documentation and identification of specimens.

References

Abu-Rabia, A., 2005. Urinary diseases and ethno botany among pastoral nomads in the Middle East. [http://www.ethnobiomed.com/content/1/14]. *Journal of Ethnobiology and Ethnomedicine,* 1: 4.

Ahmad, I. and Beg, A.Z., 2001. Antimicrobial and phytochemical studies on 45 Indian medicinal plants against multi-drug resistant human pathogens. *Journal of Ethnopharmacology,* 74: 113–123.

Annamalai, R., 2004. Tamil Nadu biodiversity strategy and action plan: Forest biodiversity. *Tamil Nadu Forest Department, Government of India, Chennai.*

Bhattarai, N.K., 1990. Herbal folk medicines of Kabhrepalanchok District, Central Nepal. *International Journal of Crude Drug Research,* 28(3): 225–231.

Chitra, V., 1983. *Flora of Tamil Nadu, India Ser. 1: Analysis,* (Eds.) N.C. Nair and A.N. Henry. Botanical Survey of India, Coimbatore, India.

Eddouks, M., Maghrani, M., Lemhadri, A., Ouahidi, M.L. and Jouad, H., 2002. Ethnopharmacological survey of medicinal plants used for the treatment of diabetes mellitus, hypertension and cardiac diseases in the south-east region of Morocco (Tafilalet). *J. Ethnopharmacol.,* 81(1): 81–100.

Ganesan, S., Suresh, N. and Kesaven, L., 2004. Ethnomedicinal survey of lower Palani Hills of Tamil Nadu. *I. J. Trad. Knowledge,* 3(3): 299–304.

Ganesan, S., Ramar Pand, N. and Banumathi, N., 2007. Ethnomedicinal survey of Alagarkoil hills (Reserved forest), Tamil Nadu, India. *Electronic Journal of Indian Medicine,* 1: 1–19.

Jain, S.K., 1964. The role of botanist in folklore research. *Folklore,* 5: 145–150.

Kala, C.P., 2002a. Indigenous knowledge of Bhotia tribal community on wool dying and its present status in the Garhwal Himalaya, India. *Curent Science,* 83: 814–817.

Kottaimuthu, R., 2007. Systematic studies on the Dicotyledonous flora of Karandamalai, Dindigul District, Tamil Nadu.

Maheshwari, J.K., Singh, K.K. and Saha, S., 1986. *Ethnobotany of Tribals of Mirzapur District, Uttar pradesh.* Economic Botany Information Service, NBRI, Lucknow.

Pullaiah, T., Sriramamurthy, K. and Karuppusamy, S., 2007. *Flora of Eastern Ghats, Hill Ranges of South East India, Vol. 3.* Regency Publication, New Delhi.

Ramachandran, V.S. and Nair, N.C., 1981. Ethnobotanical observations on Irulars of Tamil Nadu, India. *Journal of Economic and Taxonomic Botany*, 2: 183–190.

Samy, R.P. and Ignacimuthu, S., 2000. Antibacterial activity of some folklore medicinal plants used by tribals in Western Ghats of India. *Journal of Ethnopharmacology*, 69: 63–71.

Santhya, B., Thomas, S., Isabel, W. and Shenbagarathai, R., 2006. Ehnomedicinal plants used by the Valaiyan community of Piranmalai hills (Reserved forest), Tamil Nadu, India: A pilot study. *Afr. J. Trad. CAM,* 3(1): 101–114.

Saravanan, 1998. Impact of welfare programmes on the tribals in Tamil Nadu. *Review of Development and Change.* 3(1): Jan–June.

Thurston, E. and Rangachari, K., 1909. *Castes and Tribes of Southern India.* The Superintendent, Government Press, Madras, 1: 272–280.

2013, Perspectives in Plant Biodiversity   Pages **67–76**
Editor: **Dr. K. Muthuchelian,** *Vice Chancellor, Periyar University, Salem*
Published by: **Daya Publishing House, NEW DELHI**

Chapter 11

Biodiversity Assessment in Tropical Dry Evergreen Forest (TDEF) at Point Calimere Wildlife Sanctuary, Tamil Nadu: A Threatened Ecosystem

A. Balasubramanain, S. Radhakrishnan, R. Revathi, V. Thirunavukarasu*, R. Rajasekar and M. Govinda Rao

*Forest College and Research Institute,
Tamil Nadu Agricultural University, Mettupalayam – 641 301*

ABSTRACT

A two year research project funded by Tamil Nadu Forest Department was carried out to assess the floral diversity status of Tropical Dry Evergreen Forest (TDEF) of Point Calimere Wildlife Sanctuary, Tamil Nadu, the single largest patch of its kind in the entire country. This forests harbours unique floral composition encompassing invaluable medicinal plant wealth and is being continuously exploited since the Chola Dynasty (900 A.D). The floral diversity of this forest was assessed using Systematic random sampling technique with sample plots. The adequacy and size of sample plots were decided using species-effort –curve-technique. The result of the vegetation study revealed that highest dominance in trees was observed in *Manilkara hexandra (0.000837)*,

* Corresponding Author: E-mail: balayzz@yahoo.com

followed by *Hemicyclia sepiaria (0.000090)*, and *Maba buxifolia (0.000084)*. The highest density (11.3) and relative density (25.37) were observed in *Maba buxifolia*, where as the frequency (1.048) and relative frequency (14.379) were highest in *Manilkara hexandra*. Among the shrubs, *Atlantia monophylla, Glycosmis cochinchinensis, Memecylon umbellatum* were dominant with similar frequency and a relative frequency of 0.524 and 15.278 respectively. In herbal layer the density varied from 15.95 (Thattai pillu) to 0.048 (*Cassia auriculata*) with a relative density of 26.13 to 0.078. The high diversity value of 6.2855 for herbs, 8.8066 for shrubs and 0.7102 for trees were recorded in Margalef's index of species richness. In Simpson's index of species diversity, herbs, shrubs and trees of TDEF derived the values of 3.36576, 2.2139 and 0.7271 respectively. Pielouis evenness index for the measurement of partition of individuals of population among species elucidated the values of 4.5268 (herbs) 8.0825 (shrubs) and 0.7179 (trees).The vegetation analysis and diversity indices revealed the high species richness, high species diversity and the population is distributed evenly in Tropical Dry Evergreen Forests of Point Calimere Wildlife Sanctuary, Tamil Nadu.

Keywords: *Wildlife warden, Point Calimere Wildlife Sanctuary, Nagapattinam.*

Introduction

Forests are one of the most biologically rich terrestrial ecosystems. Together, tropical, temperate and boreal forests offer diverse sets of habitats for plants, animals and micro-organisms, and harbour the vast majority of the world's terrestrial species.. Furthermore, forest biodiversity is interlinked to a web of other socio-economic factors, providing an array of goods and services that range from timber and non-timber forest resources to mitigating climate change and genetic resources. At the same time, forests provide livelihoods for people worldwide and play important economic, social, and cultural roles in the lives of many indigenous communities.

In the tropics, changes in the quantity and distribution of rainfall along with temperature and the length of the dry season gradually alter the vegetation formation. The pronounced seasonality in rainfall distribution with several months of drought result in seasonally dry forests in tropical regions., these dry forests in the varying climatic regimes differ in their forest structure and physiognomy. The prevailing tropical dissymmetric climate regime on the Coromandel coast of southern peninsular India supports a unique type of vegetation named tropical dry evergreen forest (TDEF).

Point Calimere Wild life and Birds sanctuary is located on the Coromandel Coast in Nagapattinam district of Tamil Nadu bounded by Bay of Bengal in the East and Palk Straits in the south. Forest area is extending over an area of approx. 333 Sq. Kms. which includes 23 Sq. Kms of Tropical Dry Evergreen Forest (TDEF) on the eastern extremity. This Tropical Dry Evergreen Forest (TDEF) is the indigenous forest of the coastal seaboard of South East India. Historically the forest extended from Vishakapatanam to Ramanathapuram as a belt of vegetation between 30 and 50 km wide, bordered on one side by the sea and on the other side by a forest that becomes increasingly deciduous as one moves inland. Biodiversity assessment was carried out in this forest to assess the vegetation and floral diversity.

Methodology

The floral diversity assessment of tropical dry evergreen forest was taken up using sample plot techniques. To asses the floral diversity the following methodologies were followed.

Study Area

The Point Calimere Wildlife and Birds sanctuary is located adjacent to and east of Kodaikarai and Kodaikadu villages, is basically an Island surrounded by the Bay of Bengal to the east, the Palk Straight to the south and swampy backwaters and salt pans to the west and north. Coordinates are between 10.276 to 10.826 N and 79.399 to 79.884 E. Low sand dunes are located along the coast and along the western periphery with coastal plains, tidal mud-flats and shallow seasonal ponds in between. Sand dunes in the east are mostly now stabilized by *Prosopis* and the higher dunes in the west are stabilized by dense Tropical dry evergreen forests.

Perambulation of Vegetation

The entire forest area was perambulated before taking up the diversity study. The perambulation was made from west to east by having abandoned railway line in the west as a baseline. Similarly, another perambulation was made from north to south by having Ramar Patham to sevvarayan koil as a baseline in the north.

Vegetational Analysis

The vegetation analysis was carried out and sample plots were fixed based on species effort curve technique. Sample plots of dimensions 25m×25m for trees, 5m×5m for shrubs, and 1m×1m for herbs were laid. Density, frequency and dominance have been determined for the species using the following mathematical expressions.

Density

Density of number of individuals per unit area

$$\text{Density} = \frac{\text{Total no. of all individuals of all species in all quadrats}}{\text{Total no. of quadrats sampled}}$$

Relative Density

Relative density is the study of numerical strength of a species in relation to total number of species.

$$\text{Relative Density} = \frac{\text{Density of a species}}{\text{Density of all species}} \times 100$$

Frequency

Frequency is the dispersion of a particular species in a community. It denotes the distribution of a species.

Frequency = No. of quadrats in which a species occur

Relative Frequency

Relative frequency is the distribution of a particular species in relation to other species.

$$\text{Relative Frequency} = \frac{\text{No. of occurrence of a species}}{\text{No. of occurrence of all species}} \times 100$$

Dominance

Dominance refers to vigour of a species in a community

$$\text{Dominance} = \frac{\text{Total basal area of a species}}{\text{Area sampled}} \times 100$$

Floral Diversity Analysis

The floral diversity was estimated using the following diversity indices

Margalef's Index

Margalef's index reveals the species richness in a biota. Its value usually ranges between 0.1 and 4. More the value of D more is the species richness.

D = (S-1)/ln N

Where, D = Margalef's index, S = No. of species, N = Number of individuals

Simpson's Index

Simpson's index reveals the species diversity in a plant community. Expressed as $(1/\lambda)$. More the value of $(1/\lambda)$ more is the species diversity.

Simpson's Index $(\lambda) = \Sigma$ pi2 and often it is expressed as $(1/\lambda)$

Where p_i = probability of the ith species

Pielou's Evenness

It's a measure of partition of the individuals of population among species. In an evenly distributed population J is 1. J decreases with increasing unevenness.

J = H/ln S

Where, H = $-\Sigma P_i$ ln p_i, S = Individuals of all the species

Results

Thattai pillu is the predominant herb species having the maximum number of individuals and have the highest density (15.952), relative density (26.131) (Table 11.1) *Tinospora cordifolia* (Seendhil) and *Leucas aspera* (thumbai) were the two species of herbs with high density values (7.476 and 4.571) and relative density (12.246 and 7.488), next to Thattai pillu. *Euphatorium odoratum* and *Clerodendron inerme* have the next higher density values of 3.143 and 2.857, respectively. *Glycosmis cochinchinensis* (konji) is the herb species found in the most number of study plots with a frequency and relative frequency value of 11.0 The second most distributed herb is *Memecylon umbellatum* (casan)

Table 11.1: Densities and frequencies of the herb species.

Sl. No.	Name of the Species	Density	Relative Density	Frequency	Relative Frequency
1.	Thattai pillu	15.952	26.131	11	11.702
2.	*Tinospora cordifolia* (seendhil)	7.476	12.246	8	8.510
3.	*Leucas aspera* (thumbai)	4.571	7.488	5	5.319
4.	*Euphatorium odorantum*	3.143	5.148	5	5.319
5.	*Clerodendron inerme* (peechalathi)	2.857	4.680	3	3.191
6.	Pytham pillu	2.857	4.680	3	3.191
7.	*Glycosmis cochinchinensis* (konji)	1.857	3.042	3	3.191
8.	*Memecylon umbellatum* (casan)	1.381	2.262	3	3.191
9.	*Phoenix pusilla* (eechai)	1.238	2.028	3	3.191
10.	Kattuavarai	1.000	1.638	2	2.127
11.	*Ixora parviflora* (kaippalai)	0.952	1.560	2	2.127
12.	*Leuco pyrus* (mathanga pull)	0.810	1.326	2	2.127
13.	*Hemidesmus indicus* (nannari)	0.762	1.248	2	2.127
14.	*Cyprus corymbosus* (korai)	0.714	1.170	2	2.127
15.	*Maytenes emarginata* (vettalai)	0.571	0.936	2	2.127
16.	Aanaikorattai	0.524	0.858	2	2.127
17.	Paalarugu	0.524	0.858	2	2.127
18.	*Jasminum sessiliflorum* (kaattumullai)	0.476	0.780	2	2.127
19.	*Gmelina asiatica* (kumulai)	0.429	0.702	2	2.127
20.	*Abrus precatorius* (kundumani)	0.333	0.546	2	2.127
21.	*Borreria hispida* (Nahaisoodi)	0.333	0.546	2	2.127
22.	*Ocimum sanctum* (thulasi)	0.286	0.468	2	2.127
23.	*Atlantia monophylla* (kattunarthai)	0.190	0.312	1	1.063
24.	*Manilkara hexandra* (palamaram)	0.190	0.312	1	1.063
25.	*Cyanodon dactylon* (aruhampull)	0.190	0.312	1	1.063
26.	*Clitoria ternatea* (Sangupushpam)	0.190	0.312	1	1.063
27.	*Phyllanthus niruri* (kilanelli)	0.143	0.234	1	1.063
28.	*Fluggea leucopyrus* (verpula)	0.095	0.156	1	1.063
29.	*Ageratum coryzoides* (vaelipasi)	0.095	0.156	1	1.063
30.	*Terena asiatica* (kaattupavattai)	0.095	0.156	1	1.063
31.	*Hemicyclia sepiaria* (veeramaram)	0.095	0.156	1	1.063
32.	*Achyranthes aspera* (nayuruvi)	0.095	0.156	1	1.063
33.	Nilavarai	0.095	0.156	1	1.063
34.	*Maba buxifolia* (karunthuvarai)	0.095	0.156	1	1.063
35.	*Cissus quadrangularis* (pirandai)	0.095	0.156	1	1.063
36.	*Phyllanthus amarus*	0.095	0.156	1	1.063

Contd...

Table 11.1–*Contd...*

Sl.No.	Name of the Species	Density	Relative Density	Frequency	Relative Frequency
37.	Usimullai	0.048	0.078	1	1.063
38.	*Salacia primoides* (karukkai)	0.048	0.078	1	1.063
39.	*Zizyphus oenoplia* (suurai)	0.048	0.078	1	1.063
40.	*Dichrostachys cineria* (vedatharai)	0.048	0.078	1	1.063
41.	*Carissa spinarum* (kattukkilakkai)	0.048	0.078	1	1.063
42.	Mullukathiri	0.048	0.078	1	1.063
43.	*Blumia lacerea* (poripoondu)	0.048	0.078	1	1.063
44.	*Azadirachta indica* (vaembu)	0.048	0.078	1	1.063
45.	*Syzygium cumini* (naaval)	0.048	0.078	1	1.063
46.	*Cassia auriculata*	0.048	0.078	1	1.063

Table 11.2: Densities and frequencies of the shrub species.

Sl.No.	Name of the Species	Density	Relative Density	Frequency	Relative Frequency
1.	*Glycosmis cochinchinensis* (konji)	3.857	35.065	0.524	15.278
2.	*Memecylon umbellatum* (casan)	2.238	20.346	0.524	15.278
3.	*Atlantia monophylla* (kattunarthai)	0.905	8.225	0.524	15.278
4.	*Maytenus emarginata* (vettalai)	0.762	6.926	0.286	8.333
5.	*Randia dumetorum* (kattukkoiah)	0.667	6.061	0.238	6.944
6.	*Dichrostachys cinerea* (vedatharai)	0.476	4.329	0.190	5.556
7.	*Clerodendron inerme* (peechalathi)	0.429	3.896	0.143	4.167
8.	*Carissa spinarum* (kaattukkalakai)	0.333	3.030	0.143	4.167
9.	(Kilacheddi)	0.286	2.597	0.095	2.778
10.	*Adatoda vasica* (aadathodai)	0.238	2.165	0.095	2.778
11.	*Salacia primoides* (karukkai)	0.190	1.732	0.095	2.778
12.	*Zizyphus oenoplia* (surai)	0.190	1.732	0.095	2.778
13.	*Fluggea leucopyrus* (verpula)	0.143	1.299	0.095	2.778
14.	*Hemicyclia sepiaria* (veeramaram)	0.143	1.299	0.048	1.389
15.	*Gmelina asiatica* (kumulai)	0.143	1.299	0.048	1.389
16.	*Cassia fistula* (sarakondrai)	0.095	0.866	0.048	1.389
17.	*Tarena asiatica* (kaattuppavattai)	0.095	0.866	0.048	1.389
18.	*Cissus quadranglaris* (pirandai)	0.095	0.866	0.048	1.389
19.	Azima tetracantha (Sangalai)	0.048	0.433	0.048	1.389
20.	*Hugonia mystax* (Mothirakkanni)	0.048	0.433	0.048	1.389
21.	*Salvadora persica* (vasamaram or siruvahai)	0.048	0.433	0.048	1.389

Of all the shrubs documented in the study area, *Glycosmsis cochinchinensis* (konji), *Memecylon umbellatum* (casan), and *Atalantia monophylla* (Kattunarthai) have the highest densities in the order of 3.86, 2.24, and 0.9 respectively (Table 11.2). Nearly 35 per cent of the study area is occupied by the shrub species *Glycosmis*

Table 11.3: Densities and frequencies of the tree species.

Sl.No.	Name of the Species	Density	Relative Density	Frequency	Relative Frequency
1.	*Maba buxifolia* (Karunthuvarai)	11.333	25.373	1.048	14.379
2.	*Manilkara hexandra* (Palamaram)	7.095	15.885	0.619	8.497
3.	*Atlantia monophylla* (Kattunarthai)	4.095	9.168	0.619	8.497
4.	*Dichrostachys cinerea* (Vedatharai)	4.095	9.168	0.571	7.843
5.	*Hemicyclia sepiaria* (Veeramaram)	3.143	7.036	0.476	6.536
6.	*Memecylon umbellatum* (Casan)	2.857	6.397	0.429	5.882
7.	*Randia dumetorum* (Kattukkoiah)	2.429	5.437	0.381	5.229
8.	*Salvadora persica* (Vasamaram or Siruvahai)	1.667	3.731	0.381	5.229
9.	*Prosopis juliflora* (Karuvai)	1.429	3.198	0.381	5.229
10.	*Crataeva religiosa* (Maavilangai)	1.429	3.198	0.333	4.575
11.	*Gmelina asiatica* (Kumulai)	1.190	2.665	0.286	3.922
12.	*Zizyphus oenoplia* (Suurai)	1.143	2.559	0.238	3.268
13.	*Cassia fistula* (Sarakkondrai)	0.952	2.132	0.238	3.268
14.	*Ixora parviflora* (kaippalai)	0.810	1.812	0.190	2.614
15.	*Hugonia mystax* (mothirakkanni)	0.476	1.066	0.143	1.961
16.	Manicasan	0.333	0.746	0.143	1.961
17.	*Coccinia indica* (kovai)	0.333	0.746	0.095	1.307
18.	Peimullai	0.238	0.533	0.048	0.654
19.	*Fluggea leucopyrus* (verpula)	0.238	0.533	0.048	0.654
20.	Karukkamaram	0.190	0.426	0.048	0.654
21.	*Phoenix pusilla* (Eechai)	0.143	0.320	0.048	0.654
22.	*Jasminum sessiliflorum* (kaattumullai)	0.095	0.213	0.048	0.654
23.	*Euphorbia antiquorum* (kalli)	0.095	0.213	0.048	0.654
24.	*Azadirachta indica* (Vaembu)	0.048	0.107	0.048	0.654
25.	*Syzygium cumini* (Naval)	0.048	0.107	0.048	0.654
26.	*Scultia myrtina* (Thoradi)	0.048	0.107	0.048	0.654
27.	*Maytenes emarginatus* (vettalai)	0.048	0.107	0.048	0.654
28.	*Olax scandens* (kadalanji)	0.048	0.107	0.048	0.654
29.	*Sapindus emarginatus* (soapnut)	0.048	0.107	0.048	0.654
30.	*Odina wodier* (odiyamaram)	0.048	0.107	0.048	0.654
31.	Nunamaram	0.048	0.107	0.048	0.654
32.	Kaalalai	0.048	0.107	0.048	0.654

cochinchinensis (konji) and the *Memecylon umbellatum* (casan) occupies the 20 per cent of plot areas. These two species alone have more than 50 per cent of the relative density. *Atlantia monophylla* (Kattunarthai), *Glycosmis cochinchinensis* (konji), and *Memecylon umbellatum* (casan) are found to have the highest frequency value of 0.524 and relative frequency value of 15.278 per cent.

Dense vegetation of *Maba buxifolia* (karunthuvarai) tree is found in the study area which is about 11.3 trees in an area of 625 m^2. This is the highest value compared to the other tree species namely *Manilkara hexandra* (palamaram) (7.095), *Atlantia monophylla* (kattunarthai) (4.095) and *Dichrostachys cinerea* (vedatharai) (4.095)

Table 11.4: Dominance of the tree species.

Sl.No.	Name of the Species	Dominance
1.	*Manilkara hexandra* (palamaram)	0.0008370
2.	*Hemicyclia sepiaria* (veeramaram)	0.0000905
3.	*Maba buxifolia* (karunthuvarai)	0.0000848
4.	*Crataeva religiosa* (Mavilanagai)	0.0000529
5.	*Salvadora persica* (Vasamaram or siruvahai)	0.0000447
6.	*Dichrostachys cinerea* (Vedatharai)	0.0000380
7.	*Atlantia monophylla* (Kattunarthai)	0.0000346
8.	*Randia dumetorum* (Kattukkoiah)	0.0000175
9.	*Memecylon umbellatum* (Casan)	0.0000169
10.	*Cassia fistula* (Sarakkondrai)	0.0000145
11.	*Phoenix pusilla* (Eechai)	0.0000142
12.	*Ixora parviflora* (kaippalai)	0.0000087
13.	*Gmelina asiatica* (Kumulai)	0.0000081
14.	*Ziziphus oenoplia* (suurai)	0.0000069
15.	Kaalalai	0.0000045
16.	*Euphorbia antiquorum* (kalli)	0.0000035
17.	*Odina wodier* (Odiyamaram)	0.0000033
18.	*Hugonia mystax* (mothirakkanni)	0.0000024
19.	(Peimullai)	0.0000022
20.	*Fluggea leucopyrus* (verpula)	0.0000021
21.	Kovai	0.0000016
22.	*Azadirachta indica* (Vaembu)	0.0000010
23.	Karukkamaram	0.0000008
24.	Kaattumullai	0.0000006
25.	*Syzygium cumini* (*navalmaram*)	0.0000003
26.	*Scutia myrtina* (thoradi)	0.0000002
27.	Kadalanji	0.0000002
28.	*Sapindus emarginatus* (soap nut)	0.0000002
29.	Nunamaram	0.0000002
30.	*Maytenes emarginatus* (Vettalai)	0.0000001

Table 11.5: Biodiversity assessment keys.

Sl.No.	Plant Type	Herbs	Shrubs	Trees
1.	Margalef's Index	6.2855 (Usually the Margalef's index ranges between 0.1 and 4.0 and the number 4 means high species richness, but with regard to herbs in the Dry Evergreen Forest of Point Calimere, the value is more than 6 and it implies a very high species richness).	8.8066 (The value of Margalef's index ususally vary between 0.1 and 4. The more the value of D the higher the species richness. As the value is high close to 4 it could be concluded that the species richness of Shrubs are high).	0.7102 (The value of Margalef's index ususally vary between 0.1 and 4. The more the value of D the higher the species richness. As the value is more thN 4.5 it could be concluded that the species richness of trees is high).
2.	Simpsons Index	3.6576 (Simpson's index reveals the highest species diversity in herbs compared to Shrubs and Trees).	2.2139 (Simpson's index reveals the lowest species diversity in shrubs compared to Herbs and Trees).	0.7271 (Simpson's index reveals second highest species diversity in Trees compared to Herbs and Shrubs).
3.	Pielou's Evenness	4.5268 (In an evenly distributed population the Pielou's index is 1. J decreases with increasing unevenness. As the value 0.710275 herb species are justifiably evenly distributed).	8.1825 (In an evenly distributed population the Pielou's index is 1. J decreases with increasing uneven-ness. As the value is close to 1 it could be stated that the shrub species are more or less evenly distributed).	0.7179 (In an evenly distributed population the Pielou's index is 1. J decreases with increasing unevenness. As the value is 0.717972202, tree species are more or less evenly distributed).

Manilkara hexandra (palamaram) recorded the highest dominance value and it is calculated as 0.0008370. As dominance represents the area occupied by a species in a community, the tree species *Manilkara hexandra* can be considered to be the dominant tree species in the vegetation of Dry Evergreen Forests of Point Calimere (Table 11.4).

Conclusion

Manilkara hexandra (palamaram) is the dominant tree species. Among the shrub layer in the vegetation *Glycosmis cochinchinensis* (konji) and *Memecylon umbellatum* (casan) are the dominant species. The study results also revealed that the invasion of *Prosopis juliflora* is seriously endangering the diversity value of the TDEF vegetation.

Acknowledgements

The authors are expressing their sincere thanks and gratitude to the Tamil Nadu Agricultural University for extending all the help to carry out the study. The finance is given by the Tamil Nadu Forest Department is deeply acknowledged with thanks.

2013, Perspectives in Plant Biodiversity
Pages **77–86**
Editor: **Dr. K. Muthuchelian,** *Vice Chancellor, Periyar University, Salem*
Published by: **Daya Publishing House, NEW DELHI**

Chapter 12

Herb Diversity in Oak and Pine and Mixed Forest of Central Himalaya

Samina Usman and Kumud Dubey
Centre for Social Forestry and Eco-Rehabilitation,
3/1 Lajpat Rai Road, New Katra, Allahbad, U.P.

ABSTRACT

The present paper deals with the diversity of herb species of two predominant evergreen forests of central Himalaya *viz.* pure oak (*Quercus leucotrichophora*), pure pine (*Pinus roxburghii*) and oak pine mixed (*Quercus leucotrichophora* and *Pinus roxburghii*) along an altitudinal gradient 1700-2000m.

The herb cover was maximum in rainy season. The value of diversity was maximum in rainy season. The value of diversity was highest for oak forest followed by pine and mixed forest. Soil organic carbon and nitrogen concentration was highest for oak and lowest was recorded for pine forest. The nitrogen concentration of all three studied forest was peaked in rainy season and lowest in winter season.

A total of 21 medicinal species was screened in all three studied forest sites. Across the tree forest sites the distribution was contagious rarely random.

Keywords: Oak, Pine, Phyto-sociological study, Herb diversity.

Introduction

Species diversity is an important concept and is distributed heterogeneously across the planet. A number of quantitative indices of diversity have been proposed

(Simpson, 1949; Shannon and Weaver, 1963; Whittaker, 1972) by several workers to measure this diversity. The spatial variation in biodiversity includes species diversity in relation to size, area, relationship between local and regional species diversity and diversity along gradients across space or environmental factors (Gaston, 2000). However, diversity variation can be distinguished at several levels of organization, from genes to ecosystem. The pattern is measured in terms to ecosystem and the pattern is measured in terms of species richness (number of species per unit areas).

Knight (1975) reported that species diversity in a tropical rain forest increases most rapidly during the first 15 years of succession. Braun (1950) reported tree diversity between 1.69 and 3.4. Monk (1967) obtained diversity index values from 2 to 3 for temperate forests. Whittaker (1969) concluded that temperate deciduous forests generally have higher species diversity than evergreen needle-leaf forest and the greater number of species exhibit more stable environment. Species diversity in lacustrine phytoplankton studied by Sager and Hasler (1969) indicated that the component of relative abundance has greater importance. Diversity will be the greatest, if the individuals are distributed equally among the species. Concentration of number or some other measures of quantity in one or few species decrease diversity, which becomes zero if all individuals, belong to one species.

Beta-diversity increased from sub-alpine forests to those of lower elevation (Whittaker, 1965) and was still higher in coastal plain area in Texas Moral (1972) found that the beta-diversity increased with increasing elevation.

Study Sites and Methodology

Central Himalayan evergreen tree species were selected for the study. The different forest experimental sites were conducted at Nainital (29°28' N and 79°28' E) and its surroundings. Forest stands selected were:

1. Oak
2. Pine
3. Oak-pine mixed

Climate

The climate data was collected from the State Observatory (ARIES), Nainital The mean maximum temperature varied from 13.9 °C (February) to 23.7°C (April) and the mean minimum temperature from 4.9 °C (February) to 16.55°C (July). The monsoon strikes this area from mid-June to the mid-September, which sometimes extends to late-September and first week of October. Because of the high altitudes, the temperatures are similar to those of temperate regions but latitudinally it comes within subtropical belt. Three main seasons could be recognized, *viz.*, cold and dry winter, warm and dry summer and warm and humid rainy season. The climate is monsoon temperate, snowfall being frequent during winter months (December–February).

The values of the rainfall and temperature are the means of the two years (2003-2006). The study area received 2488 mm rainfall annually, of which more than 80 per cent occurred during rainy season, and varied from 5.0 mm (November) to 1244 mm (June, contributing 50 per cent of annual rainfall).

Soil

During the study period three soil samples (up to 30 cm depth) were randomly collected from each stand of different forest sites and analysed for physicochemical properties using standard methods following Misra (1968).

Soil-Texture

The composite soil samples were collected from three soil depths, *viz.*, 0-30 cm from each site, soil kept in polythene bags and brought to the laboratory for physical and chemical analysis. For determination of soil texture, the air-dried soil samples were passed through a series of sieves with different mesh sizes and the proportion by weight of soil particles was calculated.

Soil pH

For determination of soil pH, soil extract was assessed by digital pH meter using 1:5 proportions of soil and water.

Total Organic Carbon

Soil organic carbon was determined using the wet oxidation method (Jackson, 1958).

Total Nitrogen

Nitrogen content of soil was determined by Kjel Auto Vs-KTP Nitrogen Analyser based on a micro-Kjeldahl technique (Misra, 1968).

Species-Area Curve

A species area curve was obtained by placing larger and larger quadrats on the ground in such a way that each large quadrat encompasses all the smaller ones. This is also called the Nested Plot Technique (Muller-Dombois and Ellenberg, 1974).

The herb layer analyses for each forest site, by placing species area curve was done by randomly sampling 8m × 16m area for banj-oak, 16m × 16m area for oak pine mixed forest and 8m × 8m area for pine forest stands. Beyond these quadrat sizes the number of species remained constant.

Each tiller of grasses was considered as an individual plant (Singh, 1967). In the case of creeping plant any unit of the plant having functional roots was considered as one plant.

Vegetational data were analysed for abundance (A), density (D) and frequency (F) following Curtis and McIntosh (1950), A/F ratio following Whitford (1949), distribution pattern, relative values of density (RD) and frequency (RF) following Curtis and Cottam (1956). According to Curtis and Cottam (1956), the ratio of abundance to frequency below 0.025 indicates regular distribution, between 0.025 and 0.050 indicates random distribution and when exceeds 0.05, indicates contagious distribution. The Prominence Value (PV) was determined following Curtis (1959). The three forest sites were subjected to gradient analysis.

The species richness and diversity value of each species was double standardized before computing per cent similarity. Index of similarity (IS) was calculated (Muller-Dombois and Ellenberg, 1974) as:

Alpha-Diversity

The species diversity within-habitat/community is termed as alpha-diversity. The alpha-diversity is a balance between the actions of local biotic and abiotic elements of the environment, and immigration from other locations (Whittaker, 1960). The alpha-diversity comprises of two components: species richness and evenness or equitability.

Species Richness

Can be expressed by the total number of species in a given community or the number of species per unit area (Whittaker, 1960).

Evenness or Equitability

It represents the distribution of individuals among the species and was calculated (Whittaker, 1972) as:

Shannon-Wiener Index

The diversity index is based on information theory. it was calculated by using Shannon-Wiener information index (Shannon and Weiner *et al.,* 1963), on the basis of density:

$$H' = -\sum_{n=1}^{i} (Ni/N)\log_2 (Ni/N)$$

where, Ni is the total number of species i and N is the number of individuals of all species in that site.

Simpson's Index

Simpson (1949) proposed for the first time a widely used index, which varies inversely with species heterogeneity, and in fact measures the concentration of dominance (Cd) and was measured by using the equation:

$$Cd = \Sigma (Ni/N)^2$$

where, Ni and N were the same as for the Shannon-Weiner information function. It ranges from 1 (if all the individuals belong to one species) to 1/S (if they are equally divided among the species) (S is the total number of species). Thus, it is the inverse of diversity. Whittaker (1972) suggested that this index is appropriate for communities exhibiting strong dominance.

Beta-Diversity

Inter-community or inter-habitat diversity is termed as beta-diversity (BD). It was calculated by using the formula given by Whittaker (1972).

Results and Discussion

In oak site the *Quercus leucotrichphora* was the dominant tree species and in pine site the tree of *Pinus roxburghii* attain 30-35 m height. The canopy is open with crown density below 60 per cent. During winter season the herb species namely *Carex nubigene,Erigeron annue,Oryzopsis compositus* shows the maximum densities in oak, pine and mixed forest respectively. In rainy season the maximum densities was shown by *Carex cruciata* in oak and mixed forest and *Epipetus royleenum* in pine forest., however in summer season *Erigeron annua* showed in oak pine and mixed forest respectively.

During the study period it was observed that across the three season the highest species richness was recorded during rainy seasons in all three forest site. However, the species richness for oak and mixed forest was more or less same and it was minimum for pine forest it was minimum in winter Figure 12.1.

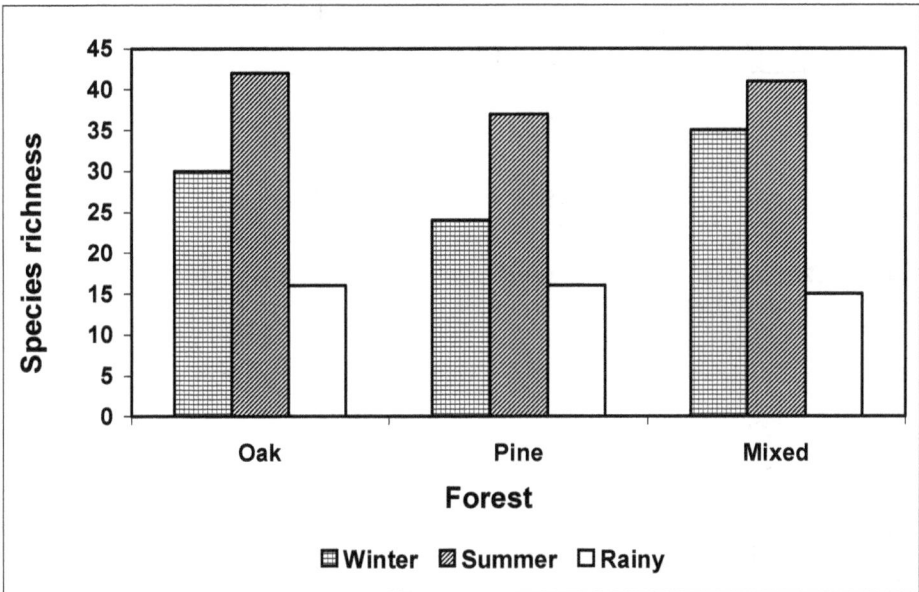

Figure 12.1: Seasonal variation in species richness in oak, pine and oak pine mixed forest in Central Himalaya.

Percent Nitrogen concentration shows similar pattern as observed for species richness *e.g.* it was peaked during rainy season and lowest in rainy season at all three forest sites (Figure 12.2).

For Soil organic carbon there is no clear cut demarcation line was observed for three different season. It was slightly higher for mixed forest as compare to oak and pine forest (Figure 12.3).

In the first year of study, 42 herb species were recorded in oak forest, 36 in pine and 43 in mixed forest. Of these, 12 herb species having medicinal values were screened in oak, 12 in pine and 17 in mixed forest (Table 12.1, Figure 12.4). During rainy and summer season the α-diversity (plants m^{-2}) was highest in oak forest and lowest in mixed forest. While in winter season it was the maximum for mixed forest.

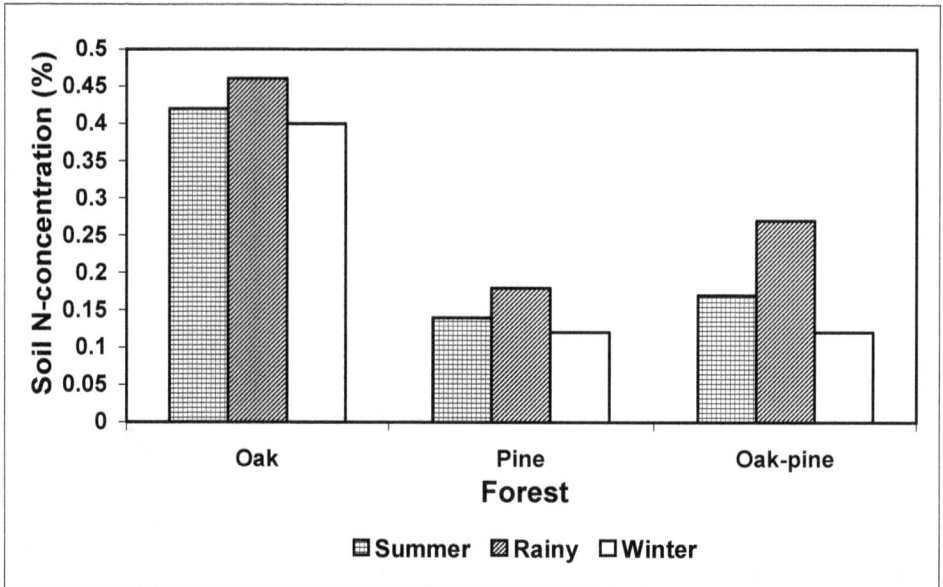

Figure 12.2: Seasonal variation in soil N-concentration (per cent) (30 cm depth) in oak, pine and oak pine mixed forest in Central Himalaya.

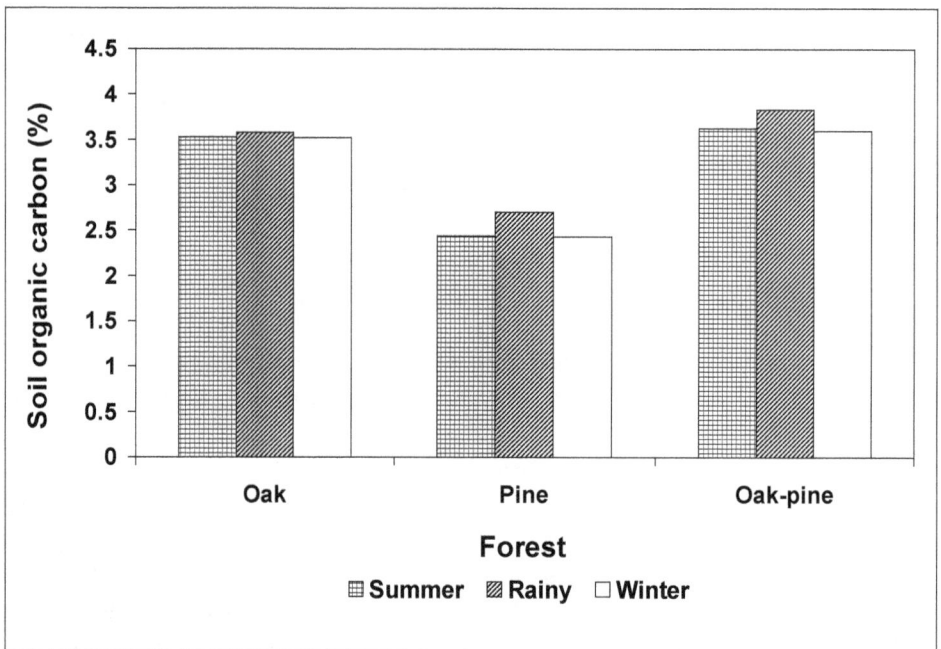

Figure 12.3: Seasonal variation in soil organic Carbon (per cent) (30 cm depth) in oak, pine and oak-pine mixed forest in Central Himalaya.

Table 12.1: Medicinal herb species of pure oak, pure pine and mixed forests.

Sl.No.	Family	Name of the Species
1.	Asteraceae	*Ainsliaea aptera* DC.
2.	Asteraceae	*Ainsliaea latifolia* (D.Don) Sch.-Bip.
3.	Asteraceae	*Anaphalis contorta*
4.	Asteraceae	*Artemisia nilagirica*
5.	Asteraceae	*Bidens pilosa* L.
6.	Commelinaceae	*Commelina benghalensis* L.
7.	Geraniaceae	*Geranium nepalense* Sw.
8.	Geraniaceae	*Geranium wallichianum*
9.	Lamiaceae	*Clinopodium umbrosum*
10.	Lamiaceae	*Leucas lanata* Benth.
11.	Lamiaceae	*Micromeria biflora*
12.	Lamiaceae	*Origanum vulgare* L.
13.	Oxalidaceae	*Oxalis corniculata* L.
14.	Polygonaceae	*Polygonum amplexicaule*
15.	Polygonaceae	*Polygonum nepalense*
16.	Ranunculaceae	*Anemone vitifolia* Buch.-Ham.
17.	Ranunculaceae	*Thalictrum foliolosum* DC.
18.	Rosaceae	*Agrimonia pilosa* Ledeb.
19.	Rubiaceae	*Galium aparine* L.
20.	Urticaceae	*Pouzolzia hirta* Roxb.
21.	Zingiberaceae	*Hedychium spicatum*

Table 12.2: Diversity and related parameters in the three forest sites H'= Shannon Winner Index, E= Equatability, Cd= Concentration of Dominance and SI = Similarity Index.

Forest	H'	E	Cd	SI
Oak	3.45	27.31	0.038	48.10
Pine	3.32	55.18	0.09	55.42
Mixed	3.26	20.5	1.0	41.02

Table 12.3: Soil parameters of three forest sites.

Soil Texture	Banj Oak Forest	Chir Pine Forest	Oak–Pine Mixed Forest
Sand	60	65	69
Silt	24	23	21
Clay	16	12	10
Bulk Density (gm^{-3})	6.3	5.7	5.9

Table 12.4: Seasonal changes in Soil parameters of the study sites.

Season	Banj Oak Forest	Chir Pine Forest	Oak–Pine Mixed Forest
Soil temperature			
Summer	19.5	23.0	20.2
Rainy	17.2	19.6	19.1
Winter	8.3	10.8	9.2
Soil moisture			
Summer	17.4	13.3	10.9
Rainy	31.2	20.0	23.0
Winter	26.0	16.8	15.0
Total nitrogen			
Summer	0.43	0.14	0.25
Rainy	0.46	0.18	0.29
Winter	0.40	0.13	0.26
Organic carbon			
Summer	3.6	2.46	1.42
Rainy	3.67	2.70	2.11
Winter	3.55	2.44	1.65

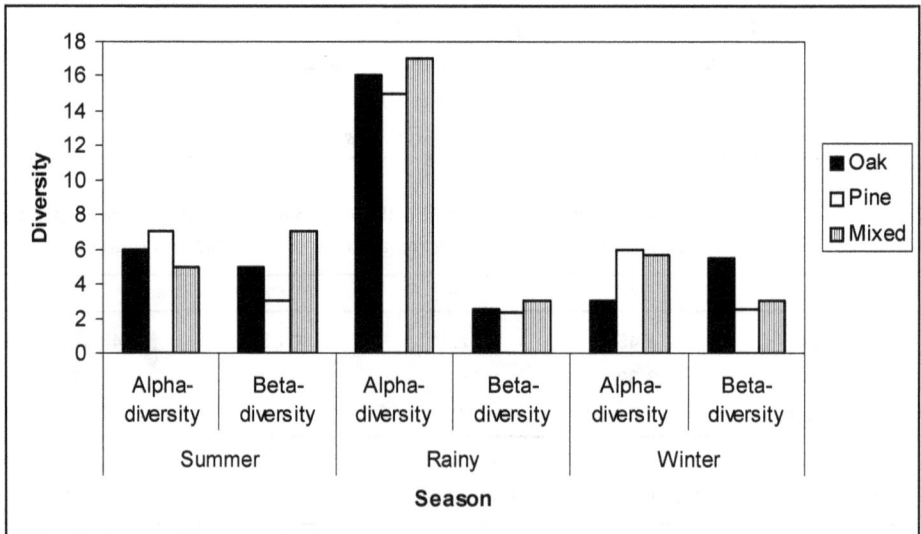

Figure 12.4: α and β-diversity in different seasons at the three sites.

The β-diversity was highest in mixed forest and minimum in oak forest in rainy and summer seasons. In winter β-diversity peaked in the pine forest and was minimum in mixed forest (Figure 12.1a). Among the three forest sites, there are twenty three herb species common in all three forest sites, eight occur only in pure oak and mixed forest sites, while only two species are common for oak and pine forest.

Among distribution pattern of herbaceous layer the contagious distribution was the most common in oak, pine and mixed forest Several workers have also observed (Grieig Smith 1952; Kershaw 1973; Singh and Yadav 1974) among regular and contagious distribution the contagious distribution is most common in the herb strata in different grassland. This may be due to majority of herb species reproduce vegetatively, however the vegetative reproduction may not be the only reason for this, as a recent researches indicate that the contagious in vegetation due to multiple factors. Kershaw 1973.. Greater diversity in rainy season in oak forest followed by pine and than mixed forest (Table 12.2).

Acknowledgements

The First is gratefully acknowledges financial support by the Department of Science and Technology, New Delhi, as Young Scientist as well as Woman Scientist.

References

Braun, E.L., 1950. The ecology of the forest of Eastern North America: Their development, composition and distribution. In: *Deciduous Forest of Eastern North America*. McGraw Hill, New York–Blakiston, Club 99: 57–64.

Curtis, J.T. and Cottom, G., 1956. *Plant Ecology Workbook: Laboratory Field Reference Manual*. Burgess Publishing, Minnesota Co.

Curtis, J.T. and McIntosh, R.P., 1950. The interrelations of certain analytic and synthetic phytosociological characters. *Ecology,* 31: 434–455.

Curtis, J.T. 1959. The vegetation of Wisconsin: An ordination of plant communities. University Wisconsin Press, Madison. Wisconsin gradient analysis of the south slopes. *Ecology,* 45: 429–452.

Greig-Smith, P., 1952. Ecological observations on degraded and secondary forest in Trinidad British West Indies. II. Structure of the communities. *Journal of Ecology,* 40: 316–330.

Knight, D.H., 1975. A phyto-sociological analysis of species rich tropical forest on Barro-Colorado Island: Panama. *Ecological Monograph,* 45: 259–289.

Monk, C.D., 1967. Tree species diversity in the eastern deciduous forest with particular reference to North Central Florida. *Amer. Natur.,* 101: 173–187.

Moral, R., 1972. Diversity patterns in forest vegetation of the Wenatchee Mountains, Washington. *Bull. Torrey Bot. Club,* 99: 57–64.

Muller Dombois, D., and Ellenberg H. Aims and methods of vegetation ecology: Relation to succession and clonal growth. Willey, New York, *Ecology,* 30: 199–288.

Sager, P.E. and Hasler, A.D., 1969. Species diversity in lacustrine phytoplankton. I. The components of the index of diversity from Shannon's formula. *Amer. Natur.,* 103: 51–59.

Shannon, C.E. and Wiener, W., 1963. *The Mathematical Theory of Communication.* University of Illinois Press, Urbana.

Simpson, E.H., 1949. Measurement of diversity. *Nature,* 163: 688.

Singh J.S., 1967. Seasonal variation in composition, plant biomass and community production in grassland. *Ph.D. Thesis*, Banaras Hindu University. Varanasi, India.

Singh J.S. and Yadav, P.S., 1974 Seasonal variation in composition, plant biomass and net primary productivity of a tropical grassland at Kurukshetra, India. *Ecological Monograph* 44L 351–376.

Whitford, P.B., 1949. Distribution of woodland plants in relation to succession and clonal growth. *Ecology,* 30: 199–288.

Whittaker, R.H. and Woodwell, G.M. 1969. Structure, production, and diversity of the oak–pine forest at Brookhaven, New York. *J. Ecol.,* 57: 155–174.

Whittaker, R.H. and Niering, W.A., 1965. Vegetation of the Santa Catalina mountains. II. A phytosociological characters. *Ecology,* 31: 434–455.

Whittaker, R.H., 1960. Vegetation of the Siskiyou Mountains, Oregon and California. *Ecol. Monogr.,* 30: 279–338.

Whittaker, R.H., 1960. Vegetation of the Siskiyou Mountains.

Whittaker, R.H., 1972. *Communication and Ecosystems.* Macmillan Publ. Co.

2013, Perspectives in Plant Biodiversity

Editor: Dr. K. Muthuchelian, *Vice Chancellor, Periyar University, Salem*

Published by: Daya Publishing House, NEW DELHI

Pages 87–95

Chapter 13

Diversity of Orchids in Koppa Taluk, Western Ghats of Karnataka, India

E.S. Kumaraswamy Udupa[1], N.R. Santhosh[2] and K. Krishnaswamy[2]*

[1]Department of Botany,
Sri JCBM College, Sringeri – 577 139, Karnataka
[2]Department of Botany,
Sahyadri Science College (Auto), Shivamogga – 577 203, Karnataka

ABSTRACT

Koppa is one of the diverse taluk in Chikamagaluru District of Karnataka, located between 75° 15' to 75° 20'N and 13° 30' to 13° 35' E with an altitude hardly about 700 to 844 msl. The documentation of orchids and their host specification in different types of selected vegetations like Evergreen, Dry deciduous, Moist deciduous, Scrubby, Acacia plantations and Grasslands by using 2x100 m belt transect reveals that the taluk has rich diversity of orchids.

A total of 68 species of orchids belonging to 37 genera were recorded within transects and out of transects in Koppa Taluk. Among these, 23 orchid species belonging to 18 genera were documented in the transect area of all types of selected vegetation including grasslands. Scrubby forest contributes the highest orchid density (72 individuals) followed by the dry deciduous forest (63). *Sarcanthus pauciflora* is the dominant orchid present in all types of vegetation except evergreen forest. The grass lands contributes 83 individuals of orchids with in the transect *viz. Satyrium nepalense* (65), *Habenaria heyneana*

* Corresponding Author: E-mail: udupa_sringeri@yahoo.co.in

(16) and *Platanthera susanae* (2). Other orchids found in the Grass land out side the transects are *Liparis wightii*, *Malaxis* sp., *Perystylis* sp., *H. longicorniculata*, *H. grandifloriformis* and *H. crinifera*.

The transect study in different forests reveals that *S.pauciflora* has highest SIV (48.28) and density (4.10). *Pholidota pallida* is the most frequently distributed orchid (0.70) and *Oberonia chandrasekharanii* is the most abundant (7.50) orchids. The recorded orchid species in different forest in Koppa taluk showed Shannon diversity value, H'= 2.16) and Simpson's species richness, C = 0.11.

The epiphytic orchids in different type of forests preferred 208 individuals of supporting trees as host plants which are belonging to 29 species and 21 families. Among the recorded host plant, *Terminalia paniculata* supports large number of orchids (37 individuals) which belongs to 14 species. 6-20 cm girth plants support the growth of highest number of orchids.

Keywords: Transects, Koppa, Western Ghats, Host specification, S. pauciflora, P. pallida, T. paniculata.

Introduction

Orchids–a precious gift of nature and our natural heritage. The family Orchidaceae is one of the largest family among flowering plants is represented by more than 25,000-35,000 known wild species in 600-800 genera in the world (Arbitti,*et.al.*,1979). In India, Orchidaceae–a second largest family, consisting of about 1,200 wild species belonging to 320 genera. Out of these nearly 287 species are endemic and they are distributed among 72 genera. The Western Ghats having many useful wild species consisting about 283 species of orchids in 76 genera. Among them 112 species in 30 genera are endemic (Rao, T.A. 1998).

Two distinctly different dangers threaten the existence of wild orchid species. One danger is over collection by amateur and professional collections. The second is deforestation, in which numerous orchids species are destroyed when the forests are cut down. The vast majorities of orchids are epiphytes and are most threatened by the loggers who make their way through the forest. It is to be noted that the decline in number of orchid species is mainly due to biotic disturbance of forest.

By considering the role and distribution of orchids in an ecosystem and threats to it, the present work is undertaken to document the diversity, host specification of orchids in different vegetations of Koppa taluk.

Materials and Methods

Koppa Taluk of Chikamagaluru District has a significant extent of forest cover, more than half of which can be described as dense. The vegetation of Koppa is a product of different lines of migration, isolation and even evolution conditioned by topography soil and bio climate. Biotic influence, especially human has modified the vegetation in varying degrees in the greater part of the taluk. The climatic condition is cold. The cold season starts from December to February, March to May is the hot season. From June to September it receives South West Monsoon and October to November it is Post Monsoon period.

Five different types of natural vegetations *viz.* Evergreen (Near Meguru), Dry deciduous (Near Kotegudda), Moist deciduous (Near Heggadde and Niluvagilu), Scrubby (Near Hulugarugudda), Grasslands (Hanathi Gudda and Meruthi Gudda) and one man made forest, Acacia plantation (Harihara pura and Niluvagilu) are selected in Koppa taluk. Frequent field visit is undertaken in this selected vegetations in the year 2009 and belt transect method is used to study orchid diversity and their host specification. In each type of selected vegetation, two 4x100m belt transects were laid at random. In the transect area the GBH of all the plants which harbor orchids, number of orchids and their height on the host plants were recorded. The data obtained are analyzed statistically by using standard formulas to calculate diversity and species richness of orchids. The flowering orchids were collected, photographed and preserved in Department of Botany, Sahyadri science college, Shivamogga and Sri J.C.B.M. College, Sringeri.

Results and Discussion

Distribution of Orchids in different Types of Vegetation

A total of 68 species of orchids belonging to 37 genera were recorded within transects and out of transects in Koppa Taluk. Among these, 23 orchid species belonging to 18 genera were documented in the transect area of all types of selected vegetation including grasslands. Scrubby forest contributes the highest orchid density (72 individuals) followed by the dry deciduous forest (63) where as Evergreen forest contributes only two orchids *i.e. Oberonia santhapaui* and *Dendrobium macrostachyum.* Moist deciduous forest contribute highest plant density with in the transect area (170 individuals) where as Acacia plantation contribute lowest number of trees (71 individuals) in the selected area (Figure 13.1). *Sarcanthus pauciflora* is the dominant orchid present in all types of vegetation except evergreen forest. The study contributes 26 orchid species addition to the Flora of Chikamagaluru where 42 species are reported (Yoganarasimhan *et al.*, 1981) and also have 24.72 per cent of orchids in the Western Ghats (Rao, T.A. 1998).

The grasslands contributes 83 individuals of orchids with in the transect *viz. Satyrium nepalense* (65), *Habenaria heyneana* (16) and *Platanthera susanae* (2). Other orchids found in the Grassland outside the transects are *Liparis wightii, Malaxis* sp., *Perystylis* sp., *H.longicorniculata, H.grandifloriformis* and *H.crinifera.*

During the course of the present survey 68 orchid species were encountered within Koppa taluk. It may be noted that Bist and Katham (1999) mention about 181 species of orchids from Buxa Tiger Reserve. Mukherjee (1972) mention about the presence of 54 species of orchids in the plans of North Bengal. Pradhan (1977, 1979) recorded as many as 255 species of orchids for the whole of West Bengal. Hegde (1990) reported 287 species (80 genera) of orchids from Darjeeling hills Yoganarasimhan *et al.* (1981) reported 42 species of orchids in Chikamagaluru District and Balakrishna Gowda (2004) recorded 41 species of orchids in Sringeri taluk.

Among the recorded 68 species of orchids, *Dendrobium* is the largest genus represents eight species followed by *Bulbophyllum* (5). 22 genera have single species (Table 13.1).

Table 13.1: Genus and species diversity of orchids in the study area.

Sl.No.	Genus	Species
1.	Acampe	A.praemorsa
2.	Aerides	A.crispa, A.ringens
3.	Aphyllorchis	A.montana
4.	Bulbophyllum	B.fimbriatum, B.tremulum, B.mysorense, B.fishcherii, B.neilgherense
5.	Chiloschista	C.pusilla
6.	Coelogyne	C.breviscapa, C.nervosa
7.	Cottonia	C.peduncularis
8.	Cymbidium	C.bicolar
9.	Dendrobium	D.aquem, D.barbatulum, D.herbacium, D.macrostachyum, D.ovatum, D.nutantiflorum, D.crepidatum, D.heterocarpum
10.	Disperis	D.zeylanica
11.	Epipogeum	E.roseum
12.	Eria	E.microchilos, E.mysorensis, E.reticosa
13.	Eulophia	E.nuda
14.	Flickingeria	F.nodosa
15.	Gastrochilus	G.pulchelllus, G.flabelliformis
16.	Geodorum	G.densiflorum
17.	Habenaria	H.grandifloriformis, H.longicorniculata, H.crinifera, H.heyneana
18.	Liparis	L.viridiflora, L.wightii
19.	Luisia	L.macrantha, L.zeylanica
20.	Malaxis	M.versicolor, Melaxis sp.
21.	Nervilia	N.crociformis, N.discolor, N.infundibulifolia, N.aragoana
22.	Oberonia	O.bruniana, O.chandrasekharanii, O.santapaui,
23.	Peristylis	P.goodyeroides, Pestylis sp.
24.	Phalaenopsis	P.decumbense
25.	Pholiodota	P.pallida
26.	Platanthera	P.susanae
27.	Polystachya	P.flevascens
28.	Porpax	P.jerdoniana, P.reticulata
29.	Rhychostylis	R.retusa,
30.	Sarcanthus	S.pauciflora
31.	Satyrium	S.nepalense
32.	Sirhookera	S.lanceolata
33.	Smithsonia	S.straminea
34.	Trias	T.stocksii
35.	Vanda	V.testacea, V.tesellata
36.	Xenikophyton	X.smeeanum
37.	Zeuxine	Z.longilabris, Z.gracilis

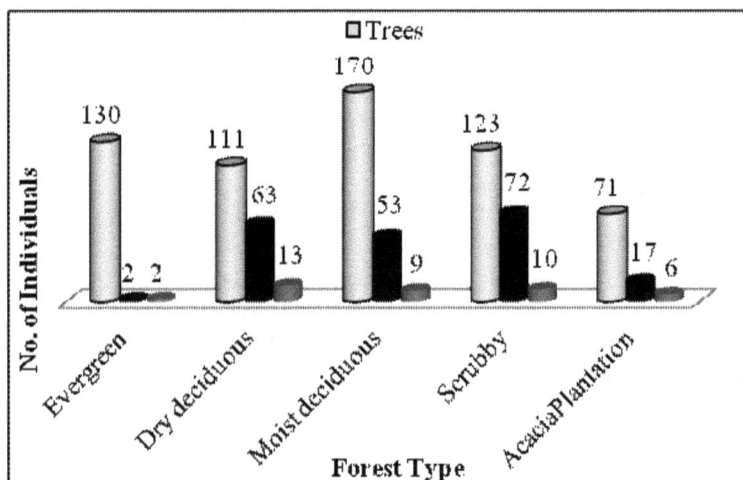

Figure 13.1: Number of trees and number of orchids in different forest with in the transect of Koppa Taluk.

Table 13.2: Species composition and their Importance Value of Orchids in selected forests of Koppa Taluk.

Sl.No.	Species	F.	RF.	D.	RD.	A.	SIV
1.	Aerides crispa	0.40	19.05	2.30	11.06	5.75	30.11
2.	Aerides ringens	0.30	14.29	2.00	9.62	6.67	23.90
3.	Chiloschista pusilla	0.10	4.76	0.20	0.96	2.00	5.72
4.	Cottonia peduncularis	0.30	14.29	0.60	2.88	2.00	17.17
5.	Cymbidium bicolar	0.20	9.52	0.30	1.44	1.50	10.97
6.	Dendrobium herbacium	0.10	4.76	0.20	0.96	2.00	5.72
7.	Dendrobiu macrostachyum	0.60	28.57	0.90	4.33	1.50	32.90
8.	Dendrobium ovatum	0.40	19.05	0.80	3.85	2.00	22.89
9.	Eria microchilos	0.20	9.52	0.20	0.96	1.00	10.49
10.	Gastrochilus pulchelllus	0.10	4.76	0.10	0.48	1.00	5.24
11.	Luisia macrantha	0.30	14.29	0.30	1.44	1.00	15.73
12.	Oberonia bruniana	0.10	4.76	0.10	0.48	1.00	5.24
13.	Oberonia chandrasekharanii	0.40	19.05	3.00	14.42	7.50	33.47
14.	Oberonia santapaui	0.20	9.52	0.20	0.96	1.00	10.49
15.	Pholiodota pallida	0.70	33.33	3.10	14.90	4.43	48.24
16.	Rhychostylis retusa	0.20	9.52	0.70	3.37	3.50	12.89
17.	Sarcanthus pauciflora	0.60	28.57	4.10	19.71	6.83	48.28
18.	Trias stocksii	0.50	23.81	1.20	5.77	2.40	29.58
19.	Vanda testacea	0.20	9.52	0.30	1.44	1.50	10.97
20.	Zeuxine longilabris	0.10	4.76	0.10	0.48	1.00	5.24

F: Frequency, RF: Relative Frequency, D: Density, RD: Relative Density, A: Abundance, SIV: Species Important Value.

The transect study in different forests reveals that *S.pauciflora* has highest SIV (48.28) and density (4.10). *P.pallida* is the most frequently distributed orchid (0.70) followed by *D.macrostachyum* (0.60). *O.chandrasekharanii* is the most abundant (7.50) orchids and is followed by *S.pauciflora* (6.83) (Table 13.2). The recorded orchid species in different forest in Koppa taluk showed Shannon diversity value, H' = 2.16) and Simpson's species richness, C = 0.11.

Host–Orchid Interaction

The recorded 207 individuals of epiphytic orchids belonging to 20 species with in the transect of different type of forests preferred 208 individuals of supporting trees as host plants which are belonging to 29 species and 21 families. *S.pauciflora* preferred 41 number of host plants followed by *P.pallida* (31).

Among the recorded host plant, *Terminalia paniculata* supports large number of orchids (37 individuals) which belongs to 14 species. Out of 37 numbers of orchids on *T.paniculata*, *A.crispa* and *P.pallida* represents highest number (7) and it is followed by *A.ringens* and *O.chandrasekharanii* (4). 13 host plants represent only single orchid species (Table 13.3).

Table 13.3: Documented host plants with in the transect which supports the growth of orchids.

Sl.No.	Host Plant	Orchid	No. of Orchids
1.	Acacia auriculiformis	A. crispa	1
2.	Aporosa lindleyana	D. macrostachyum	3
		D. ovatum	1
		G. pulchellus	1
		P. pallida	4
		S. pauciflora	9
		T. stocksii	1
3.	Buchanania lanzan	P. pallida	1
		R. retusa	1
		V. testacea	1
4.	Caralia brachiata	A. crispa	1
5.	Careya arborea	C. peduncularis	1
		R. retusa	1
		T. stocksii	1
		p. pallida	1
6.	Dillenia pentagyna	A. crispa	2
		C. peduncularis	1
		D. macrostachyum	1

Contd...

Table 13.3–*Contd...*

Sl.No.	Host Plant	Orchid	No. of Orchids
		O. chandrasekharanii	1
		P. pallida	2
		R. retusa	1
		S. pauciflora	1
		T. stocksii	1
7.	*Diospyrus Montana*	*O. chandrasekharanii*	1
8.	*Flacourtia indica*	*A. crispa*	2
		D. ovatum	1
		P. pallida	1
		S. pauciflora	1
9.	*Glochidion zeylanicum*	*D. ovatum*	1
10.	*Gmelina arborea*	*P. pallida*	1
11.	*Gnidia glauca*	*D. ovatum*	1
12.	*Gordonia obtusa*	*O. santhapu*	1
13.	*Hopea ponga*	*D. herbaceum*	2
		S. pauciflora	3
14.	*Ixora brachiata*	*S. pauciflora*	4
15.	*Lagerstroemia lanceolata*	*C. bicolor*	1
		P. pallida	1
16.	*Murraya paniculata*	*S. pauciflora*	2
17	*Madhuca nerifolia*	*P. pallida*	1
18.	*Memecylon maladaricum*	*S. pauciflora*	3
		T. stocksii	2
19.	*Memecylon umbellatum*	*P. pallida*	6
		S. pauciflora	12
		T. stocksii	1
20.	*Paramigyna monophylla*	*A. crispa*	3
		A. ringens	4
		T. stocksii	2
21.	*Psychotria dalzellii*	*A. crispa*	2
		A. ringens	7
		D. ovatum	2
		E. microchilos	2
		O. chandrasekharanii	8
		P. pallida	1
		S. pauciflora	1

Contd...

Table 13.3–*Contd...*

Sl.No.	Host Plant	Orchid	No. of Orchids
22.	Pterocarpus marsupium	P. pallida	1
23.	Randia dumetorum	A. crispa	1
		A. ringens	2
		C. pusilla	1
		C. peduncularis	2
		D. macrostachyum	1
		O. chandrasekharanii	11
		P. pallida	3
		R. retusa	2
		S. pauciflora	1
		T. stocksii	1
24.	Schefflera venulosa	O. santhapu	1
25.	Syzygium caryophyllatum	S. pauciflora	2
		T. stocksii	1
26.	Syzygium cumini	D. macrostachyum	2
		T. stocksii	1
27.	Terminalia bellerica	R. reusa	1
28.	Terminalia panuculata	A. crispa	7
		A. ringens	4
		c. pusilla	1
		C. peduncularis	2
		C. bicolor	2
		D. macrostachyum	2
		D. ovatum	2
		L. macrantha	1
		O. drunoniana	1
		O. chandrasekharanii	4
		P. pallida	7
		R. retusa	1
		T. stocksii	1
		V. testacea	2
29.	Terminalia tomentosa	A. crispa	2
		A. ringens	3
		L. macrantha	2
		O. chandrasekharanii	4

The GBH of the host plants documented in the transect area ranges between 6-200 cm. Among these 6-20 cm girth plants supports the growth of highest number of orchids (41individuals) belonging to 6 genera compared to higher girth class plants. This represents small sized host plants highly supports the growth of the orchids. The pooled data of the height of the orchids on a host plant indicates the highest growth of the orchids at 6-10 feet (67). No comprehensive works are undertaken in the field of orchids by using transect methods to compare present data.

Acknowledgements

Authors are grateful to Principal of Sahyadri Science College, Shivamogga and Sri JCBM College, Sringeri for their kind help and UGC for financial assistance.

References

Abraham, A. and Vatsala, P., 1981 *Introduction to Orchids.* St. Joseph's Press. Trivandrum, India.

Arditti, J., Michaud, J.D. and Healey, P.L., 1979. Morphometry of orchid seeds. *Paphiopedilum* and native California and related species of *Cypripedium. Amer. J. Bot.,* 66: 1128–1137.

Bist, S.S. and Kantham, Tapan, 1999. Status of orchids in Buxa tiger reserve. *Indian Forester,* p. 460–489.

Gowda, Balakrishna, 2004. *Vanaspathi Kosha: Plant Wealth of Sringeri,* Karnataka.

Hegde, S.N., 1980. Preliminary observation and list of orchids of Arunachal Pradesh. *Arunachala for News,* 3 (3): 1–11.

Mukherjee, S.K., 1972. Orchids of the Plains of North Bengal, *Bull. Bot. Surv.* India, 14(1–4): 921–1003.

Pradhan, U.C., 1977 and 1979. *Indian Orchids: Guide to Identification and Culture, Vols. I–II:* Kalimpong.

Rao, T.A., 1998. Conservation of wild orchids of Kodagu in the Western Ghats. *KAAS,* Bangalore.

Yoganarasimhan, S.N. and Razi, B.A., 1981. *Flora of Chickmagalure District, Karnataka.* International Book Distributors, Dehradun.

2013, Perspectives in Plant Biodiversity   *Pages* **96–106**
Editor: **Dr. K. Muthuchelian,** *Vice Chancellor, Periyar University, Salem*
Published by: **Daya Publishing House, NEW DELHI**

Chapter 14

Studies on Genetic Diversity in Cowpea [*Vigna unguiculata* (L). Walp] Using Molecular Markers as a Means for Germplasm Conservation

R.M. Nagalakshmi, R. Usha Kumari,
K.S. Usharani, and R. Suguna
Department of Plant Breeding and Genetics,
Agricultural College and Research Institute,
Madurai – 625 104, Tamil Nadu

ABSTRACT

Cowpea (*Vigna unguiculata* (L). Walp). is one of the most important legume crops in the semi tropics covering Asia, Europe and America. It forms an integral part of varied cropping systems and provides food, fodder, and sustains soil fertility. Recently molecular markers were used for studying the genetic diversity. Sixty six genotypes of cowpea were investigated to understand the existing pattern of variability for twelve quantitative traits and assess the extent of genetic diversity among these genotypes. Heritability estimates were high for all the characters. Genetic advance as percentage of mean was high for number of leaves per plant, plant height, grain yield per plant, number of pods per plant, number of clusters per plant, 100 seed weight, number of branches per plant, days to 50 per cent flowering and number of seeds per pod.

Mahalanobis's (1936) D² analysis established the presence of wide genetic diversity among the genotypes by the formation 23 clusters. Among the 23 clusters, cluster XXIII was found have only one genotype because of its distinct superiority over the other genotypes studied. Thirteen genotypes were selected based on diversity and other characters for RAPD marker studies. The study confirmed the diversity in Kanagamoni with COCP 7, and CP 18 with CP 222. But the diversity in PGCP 1 with ACM 05-07 and V240 with VBN 1 does not go in line with the D² studies. The most divergent genotypes selected from different clusters derived from Mahalanobis D² statistic were used for molecular marker study. The thirteen genotypes were clustered in a dendrogram and relative similarity coefficient and inter cluster distances between the clusters to which they belong are compared. Representative samples from thirteen clusters were taken to compare using molecular markers. The genotypes Kanagamoni and COCP 7 had a similarity coefficient of 79 per cent and it confirms the results of D² analysis which had an intercluster distance of 6.406. The genotypes PGCP 1 and ACM 05-07 has a similarity coeffients of 82 per cent but inter cluster distance between them is 15.93. The genotypes CP 18 and CP 222 has a similarity coefficient of 79 per cent Considering the genetic diversity using D² analysis and RAPD marker analysis the genotypes Kanagamoni, COCP 7 and CP 18 can be selected for further crossing and crop improvement.

Keywords: *Genetic diversity, Molecular markers, Germplasm conservation, Cowpea.*

Introduction

Cowpea (*Vigna unguiculata* (L). Walp). also known as lobia, black eye pea, southern pea is one of the most important legume crops. It has been grown in 14 m.ha (Pandey *et al.,* 2007) in the world. Cowpea is a major source as protein, minerals and vitamins in daily diet. As a nutritious fodder to the livestock, the cowpea haulms have a crude protein content of 17 per cent, with a high digestibility and low fibre content. It has the ability to fix atmospheric nitrogen through its nodules and it grows well even in soils with more than 85 per cent sand and with less than 0.2 per cent organic matter. Breeding is concerned with genetic variability brought by a group of genes, each having a small individual effect and fundamental nature of this variability involves the study of quantitative characters. The genetic diversity of land races is the most immediately useful part of cowpea crop improvement. Genetically diverse parents selected on the basis of diversity analysis yields hybrids with better heterosis (Ramanujam *et al.,* 1974). Genetic diversity is the best criterion to select good parent lines for hybridization programme.

Diversity analysis will be useful in choosing acceptable parents for hybridization with considerable diversity and evolve superior varieties with improved yield and desirable plant type. Recently molecular markers were used for studying the genetic diversity. These studies allow researchers to identify genotypes at the taxonomic level, assess the relative diversity within and among the species and locate diverse accession for breeding purposes. The development of randomly amplified polymorphic DNA (RAPD) markers, generated by the polymorphic chain reaction (PCR) using arbitrary primer, has provided a new tool for the detection of DNA polymorphism

(Williams *et al.,* 1990). The objectives of this study were to quantify the genetic divergence among the collected genotypes and to study the relative contribution of each trait towards the divergence at genotypic level.

Materials and Methods

A total of sixty six germplasm entries of diverse origin were used for the present study. All sixty six genotypes were used for morphological diversity studies and based on diversity results only 13 genotypes were used for molecular analysis. All germplasm entries were raised in Randomized Block Design with two replications under irrigated conditions at the research block, Department of Plant Breeding and Genetics, Agricultural College and Research Institute, Madurai, during Kharif 2008. Each entry was raised in two rows of 3 m length spaced at 30 cm between rows and 20 cm between plants in each replication. The recommended agronomic practices were adopted uniformly. Observation on twelve quantitative characters viz., Days to 50 per cent flowering, Plant height, Number of branches per plant, Number of clusters per plant, Petiole length, Peduncle length, Number of leaves per plant, Pod length, Number of pods per plant, Number of seeds per pod, Grain yield per plant, hundred seed weight were recorded on three randomly selected plants in each replication. The estimates of mean, variance and standard error were worked out by adopting the standard methods (Panes and Sukhatme, 1961).

The genetic analysis of the traits for variability, heritability, genetic advance, estimation of genetic diversity by D^2 analysis were carried out, and the results were confirmed using molecular marker studies. Heritability (h^2) in a broad sense was calculated according to Lush (1940). Genetic advance was derived according to the method of Johnson *et al.* (1955) for each character. The quantitative characters are alone taken into consideration for divergence studies. The square of Mahalanobis generalized distance between any two populations is given by the formula

$$\Delta^2 = \Sigma v_i v_j \lambda_{ji}$$

where, Δ^2 is the square of generalized distance, λ_{ji} is the reciprocal of common depression matrix, V_i is $(m_{i1}-m_{i2})$ and V_j is the $(m_{j1}-m_{j2})$.

For determining the group constellation, a relatively simple criterion by Rao (1952) was followed. The criterion of grouping was that any two populations belonging to the same cluster at least on the average show a similar D^2 than those belonging to different clusters. After establishing the clusters, the intra cluster distances were worked out taking the average of the component genotypes in that cluster. The average inter-cluster divergences were arrived at by taking into consideration, all the component D^2 values possible among the members of the two clusters considered. The square root of the averaged D^2 values gave the genetic distance between the clusters.

The genomic DNA was isolated from thirteen genotypes using the CTAB method. A total of six RAPD markers were used to access the extent of diversity across the thirteen genotypes. All the primers (OPA 01, OPA 06, OPA 10, OPA 12, OPA 14, OPA 19) exhibited polymorphic bands. The amplified samples were run in the agarose gel. The banding patterns were documented in Alpha Innotech Image Analyzer. PCR products from individual plants were scored as either present or absent. The genetic distance was calculated as percentage of total number of bands scored that were clearly different between each pair of accession. The result from each accession was analysed by the UPGMA clustering method to produce a dendrogram.

Results and Discussion

Heritability and genetic gain are the two important parameters, of which, former is used to estimate the expected genetic gain through selection. In the present study, all the characters showed high heritability along with high genetic advance (Table 14.1). Hence, all the characters appears to be more promising for improvement in a breeding programme and can be utilized for developing high yielding cowpea varieties. Venkatesan *et al.* (2003) also reported high genetic advance in cowpea for all the above characters.

Genetic diversity is the basic requirement for successful breeding programme. The mean values of the sixty six genotypes were transformed into standardized uncorrelated mean values. The D^2 values corresponding to all possible combinations among the sixty six genotypes were computed. By the application of clustering technique, the 66 genotypes were grouped into twenty three different clusters. Among the twenty three clusters, it was observed that the cluster I was the largest with twenty two genotypes followed by cluster VI with three genotypes. The clusters II, III, IV, V, VII, VIII, IX, X, XI, XII, XIII, XIV, XV, XVI, XVII, XVIII, IX, XX, XX1 and XXII had only two genotypes each. The cluster XXIII was recorded to be the smallest with one genotype. The minimum intra cluster distance was recorded in cluster II (6.40) whereas the maximum was recorded in cluster XXII (43.67). Similarly the least inter cluster distance was observed between cluster II and V (8.29) and maximum between XXII and XXIII (74.30). Cluster means (Table 14.2) and Relative contribution (Table 14.3) for twelve characters in 66 diverse accessions of Cowpea were analyzed.

The most divergent genotypes selected from different clusters derived from Mahalanobis D^2 statistic were used for molecular marker study. The thirteen genotypes were clustered in a dendrogram (Figure 14.1) and relative similarity coefficient and inter cluster distances between the clusters to which they belong are compared.. Representative samples from thirteen clusters were taken to compare using molecular markers. All the primers exhibited polymorphic bands (Figures 14.2 and Figure 14.3). The genotypes Kanagamoni and COCP 7 had a similarity coefficient of 79 per cent and it confirms the results of D^2 analysis which had an intercluster distance of 6.406. The genotypes PGCP 1 and ACM 05-07 has a similarity coeffients of 82 per cent but inter cluster distance between them is 15.93. The genotypes CP 18 and CP 222 has a similarity coefficient of 79 per cent and confirms the results of D^2 analysis which shows an inter cluster distance of 13.77 per cent as reported by Reza talebi *et al.* (2008) Saraladavi *et al.* (2008) and Roopa Lavanya *et al.* (2008). The genotypes V 240 and VBN 1 had a similarity coefficient of 79 per cent but D^2 analysis shows a greater inter cluster value of 33.39. Considering the genetic diversity using D^2 analysis and RAPD marker analysis the genotypes Kanagamoni, COCP 7 and CP 18 can be selected for further crossing and crop improvement.

Table 14.1: Range, variability, heritability and genetic advance in 66 diverse accessions of cowpea.

Sl.No.	Characters	Range	Variance		PCV per cent	GCV per cent	Heritability (Broad Sense) per cent	Genetic Advance as per cent of Mean
			Phenotypic	Genotypic				
1.	Days to 50 per cent flowering	22.50-39.50	20.13	19.60	14.31	14.12	97.36	28.71
2.	Plant height (cm)	12.3-18.78	260.18	253.17	46.51	45.88	97.30	93.24
3.	Number of branches per plant	2.88-8.20	1.60	1.04	25.36	20.45	65.02	33.97
4.	Number of leaves per plant	28.33-275.21	1596.68	1559.70	46.76	46.21	97.68	94.10
5.	Petiole length (cm)	7.98-25.00	11.95	8.72	27.63	23.60	72.98	41.54
6.	Peduncle length (cm)	9.36-41.23	68.83	66.44	37.58	36.92	96.52	74.73
7.	Number of clusters per plant	3.50-20.76	13.39	12.19	30.66	29.25	91.01	57.49
8.	Number of pods per plant	5.38-66.69	112.31	108.13	39.24	38.50	96.27	77.82
9.	Pod length (cm)	9.91-55.84	37.14	36.52	37.92	37.61	98.34	76.83
10.	Number of seeds per pod	9.52-19.68	7.18	5.01	18.85	15.76	69.90	27.15
11.	100 seed weight (g)	5.89-21.02	10.52	9.93	26.81	26.05	94.41	52.14
12.	Grain yield per plant (g)	4.73-46.55	81.04	78.63	44.98	44.31	97.03	89.91

Table 14.2: Cluster means for twelve characters in 66 diverse accessions of cowpea.

Cluster	Days to 50 per cent Flowering	Plant Height (cm)	Number of Branches per Plant	Number of Leaves per Plant	Petiole Length (cm)	Peduncle Length (cm)	Number of Clusters per Plant	Number of Pods per Plant	Pod Length (cm)	Number of Seeds per Pod	100 Seed Weight (g)	Grain Yield per Plant (g)
I.	29.47	27.09	4.50	62.19	13.02	15.58	10.74	22.30	14.37	13.44	13.47	14.70
II.	34.25	30.75	4.98	90.09	13.02	27.55	10.59	25.99	15.28	16.04	9.92	24.07
III.	36.00	44.99	6.48	155.17	12.04	31.22	9.41	24.82	17.03	12.77	9.32	24.24
IV.	28.75	28.17	4.55	78.88	11.73	20.25	12.79	27.47	16.39	13.30	13.03	21.29
V.	35.00	31.06	5.30	88.46	14.18	28.00	12.50	24.86	13.58	10.53	12.12	20.01
VI.	28.50	64.31	4.82	63.09	12.51	16.73	7.62	13.75	23.11	15.54	16.97	12.06
VII.	34.75	35.69	6.01	91.09	12.50	24.59	18.15	27.93	18.62	16.30	11.51	34.36
VIII.	27.50	31.18	4.17	65.87	10.55	28.87	10.52	26.48	17.29	15.16	11.97	15.37
IX.	35.25	34.09	5.08	103.46	12.37	13.78	11.62	27.49	11.87	15.78	8.62	14.97
X.	28.50	32.59	6.64	101.88	14.63	21.11	12.22	24.45	16.01	16.99	8.20	22.40
XI.	36.75	31.43	4.26	83.00	12.29	24.05	12.77	25.76	17.66	13.46	10.66	15.89
XII.	34.75	31.28	4.45	80.67	12.29	29.79	10.28	30.96	15.14	16.11	8.60	34.84
XIII.	35.25	35.99	4.30	55.12	14.75	23.58	10.60	22.94	15.82	16.71	12.03	18.31
XIV.	35.00	37.15	6.28	64.02	13.03	18.56	14.85	24.76	13.53	15.56	9.95	35.71
XV.	34.00	29.31	5.36	139.40	11.81	28.25	13.41	36.83	12.87	14.89	9.31	14.91
XVI.	27.25	27.32	4.46	100.65	9.40	19.81	15.29	45.16	16.22	13.36	12.73	21.51
XVII.	32.25	40.19	5.38	138.52	13.58	31.59	17.06	37.65	16.12	14.59	12.05	31.77
XVIII.	34.00	32.72	5.82	100.64	12.24	36.59	18.19	63.67	12.06	12.61	11.24	29.75
XIX.	27.50	42.80	4.59	62.60	12.38	33.86	12.00	27.62	17.67	15.71	13.92	24.89
XX.	31.25	27.18	5.28	102.10	10.22	17.28	10.70	22.02	15.90	15.24	10.85	16.76
XXI.	31.00	45.54	7.37	130.27	12.47	39.57	16.59	40.82	16.29	12.06	10.67	21.60
XXII.	28.50	38.69	4.89	183.85	10.59	20.26	10.24	30.88	14.13	14.11	10.93	32.08
XXIII.	39.50	123.00	4.50	50.43	9.36	26.45	8.78	13.30	55.84	12.73	15.92	11.54

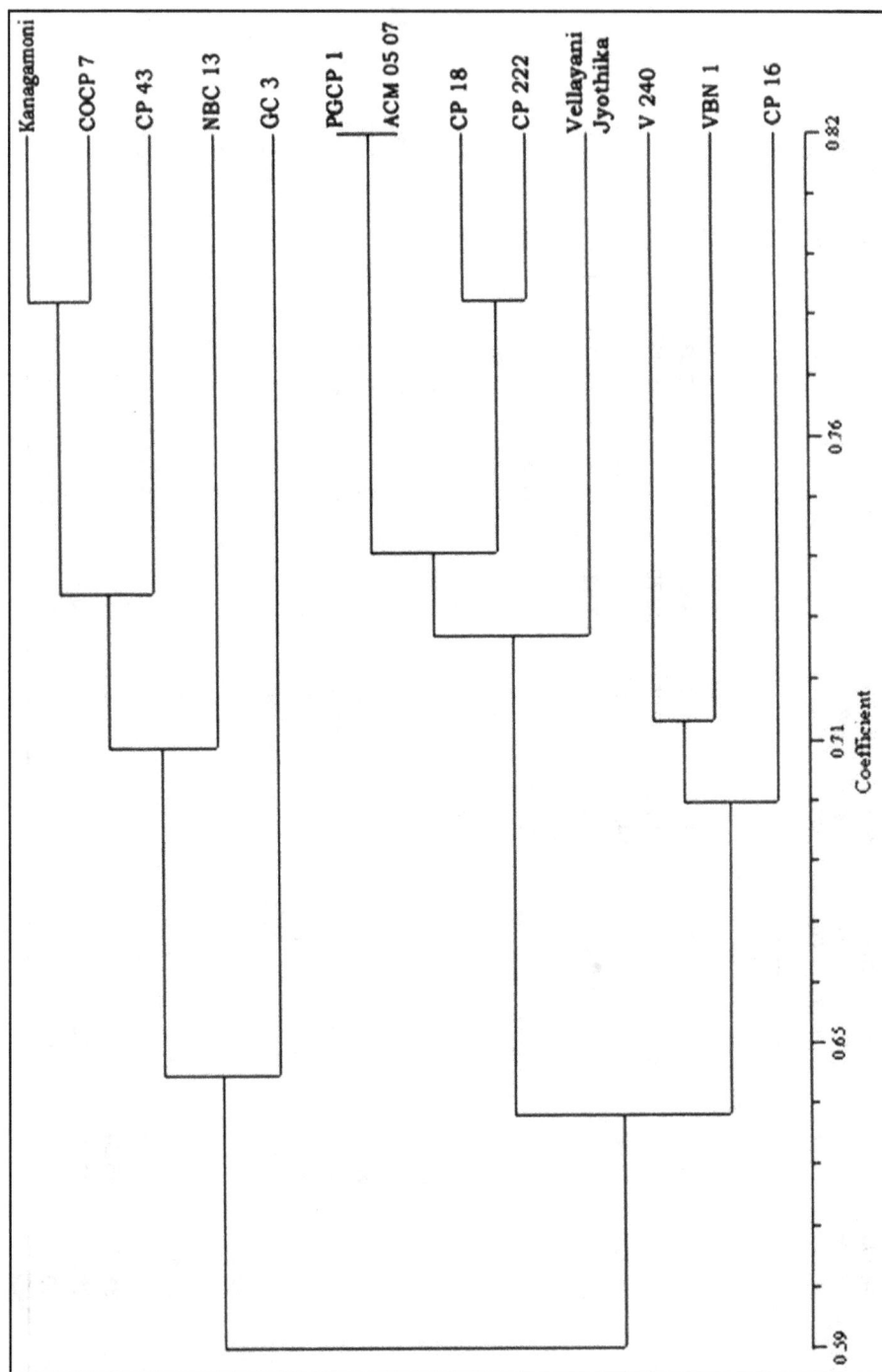

Figure 14.1: Dendrogram showing the clustering pattern of 13 cowpea genotypes.

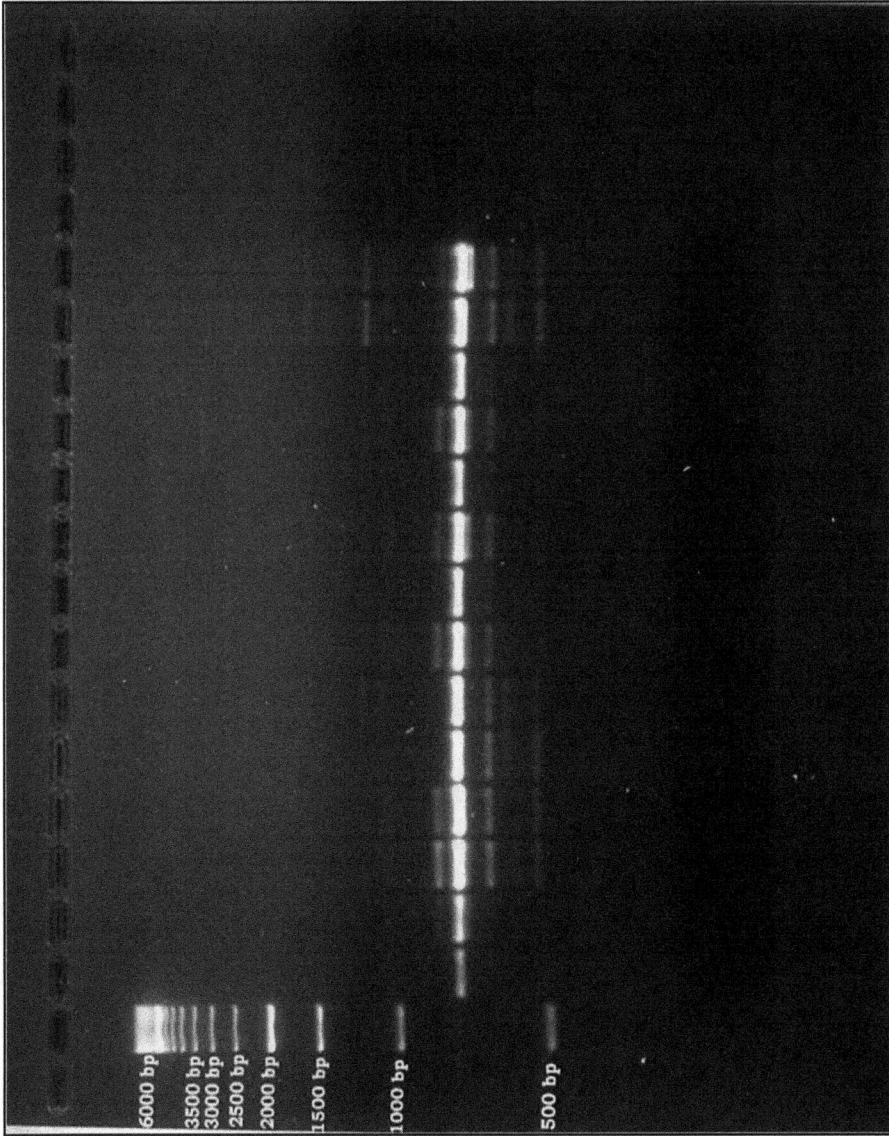

Figure 14.2: RAPD profile of 13 cowpea genotypes generated with OPA-10 primer.

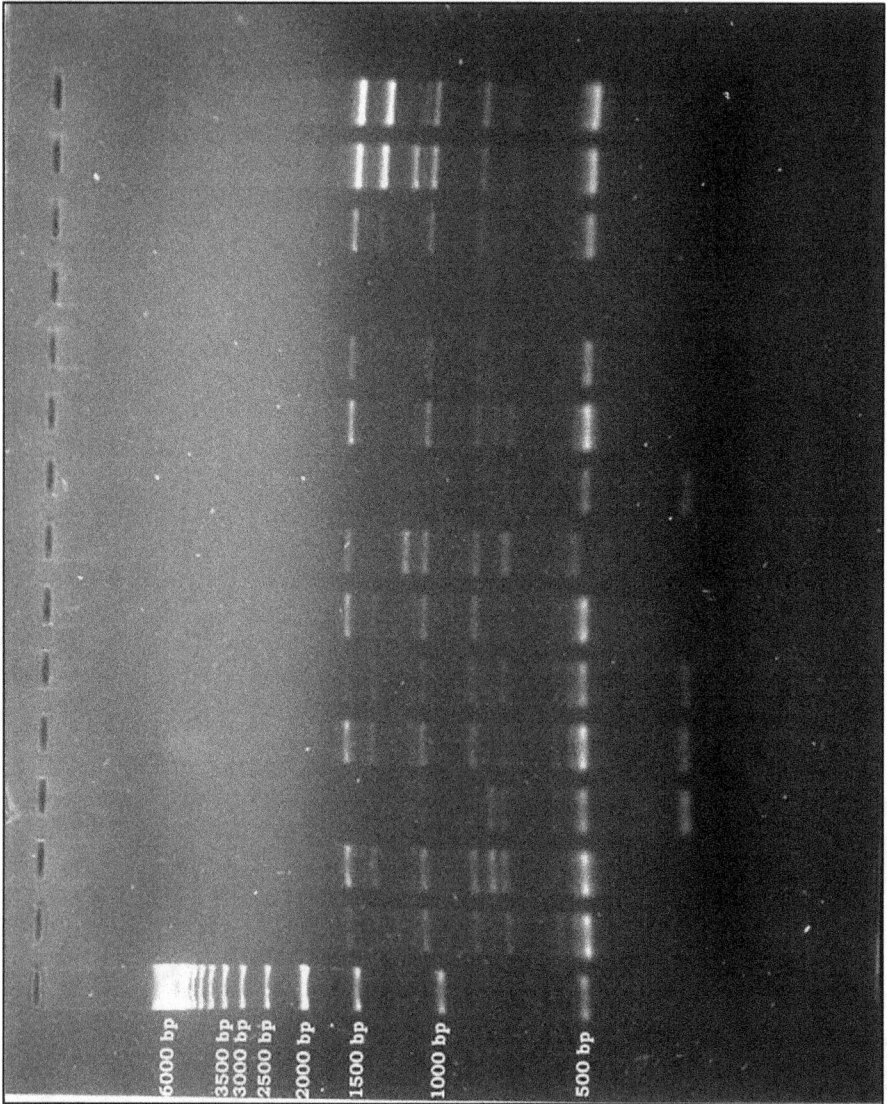

Figure 14.3: RAPD profile of 13 cowpea genotypes generated with OPA-12 primer.

Table 14.3: Relative contribution of twelve characters towards genetic divergence in 66 cowpea germplasm.

Sl.No.	Characters	Number of Times Ranked First	Contribution
1.	Days to 50% flowering	170	7.92
2.	Plant height (cm)	45	2.09
3.	Number of branches	0	0
4.	Number of leaves	138	6.43
5.	Petiole length (cm)	15	0.69
6.	Peduncle length (cm)	63	2.93
7.	Number of clusters	31	1.44
8.	Number of pods	164	7.64
9.	Pod length (cm)	108	5.03
10.	No of seeds / pod	98	4.56
11.	100 seed weight (g)	412	19.2
12.	Grain yield (g)	901	42
	Total	**2145**	**100**

References

Johnson, H.W., Robinson, H.F. and Comstock, R.E., 1955. Genotypic and phenotypic correlations in Soybeans and their implication in selection. *Agron.*, 47: 477–483.

Mahalonobis, P.C., 1936. On the generalized distance in statistics. *Proc. Nat. Istt., Sci., India*, 2: 49–55.

Muthusamy, Saraladevi, Kanagarajan, Selvaraju, Shanmugasundaram and Ponnusamy. 2008. Efficiency of RAPD and ISSR markers system in accessing genetic variation of rice bean (*Vigna umbellate*) landraces. *Electronic J. of Biotech.*, 11(3): 717–3458.

Pandey, I.D., Singh, B.B. and Kumar, D., 2007. Present status and future prospects of cowpea research in India. *Arid Legumes*, 4(1): 26–30.

Panse, V.G. and Sukhatme, P.V., 1961. *Statistical Methods for Agricultural Workers*, 2nd Edn. ICAR, New Delhi, p. 227.

Ramanujum, S., Tiwari, A.S. and Mehra, R.D., 1974. Genetic divergence and hybrid performance in mungbean. *Theor. Appl. Genet.*, 45: 211–214.

Rao, C.R., 1952. *Advanced Statistical Methods in Biometrical Research*. John Wiley and Sons Inc., New York, pp. 390.

Reza Talebi, Farzad Fayaz, Mohsen Mardi, Seyed Mostafa Pirsyedi and Amir Mohammed Nai, 2008. Genetic relationships among chickpea (*Cicer arietinum*) elite lines based on RAPD and agronomic markers. *Int. J. Agri. Biol.*, 10(3): 301–305.

Roopalavanya, G., Srivastava, Jyoti and Ranade, Shirish A., 2008. Molecular assessment of genetic diversity in mung bean germplasm. *J. of Genet.* 87 (1): 65–74.

Venkatesan, M., Prakash, M. and Ganeshan, J., 2003. Genetic variability, heritability and genetic advance analysis in Cowpea (*Vigna ungiculata* (L). Walp). *Legume Res.*, 26(2): 155–156.

Williams, J.G.K., Kubelik, A.R., Livak, K.J., Rafalski, J.A. and Tinigey, S.V., 1990. DNA polymorphism amplified by arbitrary primers are useful as genetic markers. *Nucl. Acids Res.,* 18: 6531–6535.

2013, Perspectives in Plant Biodiversity   *Pages* **107–118**
Editor: **Dr. K. Muthuchelian,** *Vice Chancellor, Periyar University, Salem*
Published by: **Daya Publishing House, NEW DELHI**

Chapter 15

Wild Relatives of Fruits in Eastern Himalayan Region of India

C.P. Suresh[1]*, K. Pradheep[2], K.D. Bhutia[1] and Sumit Chakravarty[3]
[1]Department of Forestry
[3]Department of Pomology and Post Harvest Technology
Uttar Banga Krishi Viswavidyalaya, Pundibari – 736 165
[2]Division of Plant Exploration and Germplasm Collection,
NBPGR New Delhi

ABSTRACT

Eastern Himalaya due to high rainfall, moist and cold climate coupled with altitude, longitude and latitude added to the multiplicity of habitats and thus provides a variety of microclimates and ecological niches. Hence it is agreed as a region of active speciation and designed as cradle of flowering plants. The idea behind this paper is to analyze the potentialities of those related species occurring wild in Eastern Himalaya for crop improvement from our personal observations and published literature. Those wild relatives occurring in this region are briefed with important characteristics, distribution, present and potential uses including level of nearness to cultivated counter partners, etc.

* Corresponding Author: E-mail: sureshcp2112@yahoo.com

The wild relatives of genus musa, mangifera, citrus, malus, pyrus, prunus, fragaria, vitis, rubus, actinidia, phoenix, garcinia and artocarpus are described in the paper.

Keywords: *Eastern Himalaya, Wild relative, Fruits, Crop improvement.*

Introduction

Introgression of genes from one taxon to another contributes to species evolution and in the development of modern crops. Many of the fruit crops have evolved from interspecific hybridization and polyploidy. This is particularly true in case of *Actinidia, Citrus, Fragaria, Malus, Prunus, Musa, Pyrus, Rubus* and *Vaccinium*. The evolutionary divergence within these genera into different species is often associated with polyploidy. It is a common observation that the available germplasm base of crops is not sufficient to meet out the ever ending search on desired attributes. Thus relatives occurring wild are a major for future use to develop desirable varieties. Eastern Himalaya has a wide range of habitats due to its high rainfall, moist and cold climate along with its varied altitude, longitude and latitude. This wide variation in eastern Himalayan ecosystem provides a variety of microclimates and ecological niches making it as a cradle of flowering plants supporting active speciation. The idea behind this paper is to analyze the potentialities of those related species occurring wild in the region for crop improvement from our personal observations and from those reported.

Maxted *et al.* (2006) defined crop wild relative as a wild plant taxon that has an indirect use derived from its relatively close genetic relationship to a crop. The wild relative of crop plants include wild/weedy races, the progenitors of crops, species more or less closely related and can be identified on the basis of genetic makeup which may even flow across different genera and/or on the basis of morphological similarity. In crops, where crossability and/or graft compatibility experiments were not performed, morphological similarity *i.e.* taxa within same species/series/section/subgenus/genus in descending order of importance is taken as the yardstick for finding out the wild relative. Domestication of wild species is still being done and will be continued to satisfy human needs. Human needs are likely to change with course of improvement of lifestyle as well as to solve the new problem. Consequently, the wild species of little importance today may assume great significance in days to come. Generally wild species are hardy, resistant to insect, pest and diseases and sometimes possesses noble traits, which can be introduced into the already cultivated species through convention and non-conventional breeding. Moreover, the wild species have deep tap root system and are hardy or resistant to pests. Thus compatible wild relative can be used as root stock in fruit crops. The wild relatives occurring in Indian Territory of Eastern Himalaya are briefed with their important characteristics, distribution and potential uses including level of nearness to their cultivated species.

Genus *Musa*

The cultivated banana in north eastern region belongs to AAB group (2n=33). Sections *Eumusa* (x =11) and *Rhodochlamys* (x =11) are well represented in Eastern

Himalayas whereas sections *Australimusa* and *Callimusa* are not found. *Musa cheesmanii* (*Eumusa*), *M. velutina* (*Rhodochlamys*) are found in mid hills of Arunachal Pradesh. *M. balbisiana* (*Eumusa*) is found throughout Eastern Himalaya and north eastern India is considered to be the primary centre of its origin. *M. sikkimensis* (*Eumusa*) is a rare species confined to Sikkim only. *M. rosacea* (*Rhodochlamys*) is found in higher altitudes of Arunachal Pradesh. *M. ornata* (*Rhodochlamys*) is found in Sikkim, Arunachal Pradesh while *M. sanguinea* (*Rhodochlamys*) with localized distribution in Arunachal Pradesh (Hore *et al.*, 1992). Even though *M. acuminata* (*Eumusa*) is of Malayan origin, a large number of its clones are found in all north eastern states of India. *M. aurantiaca* is reported from Arunachal Pradesh whereas *M. sapientum* var. *dubia* from northern part of West Bengal and Sikkim (Naithani, 1990). It is known that members of section *Rhodochlamys* cross freely with those of *Eumusa* producing F_1 generation that exhibits the dominance of *Rhodochlamys*. *M. balbisiana* crosses easily with *M. laterita, M. ornata, M. sanguinea, M. velutina* and *M. rosacea* producing viable hybrids. *M. ornata* is resistant to sikatoka leaf spot (Singh and Chadha, 1993a).

Genus *Mangifera*

Wild forms of *Mangifera indica* is occurring in the low hills of Eastern Himalayas up to 900 m possessing primitive traits like polyembryony and dwarf nature. *M. sylvatica* is found up to 1300 m. It bears twice in a year. Mango is closely related to this species and both have 2n = 40 (Yadav and Rajan, 1993). It is believed that the hybrid obtained from crossing of both the species was instrumental in evolution of mango. *M. khasiana* is confined to Sikkim Khasi Hills of Meghalaya.

Genus *Citrus*

Genus *Citrus* has about seven taxa of wild species in Eastern Himalaya. *C. jambhiri* found in Arunachal Pradesh is known for drought tolerance. *C. medica* which is found in Sikkim and Arunachal Pradesh can contribute to moisture stress and low soil fertility. It is the progenitor of many citron lemons having superior quality albedo (Sharma *et al.*, 2004). Taxa reported from Arunachal Pradesh are *C. limetta, C. nobilis* (subf. *Reticulata*), *Citrus aurantium* (subf. *Sinensis*) and *C. medica* f. *limon*. Latter one contributes genes for water logging and low-lying areas and also has prolific bearing tendency (Sharma *et al.*, 2004). *Citrus aurantium* is suited for stress conditions of soil and climate. *Citrus aurantifolia* is found in wild state at Sikkim. All the above taxa come under subgenera *Eucitrus*. The taxa such as *C. megaloxycarpa, C. assamensis, C. grandis* and *C. nobilis* are cultivated ones.

Citrus ichangensis (subgenera *Papeda*) distributed in Eastern Himalayas is frost tolerant (Singh and Chadha, 1993_b). This plant is successfully employed in evolving cold resistant hybrids such as ichandrins, ichang lemon and yuzu. Apart from above, Kokaya (1983) reported five new species and a subspecies from the region. They are *C. juco, C. kinosu, C. sechen, C. sechen* ssp. *sjanchen, C. serotina* and *C. tanaka*. But their validity and potential uses are unknown. Related genera/species occurring are *Hesperuthusa crenulata, Atalantia* spp., *Aegle marmelos, Feronia limonia*, etc. *Feronia limonia* is a rootstock for high density planting and has resistance to root knot nematode.

Genus *Malus*

Because of physical isolation in a range of environments occurred during comparatively recent evolutionary history, hybridization between the *Malus* species is relatively easy. The products of evolution include polyploidy and apomictic species which possess particular attributes that are helpful to apple breeders (Way *et al.,* 1990).

Malus baccata (2n = 34, 68) occurs at an altitude of 2000-3000 m showing tremendous polymorphism. It can be easily propagated through mount layering. It is crossable with apple (Cummins and Aldwinckle, 1983; Korban, 1986). It has good compatibility with 'Golden Delicious' apple as a rootstock. It possess well developed root system and adaptable to drought, poor soils of cold region and moderately tolerant to water logging. It has partial resistance to collar rot, root rot and wooly aphid. Harris *et al.* (2002) considered it as one of the close relatives of Central Asian wild apple *M. sieversii,* and considered its role in the origin of domesticated apple. It is one of the parents for inter-specific hybrids *viz. M.* x *adstringens, M.* x *hartwigii, M.* x *micromalus* and *M.* x *robusta* (Wiersema 1985; Korban 1986).

Malus sikkimensis (2n = 51) is closely related to *M. baccata,* found at 2500-2800 m height. It is apomictic; hence getting true-to-type is easy. Though it is resistant to powdery mildew and collar rot (Randhawa, 1987), it is highly sensitive to latent viruses of scion wood. *M dirangensis* is reported from Arunachal Pradesh growing at an altitude of 1600-1900 m and can serve as rootstock. *Docynia indica* is found at 1000-1800 m and is graft-compatible with apple. It is more closely related to quince. It gives semi-dwarf nature to the apple when grafted onto it. It was found resistant to collar rot (Pareek *et al.,* 1998).

Genus *Pyrus*

Very less breeding work on pear has been done in India although breeding objectives set out were high yield, increased quality, staggered harvest, ecological balance, and resistance to pests and diseases, compact and efficient plant types.

P. pashia is found in groups at an altitude of 750-2700 m and exploited as rootstock of European and Asiatic pear in India owing to its hardiness and adaptability to wide range of soil and climate conditions. Apart from seed, it is also propagated through rooted cuttings and mound layering. Tremendous variability with respect to fruit size, fruit color and leaf shape was observed. Temporal variation in flowering and fruit maturity was also noticed. It is crossable with pear and its late maturing character is of interest in breeding for late pear varieties. When used as rootstock, induces heavy bearing in scion material. It has been reported as susceptible to powdery mildew and collar rot but resistant to root rot (Ram and Randhawa, 1979). Bell (1990) reported medium resistance of this species to blossom blast, pear scab, pear decline, codling moth and blister mite.

P. arachnoidea reported from Sikkim by Naithani (1990) is not known for its potential. Related genera *Cotoneaster, Crataegus, Cydonia, Malus* and *Sorbus* found have potential as breeding material for pear (Zukovskij, 1962). Other related genera like *Sorbus insignis,* endemic to Eastern Himalaya (Jain, 1982); *S. cuspidata* (northern part of West Bengal at 2700-3700 m), *S. rhamnoides* (at 2700-3500 m), *S. microphylla* (at 3000-4200 m), *S. foliolosa* (at 2400-4000 m) and *S. ursina* (at 2900-4300 m) can be mentioned.

Genus *Prunus*

Ghora and Panigrahi (1995) recognized five subgenera of genus *Prunus* occurring in India as *Amygdalus* (almond, peach), *Prunus* (plum, apricot), *Cerasus* (sour, sweet and flowering cherries), *Padus* (Bird cherries) and *Lauro-Cerasus* (evergreen laurel cherries). Only the last three subgenera are represented in the Eastern Himalayas. Even though the region is not the centre diversity for cherry, it is sufficiently a rich home of wild relatives of cherry as is evident from presence of more than ten wild relatives. They are distributed in different altitudes hence may offer immense potential for popularizing cherry in non-traditional areas.

Subgenus *Cerasus* Pers.

Cultivated cherry belongs to section Cerasus Koehne while all the wild relatives in Eastern Himalaya come under the section *Pseudocerasus* Koehne. *P. cerasoides* (2n=16) has wider adaptability and grows between the altitude of 1000 and 2300 m. It is graft compatible with cherry but delayed incompatibility is reported. Watkins (1976) related this species with *P. campanulata* which is found up to 1500 m. The tree is small to medium size, spreading with dense and handsome canopy. It suckers freely from the exposed underground roots (Pradhan and Tamang, 2000) and also propagates from cuttings and seeds. The genetic variability is observed for fruit size, colour, pulp/ kernel ratio, sweetness and occasionally in plant type. It has several desirable attributes like gregarious bearing, very hardy and very attractive (orange red) fruit colour. Besides, it has low gestation period (of 2-3 years); very early in fruiting (March-April) when flowering just begin in cultivated cherry, sets fruits heavily; requires low chilling (250-400 hours below 7°C). Its botanical variety var. *majestica* is reported up to 3000 m in northern part of West Bengal and Sikkim whose flowers are relatively bigger sized. Another allied species found in the same area is *P. carmesiana* which is common in wet forests and river banks from 2300 to 2600 m. Potential of *P. rufa* (at 3000-3800 m) and *P. himalaica* (Sikkim) in the current context is unknown.

Subgenus *Padus* (Moench) Koehne

P. cornuta is distributed in Eastern Himalayas at altitude from 2100 to 3500 m. It bears multiple fruits/cluster. A botanical variety var. *villosa* also exists in the region (Ghora and Panigrahi, 1995). It is found susceptible to powdery mildew but resistant to root rot. It has good graft compatibility when used as a rootstock for cultivated cherry (Singh and Gupta, 1972). It is adaptable to cold and is closely related to *P. nepaulensis*, *P. padus* and *P. acuminata* (Arora and Nayar, 1984). *P. nepaulensis* is found between 1200 and-2600 m, flowers during lean season (Oct.-Nov.) and fruits take more time to mature (by July). It is compatible with cherry as rootstock. Fruits are bigger in size and pulpier than other wild species. It is related to *P. cornuta*, *P. padus* and *P. acuminata* (Arora and Nayar, 1984). It is resistant to collar rot and powdery mildew (Pareek *et al.*, 1998). It is a good bee pasturage plant and also gives green foliage in extreme winter. Another species *P. bracteopadus* reported from Sikkim and northern part of West Bengal by Ghora and Panigrahi (1995) is closely related to *P. nepaulensis* but differ mainly by the longer persisting bracts and more minute serrations in leaves. Other species coming under this subgenus are *P. glauciphylla, P. venosa* (at 1800-2000 m) and *P. wattii* but their potential is relatively unknown.

Subgenus *Lauro-Ceresus* (Tournef. Ex Duham). Rehder

P. undulata is an uncommon species growing wild between 1900 and 2700 m altitude habituating hilly forests or river valleys. It has been found incompatible as rootstock with peach and plum, however, shown resistance to powdery mildew (Kishore and Randhawa 1993; Randhawa, 1987). *P. jenkensii* (up to 900m) is also cultivated for its valuable fruits in the native areas. Both of them come under section *Lauro-Cerasus* (Duham). Miq. *P. arborea* var. *montana* coming under section *Meso-Pygeum* (Koehne) Kalkman is found in Sikkim up to 1500 m. Apart from the above, *Prunus anadaenia* was reported from Sikkim by Naithani (1990) is unknown for its potential.

Genus *Fragaria*

Cultivated strawberry (*F. x ananassa*) is an octaploid whereas related species available in India are diploids. These relatives produce sympodial runners and are self compatible in nature. The ploidy differences often make the cross more complex. Related genera are *Duchesnea* and *Potentilla*, all belonging to the tribe *Potentilleae*. Ellis (1962) obtained considerable success in hybridization between this genus and *Potentilla. Duchesnea indica* (up to 2400m) and cultivated strawberry produced many putative hybrids when former as female parent (Marta *et al.,* 2004). Three species of *Fragaria* are reported from Eastern Himalaya. *F. nubicola* commonly occurs in wild state between 1800-3800 m throughout this region along forest shrubberies and shady banks. It closely resembles *F. vesca* but with slender runners. Darrow (1966) suggested its potential for cold hardiness in breeding. Sangiacoma and Sullivan (1994) obtained synthetic octaploid by hybridization between *F. moschata* (2n=6x=42) and *F. nubicola* followed by chromosome doubling. *F. daltoniana* is distributed at an altitude from 2000 to 3600 m. It has potential for increasing the fruit size and inducing cold hardiness in strawberry (Darrow, 1966). It may be used in rock-gardens or for carpeting. *F. neilgherrensis* is also found in Eastern Himalaya at an altitude from 1800 to 2500 m). The plants are robust and fruits taste like that of banana. It has potential for increasing fruit size in cultivated strawberry (Darrow, 1966). Mochizuki *et al.* (1997) developed amphidiploids from interspecific hybrids of strawberry and this species by *in vitro* chromosome doubling.

Genus *Vitis*

Even though grape is a crop of tropics and subtropics of India, many of its relatives are found in temperate region also. India is considered to be a secondary centre of origin of *vinifera* grapes due to its resemblance with the characters of *V. lanata* found in Himalayas (Jindal and Singh, 1993). It is reported that species of section *Euvitis* are interfertile and they are separated by geographical, phonological and ecological barriers. *V. jacquemontii* (northern part of West Bengal, Sikkim) is reported to produce hybrids with *V. parvifolia* which is found in north east India. *V. latifolia* and *V. planicaulis* are also reported from this region. Intergeneric grafting between grape and *Tetrastigma lanceolarium* is reported to be successful. *V. lanata* (2n=38; subgenus *Euvitis*) is found throughout the Eastern Himalayas at an altitude ranging from 600 to 1800 m. It is resistant to fruit cracking and anthracnose (Pareek *et al.,* 1998). It is intercrossable with grapes and can be used for the development of disease tolerant rootstock. The line H-144 is the cross between grape cv. Cheema Sahebi and *V. lanata*.

Genus *Rubus*

Genus *Rubus* has around 740 species in world. All the wild species occurring in the region are more related to raspberry (*R. idaeus*) rather blackberry (*R. fruticosus*). Moore (1997) narrated the potentialities of three Eastern Himalayan species in raspberry improvement in USA. They are *R. ellipticus* (heat and drought tolerant; 600-2300 m), *R. lasiocarpus* (600-2700 m; tolerant to cane botrytis disease as mentioned by Jennings *et al.* (1990) and *R. biflorus* (1300-3300 m; heat and drought tolerant and low chilling requirement, erectness; a hybrid of it with *R. idaeus* known for heat tolerance). Another species *R. rosaefolius* (northern part of West Bengal and Sikkim at 1300-2300 m) is known for its heat tolerance and low chilling requirement. It was found cultivated in warm temperate Africa and Australasia. It was reported to be resistant to leaf spot *Sphaerulina rubi.* One selection from it namely 'Mysore' was cultivated in subtropics (Brooks and Olmo, 1972). The cross between *R. occidentalis* and 'Mysore' cultivar was found to be fertile (Ourecky and Slate, 1966). The hybrid between *R. ellipticus* and *R. rosaefolius* is known as *R. x probus.* There are many *Rubus* species found in temperate Himalaya but their taxonomical jargon has to be cleared out before attempting any breeding work. Some potential distinct species found in the region easy for exploitation other than those mentioned above includes *R. macilentus* (2000-3000 m; sweet fruit), *R. niveus* (2000-3300 m; sweet fruit) and *R. moluccanus* (up to 2300 m; large-sized fruit). Other species occurring in the region *R. splendidissimus* (West Bengal, Sikkim at 2400-3000 m), *R. fragarioides* (endemic to the region), *R. nepalensis* (northern part of West Bengal, Sikkim at 2100-3000 m), *R. paniculatus* (northern part of West Bengal, Sikkim at 1500-2900 m), *R. calycinus* (at 2100-3000 m), *R. acuminatus* (at 1000-2300 m), *R. hypargyrus* (at 2200-3300 m), *R. reticulatus* and *alpestris* (northern part of West Bengal, Sikkim at 2300-3300 m), *R. treutleri* and *hookerii* (Sikkim 2300-3300 m), *R. lineatus* (at 2000-2600 m), *R. sikkimensis* (Sikkim) and *R. ferox* (Arunachal Pradesh). Apart from above, Naithani (1990) mentioned few more species native to the region such as *R. alexeterius, calycinoides, phengodes, senchalensis* and *pectinarioides* (Sikkim); *R. burkilii, pectinaris, sumatranus* and *ghanakantus* (Arunachal Pradesh) and *R. indotibetanus* (Sikkim and Arunachal Pradesh). Potential of all these species are unknown.

Genus *Actinidia*

Kiwifruit (*A. deliciosa*) came for cultivation in India recently. Often ploidy level difference (cultivated one is hexaploid while the wild ones are diploid) makes interspecific crosses unsuccessful. *A. callosa* distributed in the region from 900 to 2700 m. Its fruits contain high vit. C (15-25mg/100g). It is propagated through cuttings. It is good source for frost tolerance in kiwi fruit. Another botanical variety var. *pubescens* is reported from Arunachal Pradesh up to a height of 1500 m (Naithani, 1990). *A. strigosa* is found in northern part of Bengal and Sikkim at an altitude of 1800-2600 m.

Genus *Phoenix*

The species of *Phoenix* are fully interfertile but more or less reproductively isolated from one another by geographical and ecological barriers (Zohary and Hoff, 1988). Interspecific crosses of *P. dactlyifera* with *P. sylvestris* (up to 1500 m); *P. rupicola* (up to 1700 m) and *P. humilis* (foothills) are successful and have meta-xenic effects. *P. sylvestris* is considered to be the progenitor of date palm. Its can provide earliness. *P. humilis* has been found to induce small seediness in several date cultivars. It is also a good source for inducing dwarf trait to date palm.

Genus *Garcinia*

G. xanthochymus and *sopsopia* occurring in the region up to an altitude of 1000 m is a successful rootstock for mangostan. *G. pedunculata* and *atroviridis* are reported from Arunachal Pradesh.

Genus *Artocarpus*

Wild form of jackfruit is found in Sikkim. *A. chama* is reported from tropical evergreen forests and *A. lakoocha* from the foothills of the region whereas *A. chaplasha* from Sikkim and Arunachal Pradesh.

Other Fruits

Pistachio Nut

Although pistachio nut (*P. vera*) is the crop of Mediterranean countries, *P. chinensis* ssp. *integerrima* is found in Arunachal Pradesh between 1000-2600 m. Cultivars budded over this species initially grow slowly but pickup growth subsequently and out yield those budded over *P. vera*. It is resistant to soil borne diseases particularly *Verticillium* wilt and hybridize with *P. atlantica* (Hartmann *et al.*, 1997).

Hazelnut

Corylus ferox is the representative of this genus found in the Eastern Himalayas from 2400 to 3000 m.

Walnut

Wild form of *Juglans regia* exists in wild form throughout Eastern Himalayas (1600-3000m) which can offer scope for extending walnut cultivation in these non-traditional areas.

Syzygium jambos

Wild forms are found in Terai regions of Sikkim.

Fig

Ficus glomerata found in low hills (up to 700m) of Eastern Himalayas can act as polliniser for *F. carica*.

Persimmon

Oriental persimmons which may have arisen from *Diospyros roxburghii* (Pareek and Sharma, 1993) occur in the region. *D. lotus* is naturalized in Arunachal Pradesh between 1700-2000 m. It is adjudged as the best rootstock for cultivated persimmon (*D. kaki*) in subtropical and temperate areas. Apart from its multiple uses, it has fibrous root system and hence easy for transplanting from nursery. It is the only cold adaptable species of *Diospyros* existing in India apart from the cultivated ones.

Currants

This fruit is not cultivated in India but its wild relatives in considerable number occurs in the region are *Ribes orientale* (2100-4000 m), *R. glaciale* (2600-4400 m), *R. alpestre* (2400-3600 m), *R. himalense* (2400-3300 m), *R. griffithii* (2700-4000 m), *R. tagare* (2200-3300 m), and *R. tenue* (Sikkim). *R. orientale* and *R. glaciale* are resistant to white pine blister rust (Brennan, 1996). *R. alpestre* related to European gooseberry *R. grossularia* L. is resistant to *Aphis grossulariae* and produced fertile hybrid with the latter species and *R. nigrum* (Keep, 1975).

Blueberries

They are also not cultivated in India but their wild relatives such as *Vaccinium donianum, V. nummularia* (2400-4000 m), *V. retusum* (1400-3600 m) and *V. vacciniaceum* (2400-3000m), *V. venosum* (alpine areas of Arunachal Pradesh) and *V. forrestii* (Arunachal Pradesh) are found in the region. Their worthiness is yet to be assessed although there are reports that there is no major breeding barrier between the cultivated and the wild species.

Ber

Wild relatives of ber are represented by *Ziziphus oenoplia* and *Z. rugosa* found in foothills of EH. They possess no graft-incompatibility with *Z. mauritiana* hence can be encouraged as rootstock.

Loquat

The genus *Eriobotrya* is represented by *E. angustissima* (1500-1800 m), *E. bengalensis* (up to 1500 m) and *E. dubia* in the region. Jain (1982) reported that *E. hookeriana* is endemic to the region.

Avocado

There are few wild relatives represented in the region such as *Machilus gammieana* (Jain, 1982), *M. edulis* and *M. gamblei*.

Chestnut

Castanea armata is the only representative of chestnut genera apart from allied genera *Castanopsis* which is well represented in Eastern Himalaya.

Hippophae rhamnoides

It is an emerging crop in Himalaya. *H. tibetana* and *H. rhamnoides* ssp. *gyantsensis* are reported from Sikkim (Naithani, 1990) may form related species.

Olive

Olea gamblei (endemic to northern part of West Bengal and Sikkim from 600 to 900 m) and *O. dioica* may offer scope for cultivating olive in non traditional areas.

Conclusion

Wild/weedy races are the close components of cultivated species and together form a genetically compatible complex. Wild forms of mango, banana, citrus, jackfruit, and persimmon exist in the Eastern Himalaya. It is well known that interspecific hybridization has been deliberately used to create new fruits in citrus (tangors, tangelos), _Prunus_ (plumcot) and _Rubus_ (tayberry). In some cases, domestication of fruit crops is aided by vegetative propagation of elite types and wild species _e.g._ apple and date palm. The tropical members of families with temperate zone representatives may represent the possibility of breeding for cold tolerance which also helps in popularization of crops in non-traditional areas. All this necessitates the correct identification of wild relatives, alleviating nomenclatural problems, biosystematic studies, pre-breeding and/or genetic enhancement, finding out ways and means of shortening long gestation period involved therein. All this efforts will lead to rapid use of wild relatives in fruit crop improvement. Priority can be given to those genera like _Malus, Mangifera, Musa, Pyrus, Prunus, Vitis, Rubus_ and _Fragaria_ which are having crossable wild relatives with known contributing traits. These traits can be incorporated into cultivated types by designing long term breeding programs. Endemic species should be given thrust for _in situ_ conservation.

References

Arora, R.K. and Nayar, E.R., 1984. _Wild Relatives of Crop Plants in India._ NBPGR Sci. Monogr. No. 7, NBPGR, New Delhi.

Bell, R.L., 1990. Pears (_Pyrus_). In: _Genetic Resources of Temperate Fruit and Nut Crops_, (Eds.) J.N. Moore and J.R. Jr. Ballington. ISHS, Wageningen, The Netherlands, pp. 657–697.

Brennan, R.M., 1996. Currants and gooseberries. In: _Fruit Breeding, Vol 2: Vine and Small Fruit Crops_, (Eds.) J. Janick and J.N. Moore. John Wiley and Sons, Inc. pp 191–295.

Brooks, R.M. and Olmo, H.P., 1972. _Registration of New Fruit and Nut Varieties_, 2nd Edn. Berkeley, Univ. California Press, pp. 528–547.

Cummins, J.N. and Aldwinckle, H.S., 1983. Breeding apple rootstocks. _Plant Breed. Rev._, 1: 294–394

Darrow, G.M., 1966. _The Strawberry: History, Breeding and Physiology._ Holt, Rinehart and Winston, New York.

Ellis, J.R., 1962. _Fragaria–Potentilla_ intergeneric hybridization and evaluation in _Fragaria. Proc. Linn. Soc. London_, 173: 99–106.

Ghora, C. and Panigrahi, G., 1995. _The Family Rosaceae in India: Revisory Studies on Six Genera (Prunus, Prinsepia, Maddenia, Rosa, Malus and Pyrus) Vol. 2._ Bishen Singh Mahendra Pal Singh, Dehradun, India, 481 p.

Harris, S.A., Robinson, J.R. and Juniper, B.E., 2002. Genetic clues to the origin of the apple. _Trend Genet.,_ 18: 426–430.

Hartmann, H.T., Kester, D. and Davis, F.T., 1997. _Plant Propagation: Principles and Practices_, 6th Edn., Indian Reprint 2002. Prentice Hall of India Pvt. Ltd., New Delhi, 770 p.

Hore, D.K., Sharma, B.D. and Pandey, G., 1992. Status of banana in northeast India. *J. Econ. Taxon. Bot.*, 16: 447–455.

Jain, S.K., 1982. Botany of Eastern Himalaya. In: *The Vegetational Wealth of Himalaya*, (Ed.) G.S. Paliwal. Puja Publishers, New Delhi.

Jennings, D.L., Daubeny, H.A. and Moore, J.N., 1990. Blackberries and raspberries (*Rubus*). In: *Genetic Resources of Temperate Fruit and Nut Crops*, (Eds.) J.N. Moore and J.R. Jr. Ballington. ISHS, Wageningen, The Netherlands, pp. 329–389.

Jindal, P.C. and Singh, R., 1993. Genetic resources of grape. In: *Advances in Horticulture, Vol. 1 Fruit Crops*, (Eds.) K.L. Chadha and O.P. Pareek. Malhotra Publishers, New Delhi, pp. 171–187.

Keep, E., 1975. Currants and gooseberries In: *Advances in Fruit Breeding*, (Eds.) J. Janick and J.N. Moore. Purdue University Press, West Lafayette, Indiana.

Kishore, D.K. and Randhawa, S.S., 1993. Wild germplasm of temperate fruits. In: *Advances in Horticulture, Vol. 1 Fruit Crops*, (Eds.) K.L. Chadha and O.P. Pareek. Malhotra Publishers, New Delhi, pp. 227–241.

Kokaya, T.D., 1983. New species of Citrus L. *Trudy. Prikl. Bot. Genet. Selek* (in Russian), 78: 98–103.

Korban, S.S., 1986. Interspecific hybridization in *Malus. Hort. Sci.*, 21: 41–48.

Marta, A.E., Camadro, E.L. and Diaz-Ricci, J.C., 2004. Breeding barriers between the cultivated strawberry *F. x ananassa* and related wild germplasm. *Euphytica* 136: 136–150.

Maxted, N., Ford-Lloyd, B.V., Jury, S.L., Kell, S.P. and Scholten, M.A., 2006. Towards a definition of crop wild relative. *Biodiversity and Conservation*, 14: 1–13.

Mochizuki, T., Noguchi, Y., Sone, K. and Morishita, M., 1997. Aroma components of amphidiploid strawberries derived from interspecific hybridization of F. x ananassa and diploid wild species. *Acta Horticulture*, 439: 75–80.

Moore, J.N., 1997. Blackberries and raspberries in the southern United States: Yesterday, Today and Tomorrow. *Fruit Var. J.*, 51: 148–157.

Naithani, H.B., 1990. *Flowering Plants of India, Nepal and Bhutan: Not Recorded in Sir J.D. Hooker's Flora of British India*. Surya Publishers, Dehradun.

Ourecky, D.K. and Slate, G.L., 1966. Hybrid vigor in *Rubus occidentalis–R.leucodermis* seedlings. *Proc. 17th Intern Hort. Cong.* 1: 277 (Abstr).

Pareek, O.P., Sharma, S. and Arora, R.K., 1998. *Underutilized Edible Fruits and Nuts*. IPGRI Office for South Asia, New Delhi, India.

Pareek, O.P. and Sharma, S., 1993. Genetic resources of underutilized fruits. In: *Advances in Horticulture, Vol. 1 Fruit Crops*, (Eds.) K.L. Chadha and O.P. Pareek. Malhotra Publishers, New Delhi, pp. 189–225.

Pradhan, M. and Tamang, S., 2000. *In situ* sucker formation in *Maesa chisia* D. Don and *Prunus cerasoides* D. Don. *J. Hill Res.*, 13: 47–48.

Ram, R.D. and Randhawa, S.S., 1979. Resistance of different species of pome and stone fruits powdery mildew incited by *Podosphaera leucotricha*. *Sci. Cult.*, 45: 256.

Randhawa, S.S., 1987. *Wild Germplasm of Pome and Stone Fruits*. IARI Regional Station, Shimla. 51p.

Randhawa, S.S. and Ram, R.D., 1979. Potentiality of the indigenous germplasm in the improvement of pome and stone fruits. In: *Fruit Breeding in India*, (Ed.) G.S. Nijjar. Oxford and IBH Publishing Company, New Delhi, pp. 131–138.

Sangiacoma, M.A. and Sullivan, J.A., 1994. Introgression of wild species into cultivated strawberry using synthetic octoploids. *Theor. Appl. Genet.*, 88: 349–354.

Sharma, B.D., Hore, D.K. and Gupta, S.G., 2004. Genetic resources of citrus of North east India. *Genetic Resources and Crop Evolution*, 51: 411–418.

Singh, H.P. and Chadha, K.L., 1993a. Genetic resources of Banana. In: *Advances in Horticulture, Vol. 1: Fruit Crops*, (Eds.) K.L. Chadha and O.P. Pareek. Malhotra Publishers, New Delhi, pp. 123–142.

Singh, H.P. and Chadha, K.L., 1993b. Genetic resources of Citrus. In: *Advances in Horticulture, Vol. 1: Fruit Crops*, (Eds.) K.L. Chadha and O.P. Pareek. Malhotra Publishers, New Delhi, pp. 95–121.

Singh, R.N. and Gupta, P.N., 1972. Rootstock problem in stone fruits and potentialities of wild species of *Prunus* found in India. *Punjab Hort. J.*, 12: 157–175.

Watkins, R., 1976. Cherry, plum, peach, apricot and almonds. In: *Evolution of Crop Plants*, (Ed.) N.W. Simmonds. Longman, London, pp. 242–247.

Way, R.D., Aldwinckle, H.S., Lamb, R.C., Rejman, A., Sansavini, S., Shen, T., Watkins, R., West Wood, M. N. and Yoshida, Y., 1990. Apples (*Malus*). In: *Genetic Resources of Temperate Fruit and Nut Crops*, (Eds.) J.N. Moore and J.R. Jr. Ballington. ISHS, Wageningen, The Netherlands, pp. 1–63.

Wiersema, J.H., 1985. *Nomenclature of Malus*. Unpublished monograph, Plant exploration and Taxonomy Lab, United States Department Agriculture Beltsville, Maryland. 22p.

Yadav, I.S. and Rajan, S., 1993. Genetic Resources of *Mangifera*. In: *Advances in Horticulture, Vol. 1: Fruit Crops*, (Eds.) K.L. Chadha and O.P. Pareek. Malhotra Publishers, New Delhi, pp. 77–93.

Zohary, D. and Hoff, M. 1988. *Domestication of Plants in the Old World*. Oxford Science Publications. Claredon Press, Oxford, London.

Zukovskij, P.M., 1962. *Cultivated Plants and their Wild Relatives* (Abridged translation from Russian text by Hudson, P.S). Commonwealth Agriculture Bureaux, England. 107p.

2013, Perspectives in Plant Biodiversity   *Pages* **119–128**
Editor: **Dr. K. Muthuchelian,** *Vice Chancellor, Periyar University, Salem*
Published by: Daya Publishing House, NEW DELHI

Chapter 16

Isolation and Analyzing the Aromatic Compounds from *Vetiver* sp. (*Vetiveria zizanioides* L. Nash) Plant Root System by Using FT-IR Method

*A. Sekaran[1], K. Muthuchelian[2], B. Pavendan[3],
N. Hariram[4] and A.Sundaram[5] ***
[1]Department of Futures Studies,
[2]Department of Bioenergy,
[5]Department of Solar Energy,
School of Energy, Environment and Natural Resources,
Madurai Kamaraj University, Madurai – 625 021
[3]Dynamic Research Centre, Mathur – 622 515
[4]Supreme Bio Energies A. Pudupatti, Alanganallur, Madurai – 625 501

ABSTRACT

Vetiveria zizanioides, an aromatic plant commonly known as Vetiver is used for its variety of ingredients. Vetiver grass (*Vetiveria zizanioides*) has potential medical benefits and is adaptable to a wide range of climatic environments. Although the grass is endemic in Sirumalai hills a part of Palani hills ecosystem, its potential for non-exploitative forest produce for local population, soil and water conservation by virtue of its fibrous root system has

* Corresponding Author: E-mail: sunmdfs@yahoo.co.in

not been fully realized. Vetiver, *Vetiveria zizanioides* L. Nash, a native grass of India has traditionally been in use in India for contour protection and essential oil production. The essential oil of Vetiver root is known to possess antioxidant properties but antioxidant potential of spent root extract has not been reported. Hence, in the present study, analysis of root compound from Vetiver plant occurring in Sirumalai hills at different locations is attempted. Anatomical studies (Cross Section of root) of root system was carried out followed by purification of Vetiver root compound by analysis using Fourier transformed infra red FT-IR method. C=C compound presence in the Vetiver root system was investigated.

Keywords: FT-IR (Fourier transformed infra red), CS: Cross section, C=C–C=C stretching within ring.

Introduction

Vetiver grass–*Vetiveria zizanioides*, L.Nash recently reclassified as *Chrysopogon zizanioides*, can be grown over a very wide range of climatic and soil conditions, and is suitable for tropical, semi-tropical, and Mediterranean climates. When Vetiver grass is grown in the form of a narrow self-sustaining hedgerow it exhibits special characteristics essential to many of the different applications that comprise the Vetiver System.

Highlighting that vetiver (i) is native to hygro-environment such as wetland, lagoon and bog, (ii) is extremely tolerant to drought as well as waterlogged/submergence conditions, (iii) is effective for soil and water conservation and (iv) is endowed with excellent biological features to suitable to wastewater reclamation and pollution mitigation.

By virtue of its long use by natives Vetiver has an ethno-botanic significance. In the last two decades use of variety of tropical plants for aromatic and medicinal oil extraction on a commercial scale has seen an increase.

Several plants have been termed as Miracle Grass, Wonder Grass and the alike for their capacity to create a living wall, a living filter strip and "live nail" reinforcement.

Vetiveria zizanioides (L). Nash (vetiver) is a perennial graminaceous plant growing wild, half-wild or cultivated in many tropical and subtropical areas (Taylor and Francis 2002). It is cultivated for its unique ability among grasses to produce in the root an essential oil, a complex mixture of sesquiterpene alcohols and hydrocarbons, which are mostly used as a basic material for perfumery and cosmetics. Because of this complexity, the oil is difficult to reproduce with synthetic aromatic chemical formulations. Moreover, differences in the quality of the oil may depend on genetic, environmental and technological factors (Taylor and Francis 2002; Champagnat *et al.*, 2006). The biological activity of the vetiver oil is also important. Termiticidal, insecticidal, antimicrobial and antioxidant activities of vetiver oil have been described (K. E. Nix *et al.*, 2006 and R. G. Grimshaw 1989). Vetiver has a cattle feed value (Grimshaw 1989) and is extensively used for land protection purposes as a barrier against erosion and for the restoration of contaminated land (R. Antiochia *et al.*, 2007) Vetiver oil is produced in secretory cells localized in the first cortical layer outside the endodermis of mature roots (Taylor and Francis 2002; J. Viano 1991). Thus, while most essential oils are extracted from aerial tissues of dicotyledonous plants, vetiver oil is distilled from the roots of this monocotyledonous plant. The terpene oils in aerial tissues are often found as a complex mixture of different terpene compounds, including monoterpenes and sesquiterpenes, arising from complex interactions between the action of the cytosolic (mevalonate) and plastidic (2-C-methylerythritol-4-phosphate) pathways, and the oils accumulate as extracellular exudates or in specialized glands (lactifiers) or oil bodies (associated with trichomes) (W. Eisenreich *et al.,* 2004). Much less is known about the biosynthesis, regulation and localization of terpenes synthesized in roots. Hence, in the present study, analysis of root compound from Vetiver plant occurring in Sirumalai hills at different locations is attempted. Anatomical studies (Cross Section of root) of root system was carried out followed by purification of Vetiver root compound by analysis using Fourier transformed infra red FT-IR method. C=C compound rich in the Vetiver root system was investigated.

Materials and Methods

Sample Collection

The sample collection (roots) of *Vetiveria zizanioides* were collected from different localities of Sirumalai hills, Dindugul, during January and July 2010. The samples of root (vetiver sp). were collected in clean plastic bag and labeled with date, number and location of samples.

Extractives

Extractives from the Vetiver grasses 35g were extracted using chloroform: methanol 2:1 ratio for 1hr. The extractives were scanned using a UV/VIS spectrophotometer at wavelengths ranging from 200-800nm. Then, solvent was removed with vacuum rotary evaporator. The dried extractives were weighed. The extractive-free samples were dried in an oven at 60'C for 2hrs. Further dried material was analysed by FT-IR for identification of present groups.

Cross Sectional Study

Morphological and physiological attributes of vetiver grass, were studied by ordinary hand section CS (cross section) by blade methods.

Results and Discussion

Agro Technology

Vetiver prefers tropical and subtropical climate for its proper growth, development and essential oil yield. It grows luxuriantly in places up to an altitude of 600 m, with an annual rainfall of 1000-2000 mm, temperature ranging from 21 to 44 C and in moderately humid climate. In places of scanty rainfall, which are otherwise suitable, it can be grown as an irrigated crop. The plant is sufficiently hardy and grows on almost all types of soils. Light soils, however, should be avoided as the roots obtained, produce very low percentage of oil. Red lateritic soils with abundant organic matter are considered ideal as the roots produced in such soils are thick and contain more essential oil. Heavy soils make harvesting of the roots difficult, with a loss of the finer roots which contain most of the oil. It can be grown even on saline and alkaline soils with a pH range of 8.5-10.0, being suitable otherwise. However, the oil contains more than 150 complex compounds including elemol 0.4-2.3 per cent, 10-epi-eudesmol 1.1-2.2 per cent, ß-eudesmol 5.5-8.5 per cent, vetiverol + cyclocopacamphenol 6.1-7.5 per cent, vetiselinenol 11-20 per cent, khusimol 13-28 per cent, ß-vetivone 2-5 per cent, -vetivone 1.5-5.8 per cent, among others (Joy *et al.*, 2001).

Study Site

Frequent field trips were carried out to the Sirumalai hill range (off-shoots) of Western Ghats and situated about 28 kilometers on the south of Dindigul District, Tamil Nadu. Information regarding the medicinal plants usage by Paliyars was gathered after developing a good rapport with the community and winning their confidence. It is interesting to note that the Paliyar community is opening up gradually after the interventions made by scientists and non-governmental organizations.

Morphological Characteristic of Vetiver Plant Species

Morphological and physiological attributes of vetiver grass, its strong, deep penetrating, aerenchymatous (Figures 16.1a–f) root system and gland cells, unusual ability to absorb and tolerate extreme levels of nutrients, agrochemicals and heavy metals make it an ideal system for environmental enhancement through appropriate interventions in the treatment regime (Truong, P. and Xia, H. 2003).

Table 16.1: Attributions of the main infrared peaks.

Frequence cm^{-1}	Assignment
3470-3420	O-H stretching highly dominant
3400-3300	O-H stretching highly dominant, N-H stretching (trace)
2940-2840	Aliphatic C-H stretching
1725-1718	C=O Stretching of COOH and Ketones (trace)
1660-1630	C=O stretching of amide groups (amide 1 band), quinone C=O and/or C=O of H bonded conjugated ketones
1620-1600	Aromatic C=C, stongly H-bonded C=O of conjugated ketones
1600-1585	C=C, stongly H-bonded C=O of conjugated ketones
1590-1520	COO⁻ symmetric stretching, N-H bending, C=N stretching (amide II bond)
1500-1400	C=C stretching within the ring*
1400-1390	OH deformation and C-O stretching of phenolic OH, C-H deformation of CH2 and CH3 groups, COO⁻ antisymmetric stretching
1280-1200	C-O stretching and OH deformation of COOH
1170-950	C-O stretcing of polysaccharide-like substances, Si-O of silicate impurities
950-500	S-O stretching, aromatic groups may regarded as insignificant
500-400	Aromatic amines
400-50	Carboxyl, amides, esters; OH, aromatic; C, H aromatic; Acetals, ketals; C=O, Carobn; Alkyl carbon

Source: Stevenson, 1994; Silverstein *et al.,* 1991.

Figures 16.1a–f: Cross sectional view of Vetiver root system

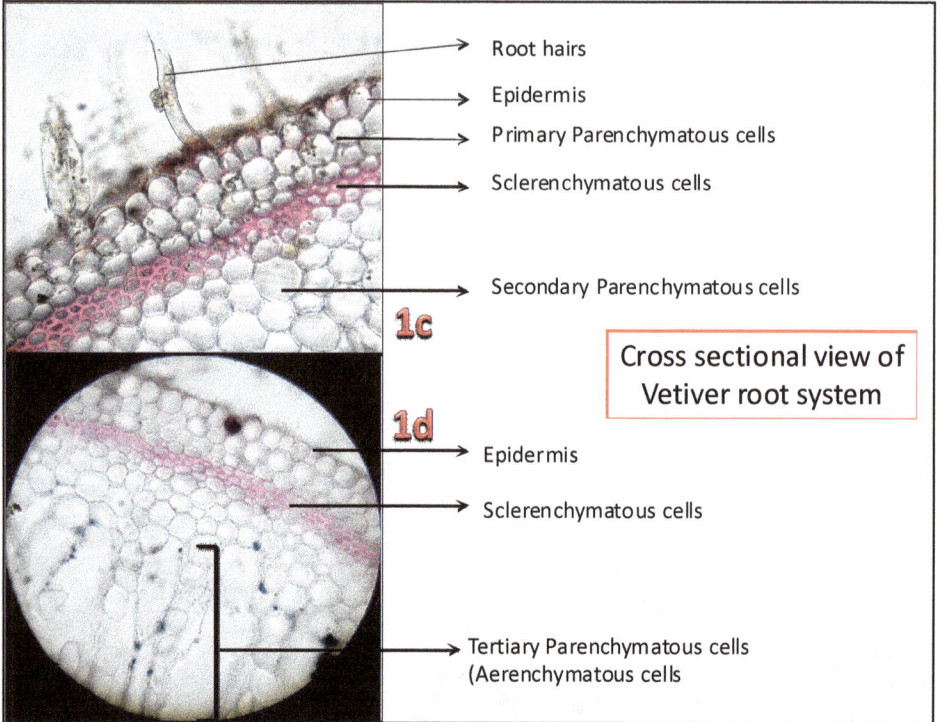

Root hairs

Epidermis

Primary Parenchymatous cells

Sclerenchymatous cells

Secondary Parenchymatous cells

1c

Cross sectional view of Vetiver root system

1d

Epidermis

Sclerenchymatous cells

Tertiary Parenchymatous cells (Aerenchymatous cells

Cross sectional view of Vetiver root system

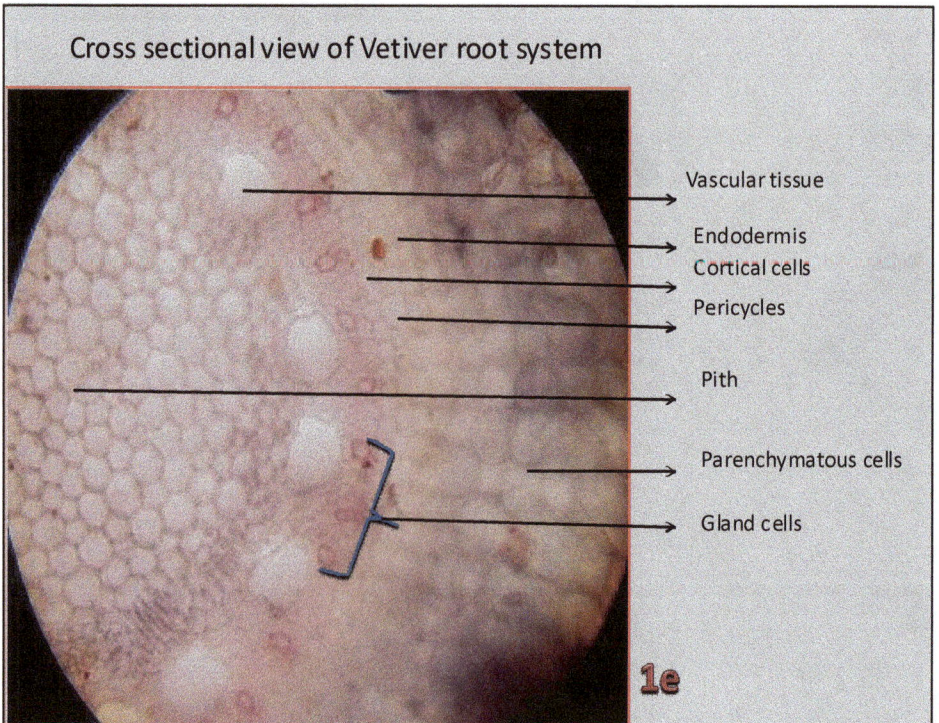

Vascular tissue

Endodermis
Cortical cells

Pericycles

Pith

Parenchymatous cells

Gland cells

1e

Cross sectional view of mature Vetiver root system

Endodermis
Perichycle

Cortical cells

Glands
Xylum

Phloem

Pith

1f

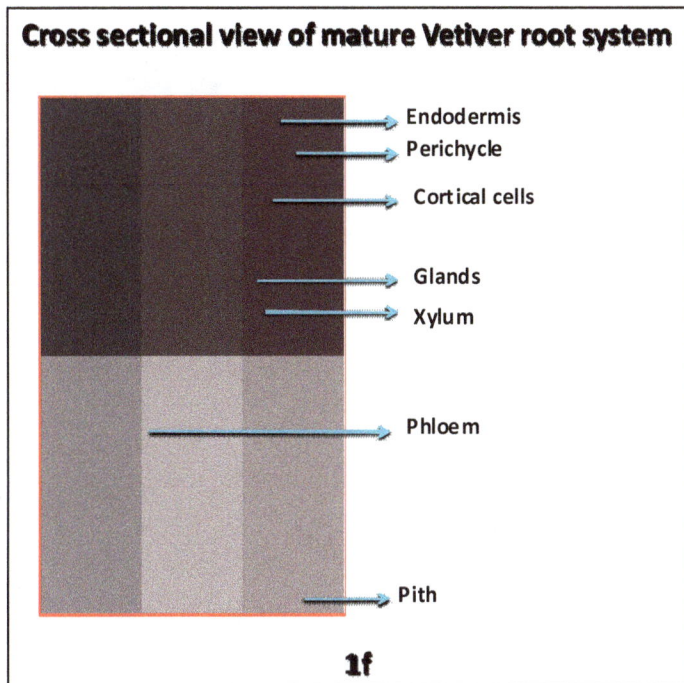

Infrared Spectra (FT-IR)

The FT-IR spectra showed features typically attributed and difference in leaf and root compound demonstrated the uniformity of the extracted material (Figures 16.2a–c and Table 16.1). In order to produce a clear presentation of the results, only one sample from each site is presented in these figures. Samples from the same sites presented identical FT-IR spectra. However results indicate that two peaks were observed in this region *Cympapogan citrate* plant (around 2920 and around 2850 cm^{-1}) (Arrow mark).

These absorption peaks are attributed, respectively, to the asymmetrical and symmetrical stretching of methylene (-CH2-) groups this being characteristic of aliphatic and nonstrained cyclic hydrocarbons (Silverstein *et al.,* 1991). In contrast, negative in (-CH2-) groups in vetiver root. Interestingly (1508.23, 1458.08 and 1421.44 cm–1) (Arrow mark) indicate that three peaks Being a terrestrial material highly unsaturated and with low nitrogen content (see its N/C and H/C ratios) the conjugated C=C stretching might be responsible for this study. The bands in this range of the infrared spectrum have several assignments (Table 16.1) including aromatic C=C stretching, amide group (RCONH2) for such an absorption signal found major in Vetiver root.

The present study has been done indicate that aromatic C=C stretching with in ring and very similar to *Chympapogan citrate* and *Vetiveria zizanioides* (1633.59 and 457.10 cm^{-1}) were found in Aromatic amines and S-O stretching, aromatic groups

Figure 16.2a-b

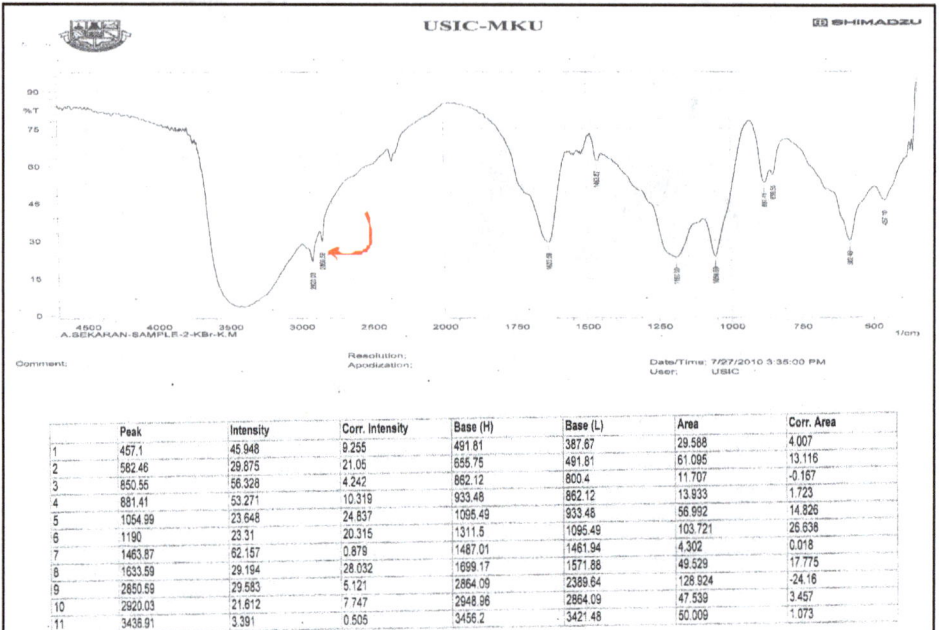

	Peak	Intensity	Corr. Intensity	Base (H)	Base (L)	Area	Corr. Area
1	457.1	45.948	9.255	491.81	387.67	29.588	4.007
2	582.46	29.875	21.05	655.75	491.81	61.095	13.116
3	850.55	56.328	4.242	862.12	800.4	11.707	-0.167
4	881.41	53.271	10.319	933.48	862.12	13.933	1.723
5	1054.99	23.648	24.837	1095.49	933.48	56.992	14.826
6	1190	23.31	20.315	1311.5	1095.49	103.721	25.638
7	1463.87	62.157	0.879	1487.01	1461.94	4.302	0.018
8	1633.59	29.194	28.032	1699.17	1571.88	49.529	17.775
9	2650.59	29.583	5.121	2864.09	2389.64	126.924	-24.16
10	2920.03	21.612	7.747	2948.96	2864.09	47.539	3.457
11	3436.91	3.391	0.505	3456.2	3421.48	50.009	1.073

2a

Control: Cympapogan citrate (leaf compound)

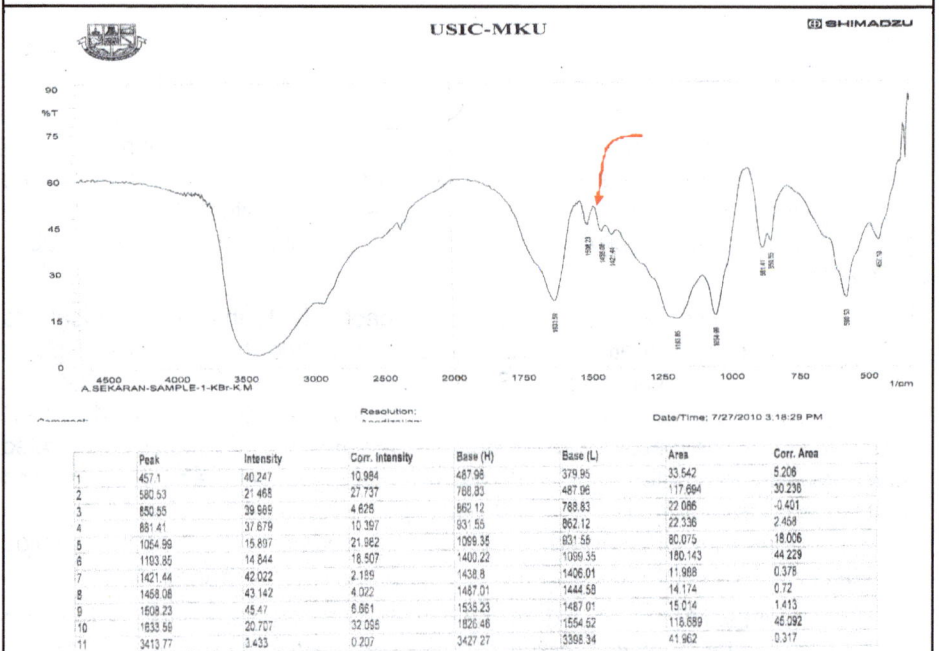

	Peak	Intensity	Corr. Intensity	Base (H)	Base (L)	Area	Corr. Area
1	457.1	40.247	10.984	487.98	379.95	33.542	5.206
2	580.53	21.465	27.737	788.83	487.98	117.694	30.238
3	850.55	39.969	4.626	862.12	788.83	22.086	-0.401
4	881.41	37.679	10.397	931.55	862.12	22.336	2.468
5	1054.99	15.897	21.962	1099.35	931.55	80.075	18.006
6	1193.85	14.844	18.507	1400.22	1099.35	180.143	44.229
7	1421.44	42.022	2.189	1438.8	1406.01	11.988	0.378
8	1458.08	43.142	4.022	1487.01	1444.58	14.174	0.72
9	1508.23	45.47	6.661	1535.23	1487.01	15.014	1.413
10	1633.59	20.707	32.098	1826.46	1554.52	118.689	46.092
11	3413.77	3.433	0.207	3427.27	3398.34	41.962	0.317

2b

Experiment: Vetiver root (root compound)

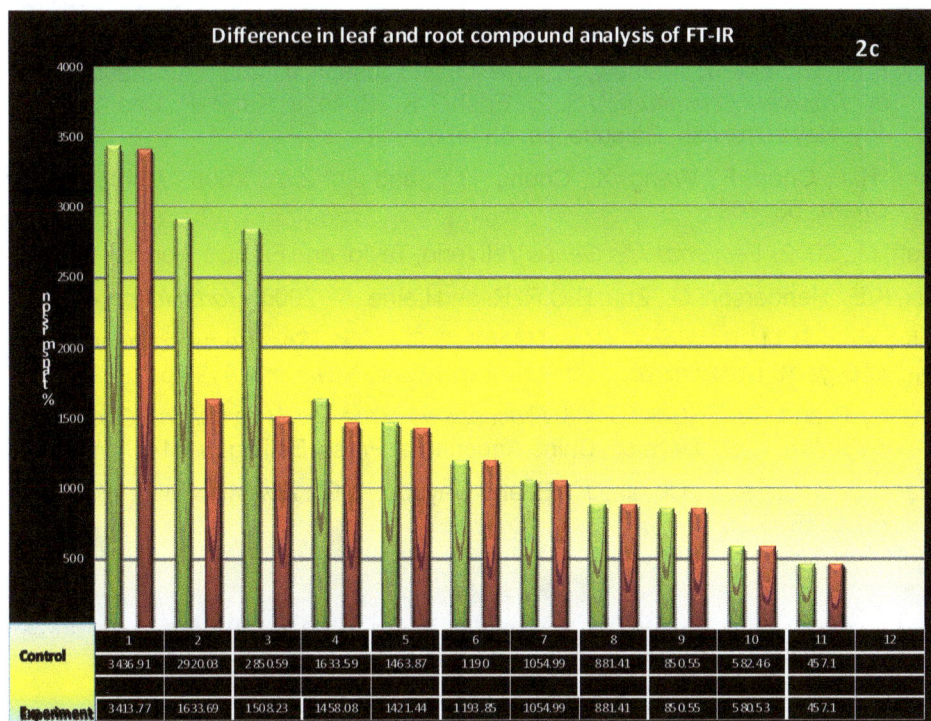

Figure 16.2c: Showing that per cent transmission of FT-IR analysis for difference in *Cympabogan citrate* leaf compound and *Vetiveria zizanoides* root compounds.

may regarded as insignificant, C-O stretching of polysaccharide-like substances, Si-O of silicate impurities (881.41 and 1054.99 cm^{-1}) were found in both plants.

Vetiver has traditionally been used as medicinal and aromatic plants in many countries, especially in Asia. Recently it has received widespread recognition as being an ideal plant for oil and water conservation as well as environmental protection. However, promoting vetiver hedgerows for soil and water conservation as the farmers complain that they do not obtain any direct benefit (*i.e.* cash return). However, it is argued that the indirect benefits the farmers could obtain are enormous.

Acknowledgements

We wish to thank The Director, Department of USIC for providing the support for FT-IR experiment, Madurai Kamaraj University, Madurai.

References

Antiochia, R., Campanella, L., Ghezzi, P. and Movassaghi, K., 2007. *Anal. Bioanal. Chem.*, 388: 947.

Champagnat, P., Figueredo, G., Chalchat, J.C. Carnat, A.P. and Bessiere, J.M., 2006. *J. Essent. Oil Res.,* 18: 416.

Eisenreich, W., Bacher, A., Arigoni, D. and Rohdich, F., *Cell Mol. Life Sci.* 61: 1401.

Grimshaw, R.G., 1989. *TVN Newslett.,* 2: 1.

Joy, P.P., Thomas, J., Mathew, S., Jose, G. and Joseph, J., 2001. Aromatic plants. In: *Tropical Horticulture,* Vol. 2, (Eds.) T.K., Bose, J. Kabir, P. Das and P.P. Joy. Naya Prokash, Calcutta, pp. 633–733.

Kim, H.J., Chen, F., Wang, X., Chung, H.Y. and Jin, Z.Y., 2005. *J. Agric. Food Chem.,* 53: 7691.

Maff, M., 2002. *Vetiveria: The Genus* Vetiveria. Taylor and Francis, London.

Nix, K.E., Henderson, G., Zhu, B.C.R. R. and Laine, A., 2006. *Hortscience,* 41: 167.

Silverstein, R.M., Bassler, G.C. and Morrill, T.C., 1991. *Spectrometric Identification of Organic Compounds.* John Wiley and Sons, New York, 419 pp.

Truong, P. and Xia, H. (Eds.), 2003. *Proceedings of the Third International Conference on Vetiver and Exhibition,* China Agricultural Press, Beijing, p. 614.

Viano, J., Smadja, J. , Conan, J.Y.E. and Gaydou, 1991. *Bull. Mus. Natl. Hist. Nat., Paris,* 13: 61.

2013, Perspectives in Plant Biodiversity   *Pages* 129–132
Editor: Dr. K. Muthuchelian, *Vice Chancellor, Periyar University, Salem*
Published by: Daya Publishing House, NEW DELHI

Chapter 17

A Comparative Survey on Water Sources in Ninety Villages in Virudhunagar District

K. Selvaraj, V. Ramasubramanian
and S. Baskaran

Centre for Research on the Effect of Global Warming and Environmental Awareness in Virudhunagar District (CREGWAV), Ayya Nadar Janaki Ammal College (Autonomous), Sivakasi

ABSTRACT

A survey was made in ninety villages during April 2010 to May 2010 on the water sources available now and in the past. From this survey, it was observed that the tanks and ponds remained filled in the past, but now in most of the villages they have become dry. Similarly, the well water level is also very less than in the past. In all the villages ponds, wells and rivers were used, as drinking water sources in the past, but now the borewell is the only water source used for drinking. In the past, ponds or rivers, which were near the villages, were full of water in all seasons but today most of the ponds are dry and have been disfigured. The amount of rainfall too has decreased. This survey clearly indicates that the water sources of Virudhunagar District have decreased rapidly. In some villages, the water bodies have vanished in the thin air, because of urbanization and industrialization, which are inferred as the main causes for water scarcity.

Introduction

Water is aptly described as the "Mother of life". It is the liquid-gold. Water is renewable natural resource of earth and is essential for all living organisms in the world. Water is not only the most important essential constituent of all animals, plants and other organisms but also the pivotal for the survivability of the mankind in the biosphere (Sharma 20005). Ground water accounts for more than 90 per cent of water supply resources for many of the developing countries (Kolraja *et al.,* 1986). In the past, the water sources for agriculture, drinking *etc.,* in Virudhunagar district were ponds, wells and rivers. Recently, frequent and indiscriminate exploitation and destruction of natural resources of water have largely disturbed the water Cycle (Kulkarni and Pawar 2006).

Human activities that involve industrial and agricultural development, over use of fertilizers, inadequate management of land, urbanization, land use patterns have directly or indirectly degrade the natural resources of surface water as wells as ground water. The impact of quality and quantity of water resources has lead to the infertility of land. The impact of ground water and surface water is significantly increasing since two decades in India, because of uncertainty of surface water resources, population growth and industrial development (Saboo, 2003). These have lead to the scarcity of drinking water. A large-scale mining and allied industrial activity in this district have resulted in vast degradation of the environment. The water bodies of this area are the greatest victims of such operations. Many, ponds and rivers were destroyed in Virudhunagar District.

Methodology

In Virudhunagar District, ninety villages were selected randomly for this survey. A survey was made during April 2010 to May 2010. Twenty five to forty people who are aged above fifty were selected in each village and asked questions from our prepared questionnaire.

Results and Discussion

The survey results have been tabulated in Tables 17.1–17.4.

Table 17.1: Tanks and ponds (Total surveyed 90 villages)

Tanks and Ponds	Filled	Half Filled	Low Level	Dried
Past	83	7	–	–
Present	–	9	13	68

In the past, maximum ponds and tanks remained filled always, but now none of the ponds and tanks is so. Ponds in Achankulam, Viswanatham, Muthalipatti, S.Paraipatti and Puthur were filled in the past but there are dry and even in rainy season only 10 per cent of its capacity has water. Anuppankulam and Amathur ponds are the two biggest ponds in the district and got filled once in two years. The water is used for agricultural, drinking and fish cultivation but now these ponds remain dry for the last fifteen years. In Ayyampatti the ponds remained filled once are, dry now–because of industrialization and urbanization, which block flow of water in to the ponds and also construction of houses in the catchments areas.

The well water level is very high in the past but now the level is medium. In Pusaripatti, the condition is worst with only 10 per cent of the capacity of well water. The main reason is digging of more bore wells.

Table 17.2: Well water level (Total surveyed 90 villages)

Well water	High	Medium	Low	Dried
Past	85	5	–	–
Present	–	6	72	12

In the past drinking water, sources in all villages were only ponds or well water but now bore well water alone is used for drinking because of the wells and ponds are dry and destroyed. Natchiarpuram and Periyapottalpatti had ponds in the past, which were disfigured because of poor rainfall resulting in conversion of them in to housing plots.

Table 17.3: Drinking water sources (Total surveyed 90 villages)

Drinking Water	Pond and Well	Well	River	Bore Well
Past	36	50	4	–
Present	–	–	–	90

Many villages had water with good taste and used for drinking but now water become tasteless and not used for drinking. For example, Vishwanatham and Sivakasi town depend on water source of Vembakottai dam. The water become tasteless because of the industrialization of this district. In Sivakasi and Mamsapuram, water has been contaminated by fluoride. This is also reported by Manimegalai and Muthulakshmi (2006).

Table 17.4: Nature of near by water bodies (Total surveyed 90 villages)

Water Bodies	Filled	Dried	Disfigured	No Water Bodies
Past	56	–	–	34
Present	–	40	16	34

In Virudhunagar District two dams are present one is Anaikuttam dam built across the river Arjuna another one is Vembakottai dam built across the river Vaipar. The two dams remained always filled in all seasons but now the two dams are dry because of poor precipitation.

Conclusion

In general there is an increasing trend in the global temperature (global warming) which has caused climatic changes throughout the world and resulted in poor rainfall, which is the major cause for the depletion in the water sources. Further, the water bodies were depleted due to urbanization and industrialization. This has also resulted up water scarcity occurred in many villages have been well understood This preliminary work has also highlighted the fact that the main sources of water had been disappeared slowly due to urbanization and industrialization resulting in water scarcity in Virudhunagar district.

References

Kolaraja,V., Vrba, J. and Zwirnmann, K.H., 1986. *Control and Management of Agricultural Impact on Groundwater,* pp. 197–228.

Kulkarni, M.K. and Pawar, N.J., 2006. Impact of urbanization on the quality of groundwater in the Ramadi Basin, Pune. *Indian J. Environmental Protection,* 26(10): 877–884.

Nimegalai, S. and Muthulakshmi, L., 2006. A survey on the levels of fluoride in ground water and prevalence of dental fluorosis in certain areas of Virudhunagar District. *Indian J. Environmental Production,* 26(6): 546–549.

Saboo, H.K., 2003. Hydro chemistry of ground water of the area around Pallahara, Angul district, Orissa. *J. Sci. Tec.,* 14 (15): 36–42.

Sharma, P.D., 2005. *Ecology and Environment.* Rastogi Publications, Meerut, India, pp. 65.

2013, Perspectives in Plant Biodiversity   *Pages* **133–139**
Editor: **Dr. K. Muthuchelian,** *Vice Chancellor, Periyar University, Salem*
Published by: **Daya Publishing House, NEW DELHI**

Chapter 18

Herbaceous Plants Diversity in Mottaipaarai of High Wavy's Moutain in Western Ghats, Tamil Nadu

S. Shanmugam* and K. Rajendran

Post Graduate and Research Department of Botany,
Thiagarajar College, Madurai – 625 009. Tamil Nadu

ABSTRACT

Natural forests are cornerstones of biodiversity conservation of any country. Species assessment provides comparable and quantifiable biological data related to change in biodiversity and number of species over a period of time and is prerequisite for making management objectives. The number of plants including endemic and rare species in an area determines the quality of its phytogeography, there by being a pointer in the direction of extinction and adaptive evolution of the occurring in it. India being a megabiodiversity nation has a lot of potential bioresources to the welfare of mankind. However, lack of awareness about biodiversity, inadequate cultivation technologies for many medicinal plants and other economic plants, massive destruction of forests, less protected area for biodiversity conservation etc., are not taking the utility value of biodiversity in a proper direction. Therefore, some of the measures like documentation of local biodiversity must be under taken to make use of the biodiversity in a sustainable manner and for its effective cultivation as well.

The High Wavy's Mountain is located in Western Ghats of Theni district, approximately between 9° 35' 54"–9° 39' 52" N longitude and 78° 18' 12"–77°

---

* Corresponding Author: E-mail: shanmugambotany@gmail.com

21' 24" E latitude. Present paper focuses on herbaceous wealth of Mottaipaarai in High Wavy's Mountain of Western Ghats in Theni district. The study reveals that the herbaceous vegetation of this zone is identified with scattered low bushes, sparsely covered grasslands, herbaceous formation and stony places. In all 81 herbaceous plant species were identified and recorded, which belongs to 76 genera and 34 families. Lamiaceae was found as dominant family which consists of 9 species, followed by Asteraceae (8 species) and Orchidaceae (5 species). Some of these plants including *Cerapegia pusilla* Wigth and Arn., *Cayratia pedata* (Lam). Juss. ex Gagnep and *Anisochilus argenteus* Gamble are under the list of Rare, Endangered and Threatened species (RET). The major plant resources in the study area are Medicinal and Aromatic Plants (MAPs). Conservation strategies must be carried out on these RET list and MAPs plants through ex-situ and in-situ conservations. The conservation of biodiversity is not only important for preservation by way of a gene bank, but also for the protection of existing forests from further degradation caused by all factors. It is therefore considering very important to conserve existing biodiversity in terms of increasing and protecting the natural population and density.

Keywords*: Herbaceous diversity, Mottaipaarai, RET list plants, Conservation need.*

Introduction

The comprehensive studies of the plants growing in a particular area are known as flora. A flora may cover any suitable area from a small patch of forest to a Taluk, City, District, State, Country or even a Continent. Floristic studies may be a simple complied check list or an elaborate analysis of the taxa of that area. Several botanists have studied the botanical wealth of Tamil Nadu and have been able to discover many taxa new to science and also several new records (Tadulingam and Venkatanarayana, 1955; Gamble and Fischer, 1957; Meher-Homji, 1969; Matthew, 1969; Matthew, 1981–83; Nair and Henry, 1983; Henry *et al.,* 1987; 1989; Subramaniyam and Henry, 1959; Krishnan and Sri Ganesan, 1971; Sri Ganesan, 1984; Matthew, 1991; Karuppusamy *et al.,* 2001; Kottaimuthu, 2008; Shanmugam, 2008).

The investigation about the floristic diversity of a particular area will definitely useful in the utilization of information by non-taxonomists like, ecological consultant, environmental engineers, forest managers, silviculturists, farmers, real estate appraisers, agricultural consultants, landscape architects, plant breeders, plant pathologists, toxicologists, forensic scientists, elementary teachers and others. Having the above facts in mind, an attempt was made to enumerate the herbaceous wealth of Mottaipaarai in High Wavy's Mountain of Western Ghats in Theni district, Tamil Nadu.

Materials and Methods

High Wavy's Mountain is located in Western Ghats of Theni District in Tamil Nadu, India. The area is lies approximately between 9° 35' 54"–9° 39' 52" N longitude and 77° 18' 12"–77° 21' 24" E latitude. The altitude of the study area varies from 800–1650m above mean sea level (M.S.L). The temperature ranges from 15° C to 30° C. Vegetation of this area ranges from scrub forests at the foothills, up to ubiquitous expanses of tea and coffee estates, to spice (pepper, cardamom, cinnamon) plantations and finally to the dense evergreen forests at the top. The entire hill ranges are consisting of Tea plantations, Coffee estates, some houses of plantation workers, artificial lakes and about six dams.

A Comprehensive and exhaustive studies about the herbaceous plants present in one of the important places of High Wavy's Mountain namely called Mottaiparai was undertaken for a period of 6 months from January, 2010 to June, 2010. The known and familiar plants were recorded on the spot in the collection site itself. The unknown and doubtful plants were collected and brought to the laboratory for identification. The nomenclature of the plants was identified with the help of the regional floras like Flora of Presidency of Madras (I–III Vols). by Gamble and Fischer (1957), Flora of Tamil Nadu Carnatic by Matthew (1983), Flora of Tamil Nadu (Vol. I) by Nair and Henry (1983), Flora of Tamil Nadu (Vol. II) by Henry *et al.* (1987) and (Vol. III) Henry *et al.* (1989) and An Excursion Flora of Central Tamil Nadu, India by Matthew (1991).

Results and Discussion

The study reveals that the herbaceous vegetation of this zone is identified with scattered low bushes, sparsely covered grasslands, herbaceous formation and stony places. Totally 81 herbaceous species belonging to 76 genera distributed among 34 families were recorded. The floristic list is presented in the Table 18.1. Among 81 species recorded from the study area, dicotyledons were represented by 72 species of 70 genera belonging to 30 families and monocotyledons were 9 species of 8 genera belonging to 4 families. Among the class dicotyledons, 25 species belonging to 23 genera distributed among 15 families were grounded under polypetalae, 40 species belonging to 38 genera distributed among 11 families under gamopetalae, and the remaining 7 species of 7 genera belonging to 4 families were monochlamydeae (Table 18.2). The plants those have been collected which are belonging to the families Cyperaceae and Poaceae were not included in this paper.

Among the 34 families listed 15 families (13 families of dicots and 2 of monocots) were represented by a single genus and single species. Lamiaceae was found as dominant family which consists of 9 species, followed by Asteraceae (8 species) and Orchidaceae (5 species). According in the level of habit, among 81 species recorded 53 species was found to be herbs, 19 species were shrubs and 9 species were climbers. *Dendrobium aqueum* Lindley, *Luisia zeylanica* Lindl., *Malaxis densiflora* (Roxb). Kuntze, *Malaxis rheedii* Sw., *Medinella indica* Clarke and *Satyrium nepalense* D.Don are the epiphytic plants presented in the study area.

Most of the plants recorded in the present study were economically important and were used for various purposes by local people living in the study area, Mottaiparai. Some of these plants including *Cerapegia pusilla* Wigth and Arn., *Cayratia pedata* (Lam). Juss. ex Gagnep and *Anisochilus argenteus* Gamble are under the list of Rare, Endangered and Threatened species (RET). The major plant resources in the study area are Medicinal and Aromatic Plants (MAPs). Conservation strategies must be carried out on these RET list and MAPs plants through ex-situ and in-situ conservations. The conservation of biodiversity is not only important for preservation by way of a gene bank, but also for the protection of existing forests from further degradation caused by all factors. It is therefore considering very important to conserve existing biodiversity in terms of increasing and protecting the natural population and density.

Table 18.1: Herbaceous plants recorded in the study area.

Sl.No.	Botanical Name	Family
1.	*Achyranthes bidentata* Blume	Amaranthaceae
2.	*Ageratum conyzoides* L.	Asteraceae
3.	*Alternanthera paronychioides* A.St.Hil.	Amaranthaceae
4.	*Anaphalis aristata* DC.	Asteraceae
5.	*Andrographis affinis* Nees	Acanthaceae
6.	*Anisochilus argenteus* Gamble	Lamiaceae
7.	*Anisochilus carnosus* (L.f).Wall. ex Benth.	Lamiaceae
8.	*Anisomeles malabarica* (L). R.Br. ex Sims	Lamiaceae
9.	*Anotis indica* (DC). W.H.Lewis	Rubiaceae
10.	*Asclepias curassavica* L.	Asclepiadaceae
11.	*Asystasia chelonoides* Nees	Acanthaceae
12.	*Blumea memranacea* wall. ex DC.	Asteraceae
13.	*Calceolaria gracilis* Kunth.	Scrophulariaceae
14.	*Cardamine hirsuta* L.	Brassicaceae
15.	*Cayratia pedata* (Lam). Juss. ex Gagnep.	Vitaceae
16.	*Cerapegia pusilla* Wight and Arn.	Asclepiadaceae
17.	*Chromolaena odorata* (*L).King and Robinson*	Asteraceae
18.	*Clematis gouriana* Roxb. Ex DC.	Rannunculaceae
19.	*Coleus barbatus* (Andr). Benth.	Lamiaceae
20.	*Conysa bonariensis* (L). Cronquist	Asteraceae
21.	*Crotalaria cordata* L.	Fabaceae
22.	*Crotalaria umbellata* Wight ex Wight and Arn.	Fabaceae
23.	*Cyananchum callialatum* Wight and Arn.	Asclepiadaceae
24.	*Cyanotis axillaris* Roem. and Schultes.	Commelinaceae
25.	*Dendrobium aqueum* Lindley	Orchidaceae
26.	*Didymocarpus tomentosa* Wight	Gesneriaceae
27.	*Diplocychlos palmatas* L.	Euphorbiaceae
28.	*Drosera peltata* Sm.	Droseraceae
29.	*Erigeron mucronatus* DC.	Asteraceae
30.	*Eriocaulon thwaitesii* Koern.	Eriocaulaceae
31.	*Euphorbia rothiana* Spreng.	Euphorbiaceae
32.	*Flemingia wightiana* Graham ex Wight and Arn.	Fabaceae
33.	*Girardinia diversifolia* (Link). I. Friis	Urticaceae

Contd...

Table 18.1–*Contd...*

Sl.No.	Botanical Name	Family
34.	*Hibiscus lunariifolius* Willd.	Malvaceae
35.	*Hibiscus sabdariffa* L.	Malvaceae
36.	*Hydrocotyl conferta* Wight.	Apiaceae
37.	*Hydrocotyl javanica* Thunb.	Apiaceae
38.	*Hypericum mysurense* Heyne ex Wight and Arn.	Hypericaceae
39.	*Impatiens balsamina* L. var. arenata Hook. f	Balsaminaceae
40.	*Isodon wightii* Benth.	Lamiaceae
41.	*Justicia procumbens* L.	Acanthaceae
42.	*Lantana camara* L.	Verbenaceae
43.	*Lantana indica* L.	Verbenaceae
44.	*Lepidium sativum* L.	Brassicaceae
45.	*Leucas lamifolia* Desf.	Lamiaceae
46.	*Leucas linifolia* (Roth). Spreng.	Lamiaceae
47.	*Lindernia pusilla* (Thunb). Merr.	Scrophulariaceae
48.	*Linum mysurense* Heyne ex Benth.	Linaceae
49.	*Lobelia nicotianifolia* Roth. ex Schultes var. *nicotianifolia* Wight	Lobeliaceae
50.	*Lobularia maritima* (L). Desv.	Brassicaceae
51.	*Luisia zeylanica* Lindl.	Orchidaceae
52.	*Malaxis densiflora* (Roxb). Kuntze	Orchidaceae
53.	*Malaxis rheedii* Sw.	Orchidaceae
54.	*Medinella indica* Clarke	Melastomataceae
55.	*Mukia maderaspatana* (L). M.Roem.	Cucurbitaceae
56.	*Murdania esculenta* (C.B.Clarke) R.Rao ex Kamm.	Commelinaceae
57.	*Naravelia zeylanica* (L). DC.	Rannunculaceae
58.	*Nicandra physalodes* (L). Geartn.	Solanaceae
59.	*Ospeckia rostrata* D. Don	Melastomataceae
60.	*Peparomia tetrahylla* (Forst.f). Hook and Arn.	Piperaceae
61.	*Pilea microphylla* (L). liebm.	Urticaceae
62.	*Plectranthus bourneae* Gamble	Lamiaceae
63.	*Plectranthus wightii* Benth.	Lamiaceae
64.	*Polycarpon tetraphyllum* L.	Caryophyllaceae
65.	*Satyrium nepalense* D.Don	Orchidaceae
66.	*Scoparia dulcis* L.	Scrophulariaceae
67.	*Sida rhomboidea* Roxb. ex Fleming	Malvaceae
68.	*Sigesbeckia orientalis* L.	Asteraceae
69.	*Smilax zeylanica* L.	Liliaceae

Contd...

Table 18.1–*Contd...*

Sl.No.	Botanical Name	Family
70.	*Smithia blanda* Wall. ex Wight and Arn.	Fabaceae
71.	*Solanum nigrum* L.	Solanaceae
72.	*Sopubia delphiniifolia* (L). G.Don	Scrophulariaceae
73.	*Spilanthues calca* DC.	Asteraceae
74.	*Stachytarpeta indica* Clark	Verbenaceae
75.	*Strobilanthes kunthiana* Nees	Acanthaceae
76.	*Tithonia diversifolia* (Hemsl). A.Gray	Asteraceae
77.	*Trichosanthes palmata* Lour.	Cucurbitaceae
78.	*Triumfetta pilosa* Roth.	Tiliaceae
79.	*Tylophora mollissima* Wight and Arn.	Asclepiadaceae
80.	*Utricularia caerulea* Dalz.	Lentibulariaceae
81.	*Viola partinii* auct. non Ging.	Violaceae

Table 18.2: Different taxon with their number of families, genera and species recorded in the study area.

Sl.No.	Taxon		Number of Families	Number of Genera	Number of Species
	Class	Subclass			
1	Dicotyledons	Polypetalae	15	23	25
		Gamopetalae	11	38	40
		Monochlamydeae	4	7	7
2	Monocotyledons	—-	4	8	9

References

Gamble, J.S. and Fischer, C.E.C., 1957. *The Flora of the Presidency of Madras.* Vols. I–III. Botanical Survey of India, Kolkata.

Henry, A.N., Kumari, G.R. and Chitra, V., 1987. *Flora of Tamil Nadu, India*, Series–I, Analysis Volume II, Botanical Survey of India, Southern Circle, Coimbatore.

Henry, A.N., Chitra, V. and Balakrishnan, N.P., 1989. *Flora of Tamil Nadu, India*, Series–I, Analysis Volume III, Botanical Survey of India, Southern Circle, Coimbatore.

Karuppusamy, S., Rajasekaran, K.M. and Karmegam, S., 2001. Endemic flora of Sirumalai Hills (Eastern Ghats), South India. *J. Econ. Taxon. Bot.,* 25(2): 367–373.

Kottaimuthu, R., Ganesan, R., Natarajan, K., Brabhu, J. and Vimala, M., 2008. Additions to the Flora of Eastern Ghats, Tamil Nadu, India. *Ethnobotanical Leaflets,* 12: 299–304.

Krishnan, S. and Sri Ganesan, T., 1971. Flora of Alagar Hills. *Suppl. List. J. Madurai University,* 3(1): 50–53.

Matthew, K.M., 1969. A Botanical Exploration of Kurseong in Darjeeling District, West Bengal. *J. Indian Bot. Soc.*, 48 (3–4): 289–295.

Matthew, K.M., 1981–83. *Flora of Tamil Nadu Carnatic.* The Rapinat Herbarium, St. Joseph's College, Tiruchirapalli, India.

Matthew, K.M., 1991. *An Excursion Flora of Central Tamil Nadu.* Oxford and IBH Publishing Co. Pvt. Ltd., New Delhi.

Meher-Homji, V.M., 1969. Some considerations on the succession on vegetation around Kodaikanal. *J. Indian Bot. Soc.*, 48: 43–52.

Nair, N.C. and Henry, A.N., 1983. *Flora of Tamil Nadu, India*, Series–I, Analysis Volume 1, Botanical Survey of India, Southern Circle, Coimbatore.

Shanmugam, S. 2008. Flora of Angiosperms and Ethnobotanical studies on Paliyar tribes of Pachalur in Dindigul district of Tamil Nadu. *M.Phil., Dissertation*, Thiagarajar College (Autonomous), Madurai, Tamil Nadu, India.

Sri Ganesan, T., 1984. *Flora of Alagar Hills. Ph.D. Thesis*, Madurai Kamaraj University, Madurai, Tamil Nadu, India.

Subramaniyam, K. and Henry, A.N., 1959. A Contribution to the flora of Alagar hills, Karandamalais and surrounding regions in Madurai District, Madras State. *J. Indian Bot. Soc.*, 38: 492–527.

Tadulingam, C. and Venkatanarayana, G., 1955. *A Handbook of Some South Indian Weeds.* Govt. Press, Madras.

2013, Perspectives in Plant Biodiversity   *Pages* **140–154**
Editor: **Dr. K. Muthuchelian,** *Vice Chancellor, Periyar University, Salem*
Published by: **Daya Publishing House, NEW DELHI**

Chapter 19

Importance of Soil Properties on Growth Response of *Triticum aestivum* L. var. Inoculated with *Glomus fasciculatum*

V.S. Bheemareddy and H.C. Lakshman

*P.G. Department of Botany, Microbiology Laboratory,
Karnatak University, Dharwad – 580 003*

Introduction

Arbuscular mycorrhizal Fungi (AMF) are known to improve water and mineral uptake in host plants. AM Fungi are helpful in reclamation of saline soils, they mobilize and fix nutrients in the soil. Soils are the most important factors in deciding the crop yield. One of the most overlooked problem areas in crop production today is soil pH. It has direct effect on soil chemistry and soil plant interactions. Nearly 7 per cent of the world's land is saline due to the presence of excess soluble salts in soil. (Dudal and Purnell 1986) Change or variation in pH of soils becomes toxic to plants due to increased availability of certain nutrients. Soil moisture, soil texture, soil temperature affect the plant growth as well as mycorrhizal efficiency. Most of Soils are generally deficient in phosphorus and Nitrogen. Deficiency of these nutrients may limit crop production.

Some soils contain adequate amount of phosphorus (P), but biologically available amount of P in soil is very low (Katznelson, 1977). Availability of P to the plant depends on the soil type. Rhizosphere acidity affects uptake of P as well as it's availability to the plant. The importance of soil as a reservoir of plant nutrients responsible for primary production is well known. Soil is a dynamic living matrix that is essential part of terrestrial ecosystem. It is a critical resource for agricultural production and food security. The employment of beneficial microorganisms has gained popularity (Perotti *et al.,* 1996).

The mutralistic association between roots and mycorrhizal fungi can improve nutritional status of plants. These Fungi are known to tolerate high alkaline conditions, low nutrient and irrigation regimes in saline soils (Rani and Bhaduria, 2001). AM fungi are present invariably in all kinds of soils and helps to improve unfavorable saline soils to promote plant survival and growth (Robert Dixon *et al.,* 1994). These fungi are known to increase yield and mineral nutrition of associated plants. Therefore efforts must be done to optimize the beneficial uses of AM fungi is sustainable agriculture (Sieverding, 1991; Lakshman and Patil, 2004). AM Fungi form essential components of sustainable soil plant systems and improve crop growth and productivity (Cavagnarao *et al.,* 2006). Now a days application of AM Fungi as biofertilizers in crop production is recommended with the aim of increasing productivity and reducing fertilizer use (Schwartz *et al.,* 2006).

The present study was undertaken to evaluate the effect of AM Fungi on *Triticum aestivum* varieties grown on different soils with pH varying between moderately acidic to moderately alkaline. Using soil samples from the fields of traditionally cropped with wheat in North Karnataka experiments were conducted in polybags under green house conditions. The four varieties of *Triticum aestivum* were selected for the experiments were DWR-162, DWR-195, DWR-225 and NI 5439. Effect of soil pH, soil texture, available phosphorous in soil and organic carbon on growth and yield of the above said four varities was undertaken.

Materials and Methods

Experiments were conducted in earthen pots under polyhouse conditions. soils collected from four sites such as Sattur, Holatikoti, Lakamanhalli and Sirsi were used for conducting the experiments.

Physical properties of the soil samples were determined using soil hydrometer according to Gee GW and Bauder JM (1986) method.

Table 19.1: Physico chemical properties of soil samples.

Soil Sample	pH	Per cent Silt	Per cent of Clay	Per cent of Sand	P kg/acre	K kg/acre	Organic Carbon
Altisol	7.20	26	34	40	14.90	120	0.64
Vertisol	7.95	20	58	22	17.29	130	0.66
Alfisol	6.75	28	42	30	21.84	167	0.77
Entisol (Guggaratti)	6.28	29	38	33	20.00	144	0.74

Root Sampling and Analysis was done according to Phillips and Hayman method. 1970. Per cent root colonization was calculated as follows.

$$\% \text{ of mycorrhizal colonization} = \frac{\text{No. of root segments colonized}}{\text{Total No. of root segment observed}} \times 100$$

Preparation of AM Inoculum

The AM spores belongs to *Glomus fasciculatum* were isolated from rhizospheric soils of wheat fields near sattur were isolated. Mass multiplication of AM fungal spores of *Glomus fasciculatum* was done by using *Sorghum vulgare* L as host plant.

Determination of Spore No.

10gm rhizospheric soil was taken from the pots. spore no was determined by wet sieving and Decanting method as suggested by Gerdemann and Nicolson (1964)

Plant Treatment and Cultivation

All the four soil samples used for the experiments were sterilized by autoclave method. Three kg sterilized soil was taken in each earthen pots. The earthen pots were arranged in following set up.

1. Sterilized Altisol (sample 1) without AM Fungal inoculum.
2. Sterilized Altisol soil (sample 1) with AM Fungal inoculum.
3. Sterilized Vertisol soil (sample 2) without AM Fungal inoculum.
4. Sterilized Vertisol (sample 2) with AM Fungal inoculum.
5. Sterilized Entosol (sample 3) without AM Fungal inoculum.
6. Sterilized Entosol (sample 3) with AM Fungal inoculum.
7. Sterilized Alfisol (sample 4) without AM Fungal inoculum.
8. Sterilized Alfisol (sample 4) with AM Fungal inoculum.

Earthen pots were placed in poly house with mean temperature of about 26°C. Pots were placed in randomized complete block design with three replicates per treatment. Plants were watered regularly on alternate day. Hogland solution minus P was given to all treatments once in 15 days.

Plant Harvest and Analysis

Plants were harvested after 60 DAS and 90 DAS. They were uprooted, parameters such as plant height, stem diameter, leaf area, fresh wt of root and Shoot were measured. The roots and shoots of experimental plants were oven dried at 70°C. for 48 hrs to determine dry weight of root and shoot. Number of grains produced per plant and 100 grains weight was also calculated from the experimental plants Absolute growth rate was calculated using following equation.

$$\text{Absolute Growth Rate} = \frac{W_2 - W_1}{T_2 - T_1}$$

where, W_1 = Dry wt. of plant in grams at time t_1;

W_2 = Dry wt. of plant in grams at time t_2.

Results

The growth response of inoculated plants belongs to four varieties was better than uninoculated plants grown in different soils. Growth responses of the experimental plants were found to be very high in Altisol with pH 7.20. The plants grown in Alfisol had shown comparatively lesser growth response than Altisol, where as the plants grown in vertisol (pH 7.90). and Entisol (6.30) had shown lesser growth both under control and inoculated conditions. Over all observation of growth parameters studied in the experimental plants grown in different soils revealed that, plants grow better and produce higher dry weight and yield in Altisol with pH 7.20. The least growth and yield was observed in experimental plants grown on Entisol with pH 6.30. Plants grown in vertisol and Entisol were shown moderate growth response.

At 60 DAS and 90 DAS plants had shown increased plant height under inoculated condition than control. In Altisol DWR-195 had shown highest plant height at 90 DAS period followed by DWR -225 under inoculated conditions. Stem diameter was found to be significantly high in DWR-225 and DWR 195 under inoculated conditions. About 15 to 20 percent increase in stem diameter was observed in inoculated plants over control in all the four varieties.

Flag leaf area was found to be maximum in NI 5439 and DWR-162 varieties under inoculated conditions. Flag leaf area observed was more in plants of 90 DAS than 60 DAS both under inoculated and control conditions. Flag leaf area was found to be more in plants grown in Altisol, Mycorrhizal inoculation helped to increase flag leaf area in all the varieties grown in different soils. But it was observed that Altisol favours to attain maximum leaf area. Plants grown vertisol and Alfisol had shown moderate increase in leaf area, it was found to be least in Entisol.

Increase in dry weight of shoot and root was observed in inoculated plants than the control. In all the four varieties dry weight increase was significantly high in inoculated plants than control plants. Further it was observed that all the four varities of *Triticum aestivum* L. had shown varied response to different soil compositions. At 60 DAS stage highest increase in dry weight of shoot and root was observed in plants grown in Altisol under inoculated conditions. The two varieties DWR-162 and NI-5439 had shown maximum increase in dry weight of shoot and root under inoculated conditions. The other two Var. DWR-195 and DWR-225 had shown moderate increase in dry weight of Shoot and root. At 60DAS stage least increase in dry weight of shoot and root was noticed in plants grown in Entisol. The two varieties DWR-195 and NI–5439 had shown more dry weight of shoot and root than the other two varieties. Plants grown in Alfisol had shown better response in terms of increase in dry wt of shoot and root, but this increase was comparatively lesser than the plants grown in Altisol. In the Alfisol it was DWR-162 and NI 5439 had shown comparatively more increase in dry weight than the other two varieties. The plants grown in Entisol had shown least growth response in terms of dry weight increase at 60DAS stage. Among the four varieties DWR-195 had shown highest dry weight than the other three varieties.

The growth response in terms of increase in dry weight of Shoot and root at 90 DAS stage was quite different than 60DAS stage. Dry wt of shoot and root was more in 90 DAS stage. Highest shoot and root dry weight was observed in plants grown in

Altisol under inoculated conditions at 90 DAS stage. Least increase was noticed in Entisol. In two varieties DWR-225 and DWR-162 highest increase in dry weight of shoot and root was observed under inoculated conditions.

Figure 19.1a–d: Absolute Growth Rate (AGR) in *Triticum aestivum* L. varieties. under inoculated and control conditions grown on different soils.

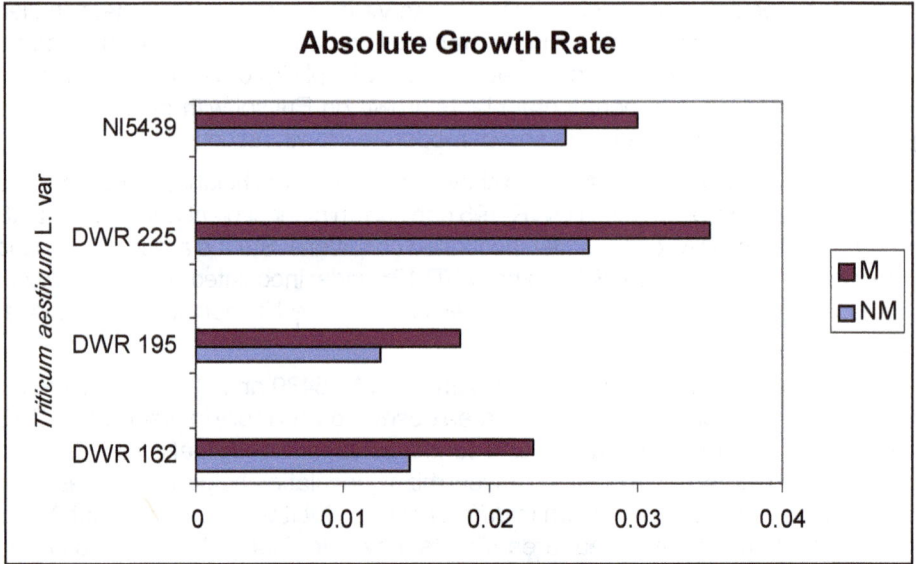

a: Alfisol

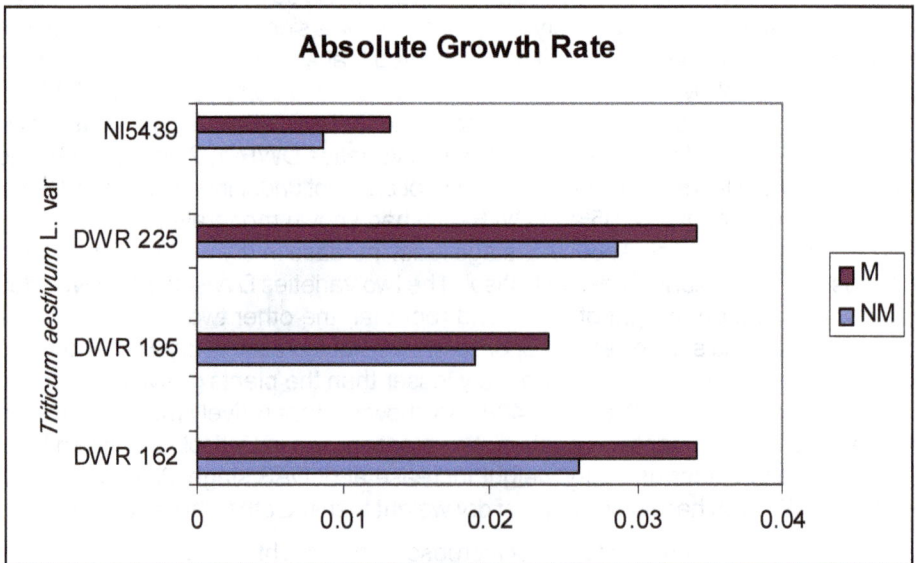

b: Entisol

Contd...

Figure 19.1a–d–*Contd...*

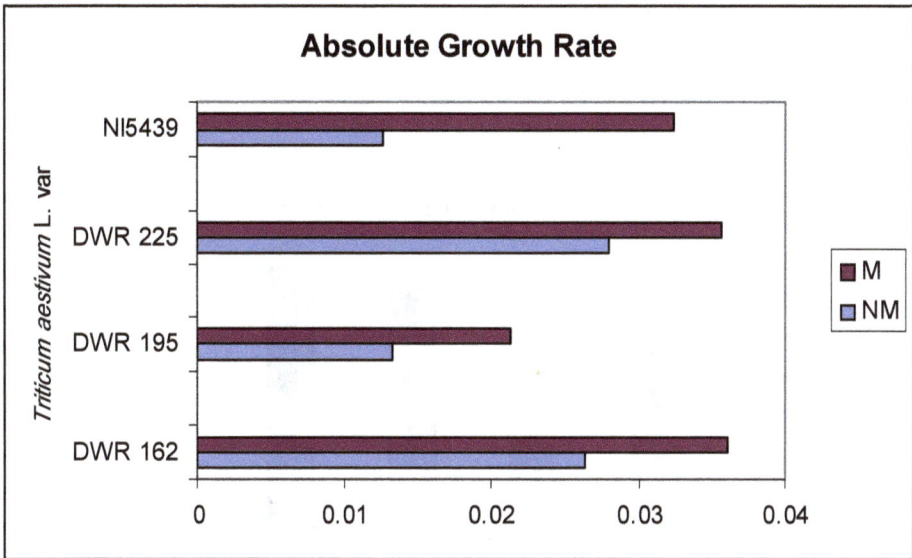

Absolute Growth Rate

c: Vertisol

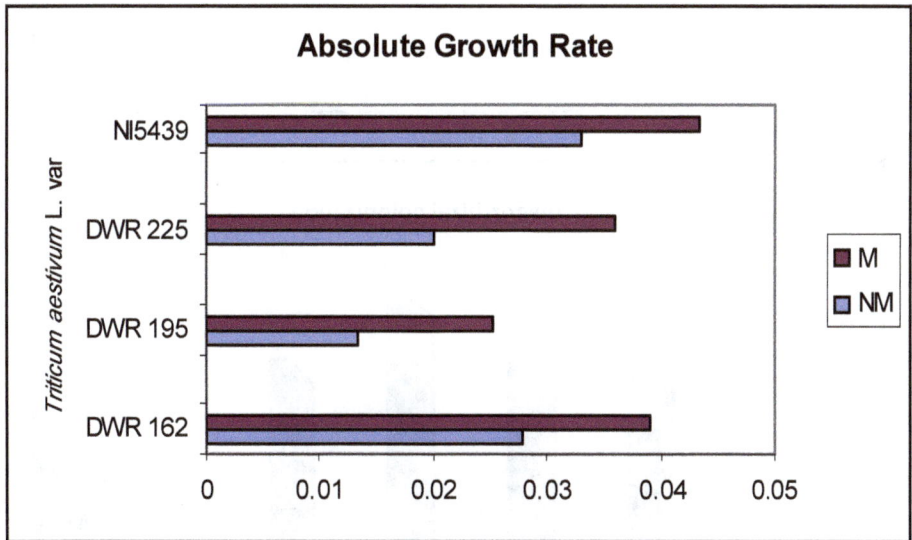

Absolute Growth Rate

d: Altisol

Absolute Growth Rate (AGR) was studied in all the experimental plants belong to four varieties. There was increase of AGR in inoculated plants over the control grown in all the soil types.

Varied AGR was observed in different soils. In Altisol, Alfisol and Entisol plants belong to DWR 162 and DWR-225 varieties had shown highest AGR than the other two varieties. Where as in Vertisol AGR was comparatively more in DWR-162 and NI

5439 varieties. AMF inoculation resulted in about 25 per cent increase in AGR of inoculated plants grown in different soils.

Figure 19.2a–d: Percentage of root colonization in *Triticum aestivum* L. varieties grown on different soils.

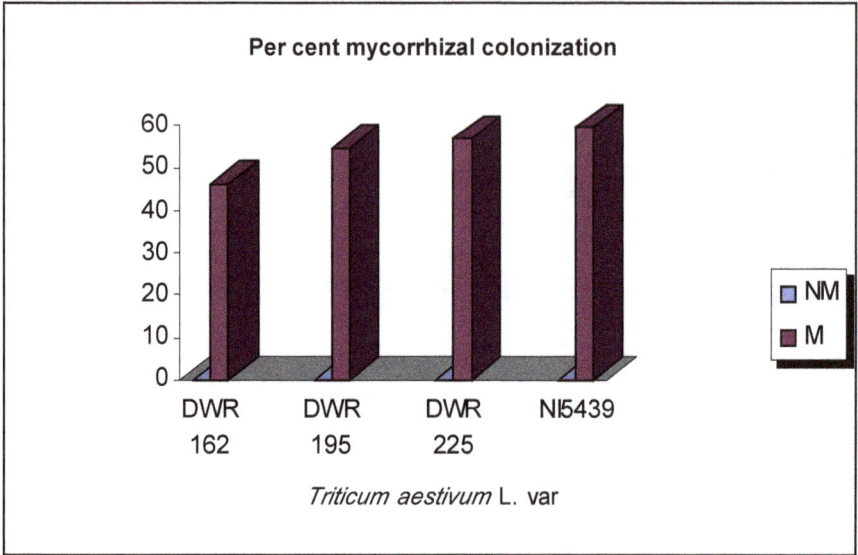

a: Vertisol

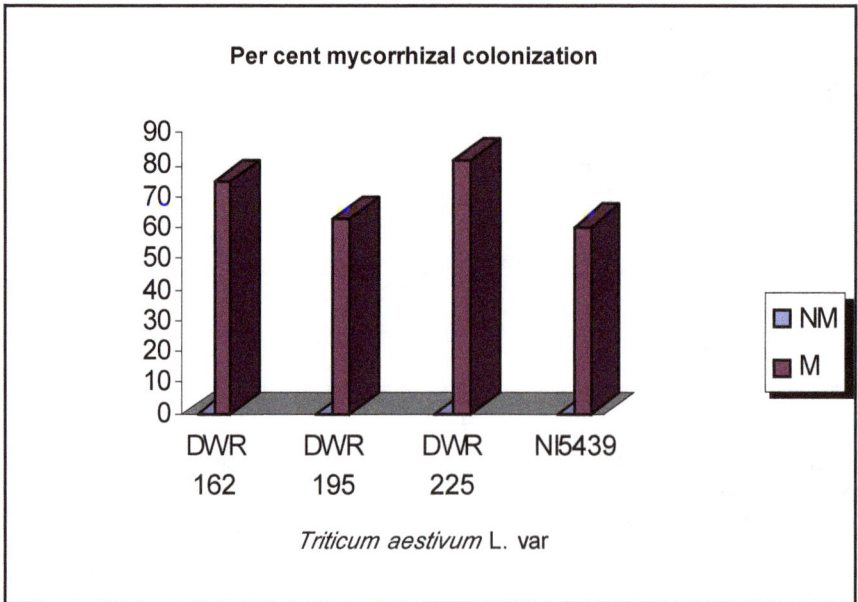

b: Altisol

Contd...

Figure 19.2a–d–*Contd...*

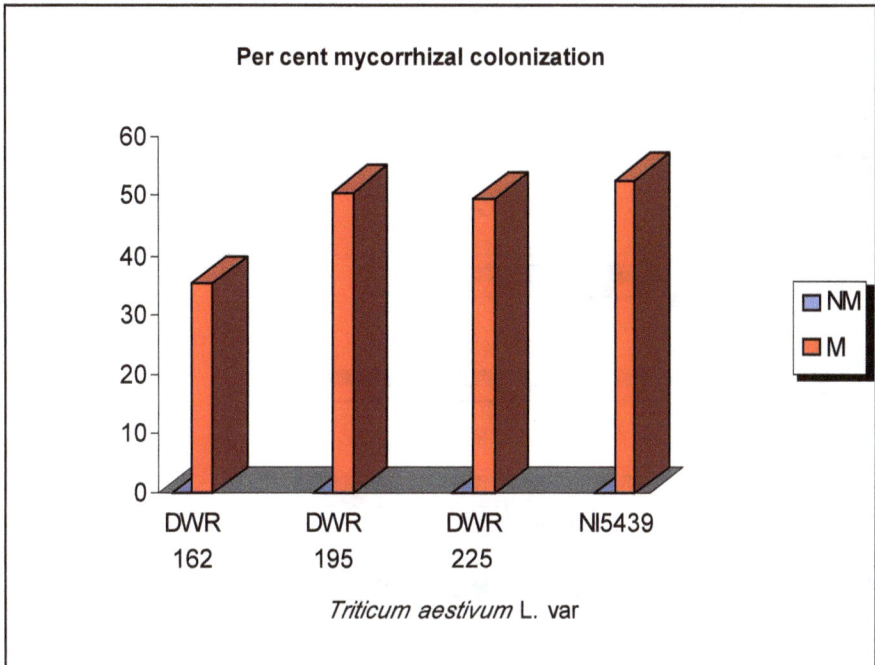

Per cent mycorrhizal colonization

Triticum aestivum L. var

NM
M

c: Entisol

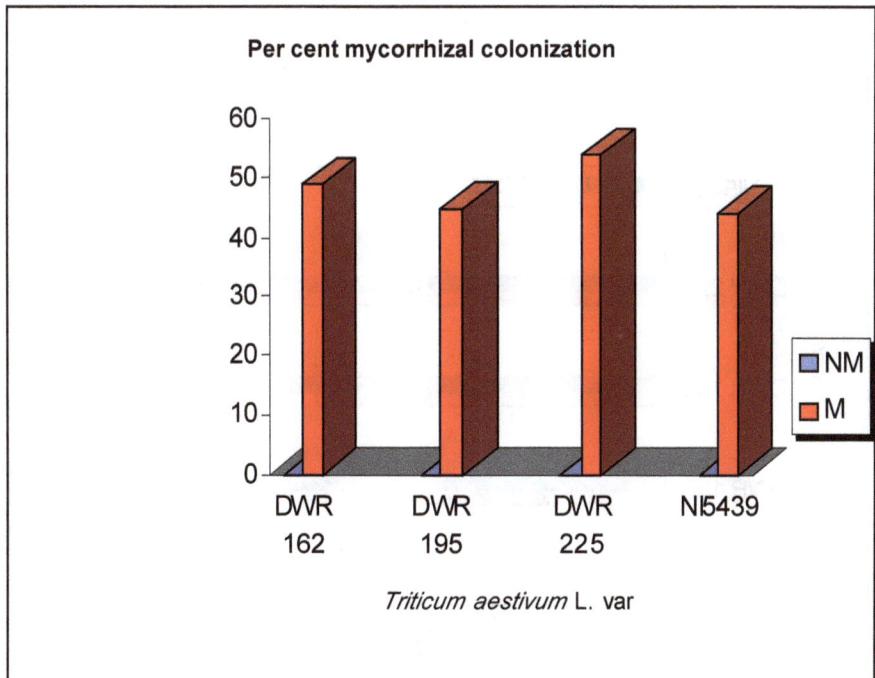

Per cent mycorrhizal colonization

Triticum aestivum L. var

NM
M

d: Alfisol

100 grains weight was studied in the experimental plants. 10 to 15 percent increase in 100 grains weight was observed in inoculated plants grown in different soils over

Figure 19.3a–d: 100 Grains Weight in *Triticum aestivum* L.varieties. under inoculated and control conditions grown on different soils.

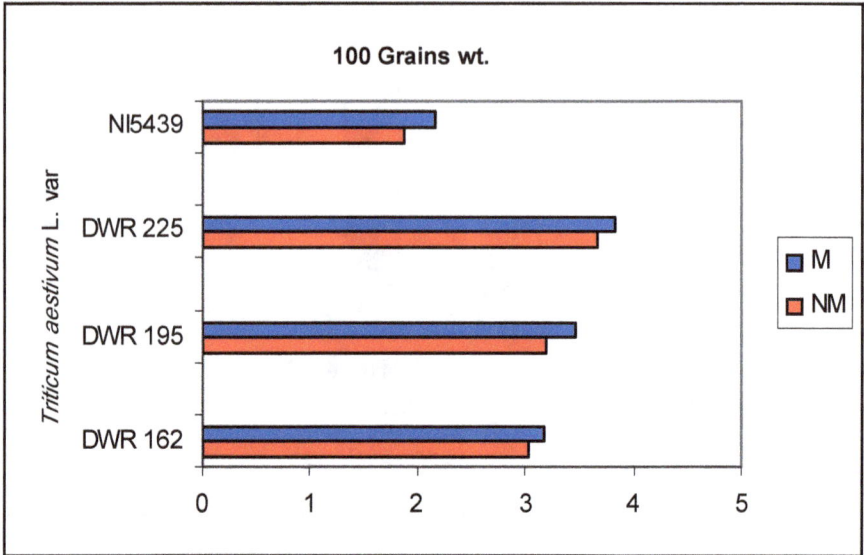

a: Vertisol

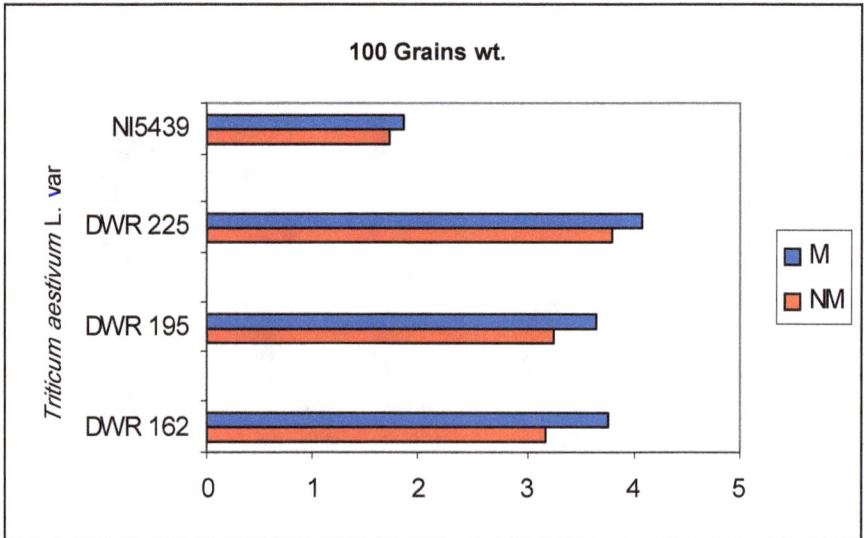

b: Alfisol

Contd...

Figure 19.3a–d–*Contd...*

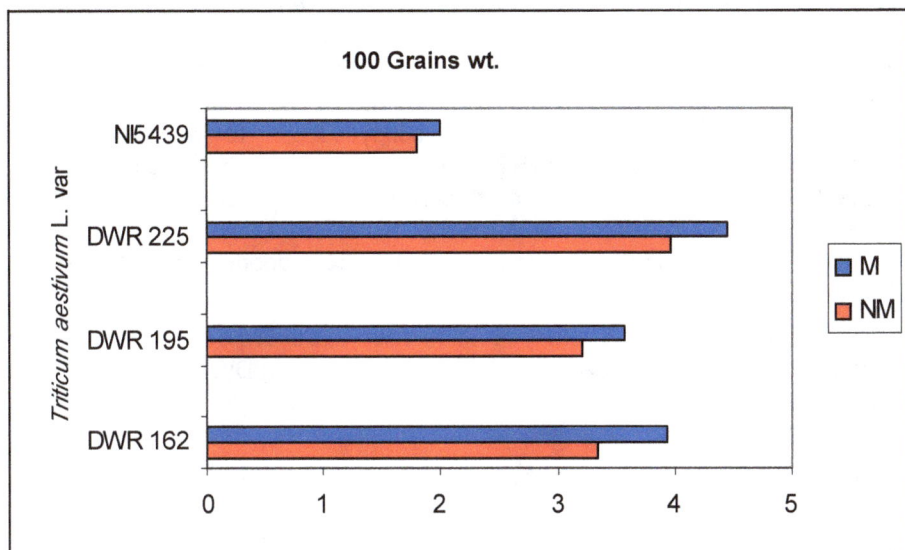

c: Artisol

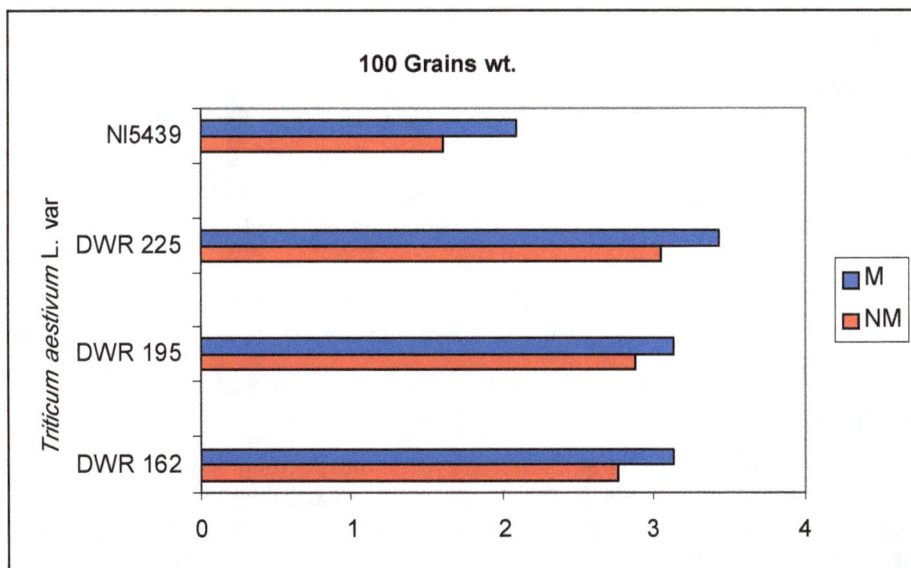

d: Entisol

control plants. Highest 100 grains weight was observed in AM inoculated plants grown in Altisol. The plants belong to DWR-162 and DWR-225 grown in Altisol, Alfisol, Entisol had shown highest 100 grains weight than other two varieties, where as in vertisol, highest 100 grain weight was observed in plants belong to DWR 195 and DWR-225 varieties.

Number of grains produced per plant was observed in control and inoculated plants. It was observed that inoculated plants had shown considerable increase in yield over control plants. There was an increase of 10 to 40 per cent yield in inoculated plants than control. Highest number of grains produced per plant was observed in DWR-225 variety grown in Altisol under inoculated Conditions. Lowest yield was observed in NI 5439 varieties grown in all the soil types taken for the study. In Altisol, maximum yield was observed in DWR225, moderate yield was observed in DWR-162 and DWR 195 varieties.Least yield was observed in NI 5439 variety than other three varieties. In vertisol, DWR-225 variety had shown maximum yield. The other varieties DWR-162, DWR 195 had shown moderate yield. NI 5439 poorly responds in terms of yield produced per plant.

In Alfisol, it was DWR 225 variety shown highest yield than the other three varieties. AM inoculation helps to increase yield up to 40 per cent in DWR-225 variety. The other two varieties DWR-162 and DWR-195 had shown moderate yield, where as NI5439 variety least yield was observed..

The poor yield response was observed in Entisol than the plants grown in other soil types. Inoculated plants had shown increased yield than control plants. About 16 to 30 per cent increase in yield was observed in inoculated plants over their control. The DWR-225 variety had produced more yield than the other three varieties. The DWR-162 and DWR 195 varieties have moderate yield and NI 5439 variety had shown lesser yield.

It was evident from the experimental results that Altisol was the best soil for the cultivation *Tritium aestivum* varieties under irrigated regimes with AM inoculation. The least preferred soil was Entisol, where plants had shown poor growth and yield.

Percent of mycorrhizal root colonization was examined in all the four varities. It was relatively high in Altisol and least percent of mycorrhizal root colonization was observed in plants grown in Alfisol and Entisol.

The spores number in 10 gm rhizospheric soil was studied. Variation in the number of spores was noticed. Highest spore number was observed in Altisol than other soil types.

Discussion

Maximum plant growth was observed in soils with pH ranging from 6 to 7. At this pH range greatest uptake of N, P and Mg was noticed. Crop production was drastically reduced in high pH due to reduced water availability to the plants. Acidic soils may not be productive due to the availability of increased amounts of toxic elements such as Aluminum and Manganese. Similar findings were reported by Pan *et al.* (1988).

There is strong correlation between cation concentration and percent root colonization, with increase in Na^+ ion concentration the percent root colonization decreases. Our findings were consistent with reports of Abds *et al.* (2003). Percent root colonization was depending on soil pH. The soil P play key role in deciding percent root colonization,addition of 'P' to the soil has resulted in the decrease of percent root colonization. AM fungal colonization was found to be more in P deficient soils. Similar results were reported by Zhu *et al.* (2001). Altisol was poor in 'P' content as well as it had favorable pH, these factors are favorable for mycorrhizal colonization. The percent root colonization was found to be lesser in vertisol and Entisol. Higher clay and P content of soil may be responsible for lesser percent root colonization. The Percent root colonization was found to be very less in Entisol, because Entisol is more fertile than Altisol and have higher capacity to adsorb ions. Higher cationic concentration in clayey soil limits the AMF development. Similar findings were also reported by Eason *et al.*,(1999). Koske and Lim (1987) ascertained another explanation for the variation in mycorrhizal colonization in soils. Finer soil particles favour the deposition of suberin on the epidermis, which increases the resistance to infection by AM fungi. Breakage of cortical layers due to mechanical stress colonization is lost. Our findings correlate with the findings of Louis and Lim (1987).

Siqueira and Saggin Junior (1995) reported that soils with low fertility limit plant development and increase the dependence of plants on Mycorrhizal colonization. Sanchez and Salinas (1981) suggested that high soil fertility reduced the dependence of plant on mycorrhizae. our experiments had shown the similar results and we strongly agree with findings of earlier workers. In Vertisol and Entisol clay percentage and P concentration were high. Plants grown in these soils are less dependant on AM fungi for the mineral acquisition, lesser difference in growth and yield parameters were observed between inoculated and control plants.

Janos (1987) reported that high acidity enhances the cation exchange which will immobilize phosphate, so that higher amount of phosphate fixed in the soil. This resulted in reducing the availability of phosphorous to the plant resulting in poor plant growth. Several mechanisms have been suggested by which the Mycorrhizal plants can enhance growth and nutrient uptake. Bethlenfalvay *et al.* (1997) explained that plants acquire more phosphorous through the production of root exudates then enhance the avartability of P.

Altisol supports maximum plant growth and yield, even the per cent of mycorrhizal colonization was found to be high. Plants grown in Altisol had shown comparatively higher growth under inoculated conditions than the plants grown in other soil types. Altisol contains relatively more sand than other soil types. Soil texture play significant role in plant growth. Our experimental results correlates with the results of earlier workers. Gupta *et al.* (2006) reported that soil texture might affect plant growth in maize plants. Addition of sand enhanced the porosity of soil and provided the nutrients. In such conditions inoculation of AM fungi helped in the enhancement of plant productivity. Raja *et al.* (2002) reported that wheat and maize plants had shown P uptake in presence of AM root colonization.

Omar (1995) reported that in sterilized pot soil, inoculation with VAM fungi dramatically increased the growth and per cent root colonization in maize plants. M.J. Mohammad *et al.* (1998) explained that AM inoculation increased grain yield more than 25 per cent in wheat plants. Our results correlates with these findings. In the four *Triticum aestivum* varieties the AM inoculation had resulted in the increase of 10 to 40 per cent grain yield per plant than control plants. and Meisinger (2002) reported that soil fertility is very important for essential plant nutrients and for soil texture, structure, organic matter anion and cation retention.

Conclusion

Conclusion drawn from the experimental results strongly confirms the beneficial effects of AM fungi or crop plants. Some of the soil factors such as soil texture and composition had significant effect on plant growth and yield produced per plant. It was observed that there was about to 10 to 40 per cent increase in the yield the of AM inoculated plants over their control. Soil properties also had significant effect on percent root colonization by AM Fungi. Out of the four soil samples taken for the study, it was Altisol that promotes maximum plant growth and yield. Out of the four Vars of *Triticum aestivum* L vars., it was DWR-225 had shown maximum growth and yield in all soil samples taken for the study. It was also observed that in presence of Altisol and AM inoculum DWR-225 produced maximum growth and yield.

Acknowledgements

First author is thankful to UGC New Delhi for sanctioning fellowship under FDP and second author is indebted to UGC New Delhi for sanctioning major research project on "Role of AM Fungi in Four rare millets of North Karnataka."

References

Abdi, R. and Dube H.C., 2003. Distribution of AM fungi along salinity gradient in a salt pan habitat. *Mycorrhiza News,* 15(2):20–22.

Al-Karaki, G.N., Al-Raddads, A. and Clark, R.B., 1998. Water stress and mycorrhizal isolate effects on growth and nutrition of wheat. *Journal of Plant Nutrition,* 21: 891–902.

Anarde, G., Mihara, K.K., Linderman, R.G. and Bethenfalvay, G.J., 1988. Soil aggregation status and Rhizobacteria in the mycorrhizosphere. *Plant and Soil,* 202(2): 89–96.

Bago, B. and Azcon-Aguilar, C., 1997. Changes in rhizosphere pH induced by arbuscular mycorrhizal formation in onion (*Allium apa*).

Bethlenfalvay, G.J., Schreiner, R.B. and Mishra, K.L., 1997. Mycorrhizal fungi effects on nutrient composition and yield of soyabean grains. *Journal of Plant Nutrition,* 20: 581–591.

Bolan, N.S., Robson, A.D., Barrow, N.J. and Aylmore, A.G., 1984. Specific activity of phosphorous in mycorrhizal and non mycorrhizal plants in relation to the availability of phosphorous to plants. *Soil Biology and Biochemistry,* 16: 299–304.

Cantrell, I.C. and Lindermann, R.G., 2001. Preinoculation of lettuce and onion with VA mycorrhizal fungi reduces deleterious effects of soil salinity. *Plant Soil,* 233: 269–281.

Covaceyich, I. and Echeverria, H.E., 2008. Receptivity of an Argentinian Pampass soil to Arbuscular mycorrhizal *Glomus* and *Aculospora* strains. *World Journal of Agricultural Sciences,* 4(6): 688–698.

Davis, E.A., Young, J.L. and Linderman, R.G., 1983. Soil lime level (pH) and VA–mycorrhiza effects on growth responses of sweetgum seedlings. *Soil Sci. Soc. Am. J.,* 47: 251–256.

Dixon, Robert K., Rao, M.V. and Garg, V.K., 1994. *In situ* and *in vitro* response of mycorrhizal fungi to salt stress. *Mycorrhiza*, 5: 6–8.

Dudal, R. and Purnell, M.F., 1986. Land resources: Salt affected soils. *Reclamation and Revegetation Research*, 5: 1–10.

Gee, G.W. and Bauder, J.M., 1986. Particle size analysis. In: *methods of Soil Analysis Part I: Physical and Mineralogical Methods*. Agronomy Monograph No. 9 (2nd edition). American society of Agronomy Madison, W.I., pp. 383–411.

Gehring, C.A. and Whitham, T.G., 1994. Comparisons of ectomycorrhizae on pinyon pines (*Pinus edulis*: Pinaceae) across extreme soil type and herbivory. *American Journal of Botany*, 81: 1509–1516.

Hetrick, B.A.D., Wilson, G.W.T., Gill, B.S. and Cox, T.S., 1995. Chromosome location of mycorrhizal responsive gene in wheat. *Canadian Journal of Botany*, 73: 981–987.

Hetrick, B.A.D., Wilson, G.W.T. and Todd, T.C., 1996. Mycorrhizal response in wheat cultivars: relationship to phosphorus. Canadian Journal of Botany. 74: 19–25.

Hoyt, P. B., Henning, A.M.I. and Dobb, J. I. (1967). Reaction of barley and luzern at liming on solonetz podzol and clay soil. *Candian Journal of Soil Science*, 47: 15–21.

Janos, D.P., 1987. VA mycorrhizae in humid tropical ecosystems. In: *Ecophysiology of VA Mycorrhizal Plants*, (Ed.) G.R. Safir. CRC Press, Boca Raton, pp. 107–134.

Khalil, S., Loynachan and McNabb, H.S., 1992. Colonization of soyabean by mycorrhizal fungi and spore populations in Iowa soils. *Agronomy Journal*, 84: 832–836.

Koske, R.E. and Gemma, J.N., 1995. Fungal reactions to plants prior to mycorrhizal formation. In: *Mycorrhizal Functioning: An Integrative Plant Fungal Processes*, (Ed.) M.F. Allen. Chapman and Hall, New York, pp. 3–36.

Kothari, S.K., Marchner, H. and Romheld, V., 1991a. Contribution of VA mycorrhizal hyphae in acquisition of phosphorous and zinc and maize grown in calcareous soil. *Plant Soil*, 131: 177–185.

Lakshman, H.C. and Geetha, B. Patil, 2004. Effect of AM fungi on finger millet in biomass production and nutrient uptake. *Bioscan.*, 11(2): 78–81.

Louis, I.M. and Lim, G., 1987. Spore density and root colonization of vesicular arbuscular mycorrhizas in tropical soil. *Transactions of British Mycorrhizal Society*, 88: 207–212.

Miller, M.A. and Jastrow, J.D., 1994. Vesicular arbuscular mycorrhizae and biogeochemical cycling. In: *Mycorrhizae in Plant Health*, (Eds.) F.L. Pfleger and R.G. Linderman. American Phytopath. Soc. St. Paul, Minnesota, pp. 189–212.

Mohammad, M.J., Pan, W.L. and Kennedy, A.C., 1998. Seasonal mycorrhizal colonization of winter wheat and its effect on wheat growth under dry land field condition. *Mycorrhiza*, 8: 139–144.

Mosse, B., 1981. Vesicular arbuscular mycorrhizae for tropical agriculture. *Research Bulletin,* 194, p. 82. University of Hawaii, Hawaii Institute of Tropical Agriculture and Human Resources.

Omar, S.A., 1995. Growth effects of VAM fungus *Glomus constrictum* on maize plants in pot trials. *Folia Microbiologica,* 40: 503–507.

Osonubi, O., Mulongoy, K., Awotoye, M.O., Atayese and Okali, D.U., 1991. Effects of ectomycorrhizal and vesicular arbuscular fungi on drought tolerance of four leguminous woody seedlings. *Plant and Soil,* 136: 131–143.

Pan, F.J. and Cheng, W.F., 1986. Effect dual inoculation on growth and nutrient uptake in *Leucaena leucocephala. Bulletin of Taiwan Forestry Research Institute,* 3(4): 209–223.

Rani, V. and Bhaduria, S., 2001. Vesicular arbuscular mycorrhizal association with some medicinal plants growing on alkaline soils of Manipuri district, Uttar Pradessh, *Mycorrhiza News,* 13(2):12–14.

Sanchez, P.A. and Salinas, J.G., 1981. Low input technology for managing oxysoils and ultisols in tropical America. *Advances in Agronomy,* 34: 279–406.

Sieverding, E., 1991. *Vesiculoarbusclar Mycorrhizal Management in Tropical Agrosystems.* Deutsche Gesellshaft Technische Zusammernarbelt (GZT), Eschborn, Germany, pp. 1–371.

Siqueira, J.O. and Saggin-Junior, O.J., 1995. The importance of mycorrhizal association in natural low fertility soils. In: *Anais do simposio International Sobre and Stress Ambiental.* Belo Horizonte 1992: EMBRAPA/CNPMS sete Lagos, pp. 240–280.

Zhu, Y.G., Smith, S.E., Barritt, A.R. and Tabatabai, 1994. Mycorrhizal dependency and nutrient uptake by improved and unimproved corn and soyabean cultivars. *Agronomy Journal,* 86: 949–958.

Zpflanzenernacher Bodenk, 160: 333–339.

2013, Perspectives in Plant Biodiversity   *Pages* **155–161**
Editor: **Dr. K. Muthuchelian,** *Vice Chancellor, Periyar University, Salem*
Published by: **Daya Publishing House, NEW DELHI**

Chapter 20

Survey on Ethnomedicinal Tree Species of Jogimatti Forest, Chitradurga District, Karnataka, India

V.T. Hiremath and T.C. Taranath*

Environmental Biology Laboratory, P.G. Dept. of Studies in Botany,
Karnatak University, Dharwad – 580 003, Karnataka

ABSTRACT

India has one of world's richest medicinal plant heritages. The wealth is not only in terms of the number of unique species documented but also in terms of the tremendous depth of traditional knowledge concerned with human and livestock healthcare systems.

Present investigation was undertaken to study medicinal tree species of Jogimatti Forest, Chitradurga district, Karnataka. Local traditional knowledge associated with the tree species of Jogimatti forest was collected through questionnaire and interaction with local health healers. The local health healers found to use 27 tree species belonging to 19 families for the treatment of 25 diseases with 47 formulations either in single/multiple applications. The study reveals that among the families, Mimosaceae claims highest number of species (04), with 22.2 per cent, followed by Euphorbiaceae and Moraceae (03 species) with 16.6 per cent, Moringaceae, Myrtaceae, Rutaceae, Fabaceae, Arecaceae (02 species) with 11.1 per cent and Santalaceae with a single species of 5.5 per cent. These medicinal tree species include *viz., Acacia arabica, Emblica*

* Corresponding Author: E-mail: hiremath2047@gmail.com

officinalis, Melia azadirachta, Eugenia jambulana, Phyllanthus acidus etc., were used to treat snake bite, antifertility, dental problems, menstrual cycle, Jaundice, etc. Analysis of the data revealed per cent contribution of different plant parts viz.,bark was frequently used (29.8 per cent), followed by leaves and fruit/seeds (21.2 per cent), root (8.5 per cent), stem and meristem (6.3 per cent) and flowers 4.2 of least percentage.

Keywords: *Ethnomedicinal, Jogimatti forest, Traditional practitioners. Hakki-pikki, Kuruba.*

Introduction

Plants have been used in traditional healthcare system of humans and livestock for several thousands of years.(Abu-Rabia, 2005). The knowledge of medicinal plants has been accumulated during the course of many centuries in the form of different medicinal systems such as Ayurveda, Unani, and Siddha. In India, it is reported that traditional healers use 2500 plant species and 100 species of plants serve as regular source of medicine (Pei, 2001). In recent years, there has been a tremendous interest in the medicinal plants which are believed to be much safer and exhibit a remarkable efficacy in the treatment of various ailments (Siddique *et al.,* 1995). The folk medicinal traditions play a vital and prominent role in human and environment interaction (Chopra *et al.,* 1956). Higher plants as source of medicinal compounds have continued to play a dominant role in maintenance of human health and livestock since ancient times (Farombi, 2003). These plants are not only used for primary health care in rural areas in developing countries, but also in developed countries where modern medicines are predominantly used. Due to the side effects of modern allopathic drugs in the present days, people are attracted towards herbal medicines and their consumption. Several workers were reported the utility of plants for the treatment of various ailments (Goel and Bhattacharya, 1981; Eddouks *et al.,* 2002; Harsha *et al.,* 2002, Hebbar *et al.,* 2004; Siva and Muthuchelian, 2004; Katz *et al.,* 2007; Leach, 2007; Alagesaboopati and Rajendran, 2009. Hiremath and Taranath 2009 and 2010.

Chitradurga district is one of the central districts of Karnataka state with much racial and socio-cultural diversity. Bedas. Besthas, Gollas, Jenu kurubas, Hakki-pikki are the tribals who are intimately associated with the plant wealth of Jogimatti forest which is located 13 kms away on the leeward side of Western Ghats at an altitude of 3,800 ft. with sparsely wooded patches of vegetation. It mainly comprises mixed and dry deciduous forests with undulating chain of hills. The present investigation was undertaken to study medicinal tree species of Jogimatti forest and their associated Traditional Knowledge from the local health practitioners.

Materials and Methods

Periodic field surveys were carried out in Jogimatti State forest of Chitradurga, Karnataka during December 2009 to June 2010. Data was collected through tribal people (Hakki-pikki, Jenu kuruba, Lambani's Golla, Kadugolla), local traditional health practitioners, and village elders through questionnaire/personal communication. Questionnaire containing information on health practices, disease, name of plant, family, part used and therapeutic use. The voucher specimens were collected and identified by referring standard flora (Hooker, 1884, Gamble 1936; Saldhana 1984).

Results and Discussion

In the present investigation 27 tree species of medicinal plants belongs to 19 families, 25 genera for the treatment of 25 diseases with 47 formulations. The major plant families used in the treatment are Mimosaceae (04 species), followed by Euphorbiaceae and Moraceae with (03) species, Moringaceae, Myrtaceae, Rutaceae, Fabaceae, Arecaceae with two species, while Meliaceae Santalaceae with a single species (Figure 20.1).

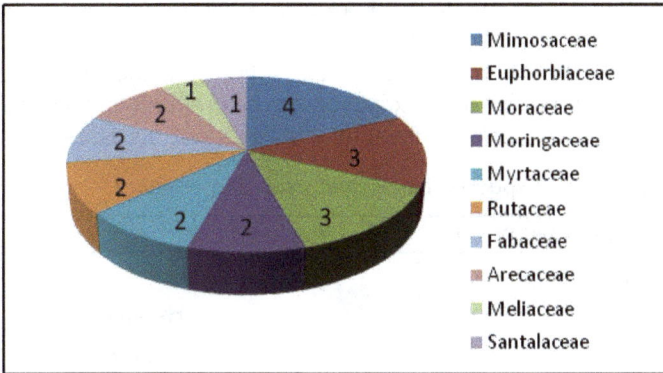

Figure 20.1: Contribution of major ten plant families for Ethnomedicine.

Major tree species used in the different formulations are *Acacia arabica, Emblica officinalis, Melia azadirachta, Eugenia jambolana, Phyllanthus acidus Mangifera indica, Annona squamosa, Areca catechu, Wrightia tinctorea, Butea monosperma, Michalia champaka, Ficus religiosa, Moringa pterygosperma, Aegle marmelos, Santalum album* etc.

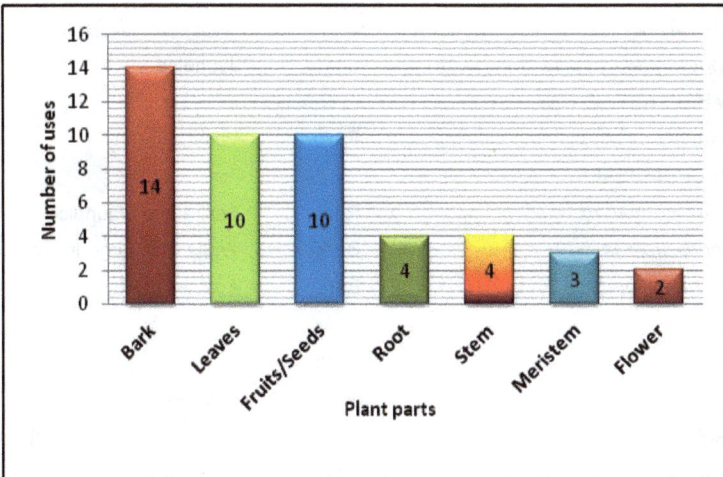

Figure 20.2: Plant part used in Ethnomedicine.

Different parts of medicinal plants were used as medicine by the traditional healers. Among the different plant parts, bark (14 uses) were most frequently used for the treatment of diseases followed by, leaves and fruits/seeds (10 uses), roots and stem (04 uses), meristem (03 uses),and flowers with two uses (Figure 20.2). These are taken internally or applied externally in the form of decoction, paste or powder. Most of the plants used in medicines are either single or mixed with ingredients. These medicinal plants are mainly used for Dandruff, Epilepsy, Jaundice, Dental problems, Worm infection, Vomit, Snakebite, White discharge in women, Against bleeding, Asthma, Scabies, Piles, Stomach ache, Sankadala, Enhancement of lactation, Improper menstrual cycle, skin diseases, Small pox, Deep wounds, Antifertility and Dysentery. Medicinal plants studied are enumerated, arranged alphabetically with their botanical name, local name, family, parts used (Table 20.1). Some important medicinal plants needs immediate conservation and their cultivation should be encouraged through which their extinction can be prevented and local village people may also get low-cost cure and treatment of live stock diseases.

Table 20.1: List of tree species, family, part used and diseases.

Sl.No.	Family/Plant Species	Part Used	Diseases
I	**ANACARDIACEAE**		
1.	*Mangifera indica*	Bark	Dandruff
		Leaf	Scabies
II	**ANNONACEAE**		
2.	*Annona squamosa*	Seed	Epilepsy
III	**APOCYANACEAE**		
3.	*Wrightia tinctorea*	Leaf	Jaundice
IV	**ARECACEAE**		
4.	*Areca catechu*	Seeds	Dental problem/Worm infection
5.	*Cocos nucifera*	Root	Worm infection
V	**ASTRACEAE**		
6.	*Grangea maderaspatensis*	Leaf	Jaundice
VI	**COMBRETACEAE**		
7.	*Terminalia chebula*	Fruit	Vomit
VII	**EUPHORBIACEAE**		
8.	*Ricinus communis*	Root	Snake bite, Jaundice
9.	*Jatropha curcas*	Bark	White discharge in women
10.	*Phyllanthus emblica*	Bark	Against bleeding
VIII	**FABACEAE**		
11.	*Pongamia pinnata*	Seeds	Asthma
12.	*Butea monosperma*	Seeds	Scabies
IX	**GRAMINAE**		
13.	*Bambusa arundinaceae*	Leaf	Piles

Contd...

Table 20.1–*Contd...*

Sl.No.	Family/Plant Species	Part Used	Diseases
X	**MAGNOLIACEAE**		
14.	*Michelia champaka*	Bark	Stomach ache, White discharge
XI	**MELIACEAE**		
15.	*Melia azadirachta*	Meristem	Stomach ache
		Leaf	Jaundice, Sankadala, Enhance lactation
		Oil	Skin disease, Worm infection
XII	**MORACEAE**		
16.	*Ficus religiosa*	Bark	Menstrual cycle, Skin disease,
		Latex	Piles
17.	*Ficus bengalensis*	Meristem	Small pox
		Bark	Stomach ache
XIII	**MORINGACEAE**		
18.	*Morinda pterygosperma*	Gum	Snake bite
		Root	Skin disease
		Leaf	Deep wounds
XIV	**MIMOSACEAE**		
19.	*Acacia arabica*	Leaf	Snake bite
		Bark	Stomach ache
20.	*Acacia feruginea*	Meristem	Stomach ache
21.	*Emblica officinalis*	Fruit	Stomach ache
22.	*Acacia nilotica*	Flowers	Antifertility
		Stem	Dental problem
XV	**MYRTACEAE**		
23.	*Eugenia jambolana*	Bark	Menstrual cycle, Stomach ache, Bleeding
		Seeds	Dysentery
XVI	**RUTACEAE**		
24.	*Aegle marmelos*	Leaf	Blood pressure
		Bark	Vomit
		Fruit	Joint pains
XVII	**SANTALACEAE**		
25.	*Santalum album*	Stem	Jaundice
XVIII	**SIMOROUBIACEAE**		
26.	*Ailanthus excelsa*	Bark	Stomach ache in animals
XIX	**SOLANACEAE**		
27.	*Withania somnifera*	Root	Stomach ache in animals
	Total 19 Families/25 Genera 27 Tree species	**39 Uses**	**25 Diseases (23 Humans/ 02 Animals)**

The tribal people life is interwoven around the forest ecology and forest resources. They were found to be the repository of accumulated experience and knowledge of indigenous vegetation, Information on some very useful medicines known to the tribal communities through experiences of ages is usually passed on from generation to generation.

As the tribal population is gradually adapting modern ways of living, their traditional knowledge on plants will soon be lost forever. At present these valuable medicinal tree species diversity are under serious threat due to their habitat destruction, overexploitation, shifting cultivation and other anthropogenic activities.

The survey indicated that, the Jogimatti forest has rich medicinal tree species diversity to treat a wide spectrum of human and livestock ailments. It is evident from the interviews conducted in different villages; knowledge of medicinal plants is limited to traditional healers, herbalists who are living in rural areas are very old. Due to lack of interest among the younger generation as well as their tendency to migrate to cities for jobs, there is a possibility of losing this wealth of knowledge in the near future. Hence there is an urgent need to assess the biodiversity of the local forest, and conserve the medicinal plants and their associated traditional knowledge by proper documentation and conservation strategies.

Acknowledgements

The authors are grateful to the local traditional practitioners in the district for sharing their knowledge on herbal medicine. We also thank the Forest department of Chitradurga for providing facilities to document the medicinal plants available in Jogimatti forest. Gratefully acknowledge the U.G.C. New Delhi for financial assistance. Thanks are due to Chairman, Department of Botany, Karnatak Univesity, Dharwad for necessary facilities.

References

Abu-Rabia, A., 2005. Urinary diseases and ethno botany among pastoral nomads in the Middle East. *Journal of Ethnobiology and Ethnomedicine*, 1: 4. http://www.ethnobiomed.com/content/1/14.

Alagesaboopati, C. and Rajendran, K., 2009. Ethnomedicinal plants of Sirumalai hills of Dindigul district, Tamil Nadu, India. *Ethnobotanical Leaflets,* 13: 159–164.

Chopra, R.N., Nayar, S.L. and Chopra, L.C., 1956. *Glossary of Indian Medicinal Plants*. Council of Scientific and Industrial Research, New Delhi.

Farombi, E.O. African indigenous plants with chemotherapeutic potentials and biotechnological approach to the production of bioactive prophylactic agents. *African J. Biotech.*, 2: 662–671.

Gamble, J.S., 1994. *Flora of Presidency of Madras, Vols. 1–3*. Bishen Singh Mahendrapal Singh, Dehradun.

Goel, A.K. and Bhattacharya, U.C., 1981. A note on some plants found effective in treatment of jaundice (Hepatitis). *J. Econ. Tax. Bot.*, 2: 157–159.

Harsha, V.H., Hebbar, S.S., Hegde, G.R. and Shripathi, V., 2002. Ethnomedical knowledge of plants used by Kunabi Tribe of Karnataka in India. *Fitoterapia,* 73: 281–287.

Hebbar, S.S., Harsha, V.H., Shripathi, V. and Hegde, G.R., 2004. Ethnomedicine of Dharwad district in Karnataka, India. *Plants Used in Oral Health.*

Hiremath, V.T. and Taranath, T.C., 2009. Ethnomedicinal plants and associated traditional knowledge of Jogimatti forest, Chitradurga district, Karnataka. *Ethnobotanical Leaflets,* 13: 1468–1475.

Hiremath, V.T. and Taranath, T.C., 2010. Traditional phytotherapy used for the snake bites by tribes of Chitradurga district, Karnataka, India. *Ethnobotanical Leaflets,* 14: 120–125.

Hooker, J.D., 1978. *Flora of British India, Vols. 1–7.* Bishen Singh Mahendrapal Singh, Dehradun.

Katz, S.R., Newman, R.A. and Llansky, E.P., 2007. *Punica granatum:* Heuristic treatment for *Diabetes mellitus. J. Med. Food,* 10(2): 213–217.

Leach, M.J., 2007. *Gymnema sylvestre* for *Diabetes mellitus:* A systematic review. *The Journal of Alternative and Complementary Medicine,* 13(9): 977–983.

Pei, S.J., 2001. Ethnobotanical approaches of traditional medicine studies: Some experiences from Asia. *Pharmaceutical Biology,* 39: 74–79.

Saldhana, C.J., 1984. *Flora of Karnataka.* Oxford and IBH Publishing Co., New Delhi.

Siddiqui, M.A.A., John, A.Q., Paul, T.M., 1995. Status of some important medicinal and aromatic plants of Kashmir Himalaya. *Advances in Plant Sciences,* 8: 134–139.

Siva, N. and Muthuchelian, K., 2004. Medicinal plant diversity and its utilization in the grizzled giant squirrel Wildlife Sanctuary, Western Ghats, South India. In: *National Workshop on Biodiversity Resources Management and Sustainable Use,* pp. 25–31.

2013, Perspectives in Plant Biodiversity   *Pages* **162–168**

Editor: **Dr. K. Muthuchelian,** *Vice Chancellor, Periyar University, Salem*

Published by: **Daya Publishing House, NEW DELHI**

Chapter 21

Studies on Medicinal Plant Diversity in Thirunelli Forest, Wayanad District, Kerala

M.V. Jeeshna, P. Sathishkumar and S. Paulsamy

PG and Research Department of Botany,
Kongunadu Arts and Science College, Coimbatore – 641 029

ABSTRACT

Plants are useful in day to day life as food or as medicine. Indigenous plants widely used for folk medicinal purposes are numerous and diverse. Medicinal uses of plants are the oldest form of health care known to mankind. The richness of wild medicinal plant species available in an ecosystem is depending upon an array of microclimatic variations. In Kerala, Thirunelli forest of Wayanad district harbours high number of medicinal plants owing to variation in climatic factors. It is that identified about 58 plants coming under 35 families are medicinally important. Among them, the plants such as *Adiantum caudatum* L., *Drosera indica* L., *Coleus aromaticus* Benth., *Exacum bicolor* Roxb., *Plumbago zeylanica* L., *Hemidesmus indicus* R. Br., are the most important species claimed to possesses pronounced medicinal activity. Thus the study ascertains the value of plants used in Ayurveda, which could be of considerable interest to the development of new drug.

Keywords: Folk medicine, Ayurveda, Thirunelli forest, Wayanad, Kerala.

Introduction

Biodiversity is dynamic in that the genetic composition of species is changing over time in response to natural and human induced selection pressures. The occurrence and relative abundance of species in ecological communities changes as a result of ecological and physical factors. India is one of the seventeen mega diversity centres of the world and harbouring about 17000 flowering plants. In India, Eastern Himalaya and Western Ghats have been recognized as two of the world's thirty two hot spots (Possingham and Wilson, 2005).

Kerala is considered as one of the most biologically rich state on the tropical Malabar Cost of South Western India. It falls under the Malabar (Western Ghats) region of the nine biogeographic regions of India. It is the region of humid equatorial climate with an average annual rainfall of 3107 mm. The forest regions in Kerala lie on the Western slope of the Western Ghats at different altitudes rising up to 2694 m above the sea level. Kerala is a home to nearly 10035 plant species, which accounts 22 per cent of total number of plant species found in India. A high number of 1272 plants and animal species, out of 3872 species are endemic. Further, 56 species out of 102 mammals, 139 species out of 169 reptiles and 86 out of 89 amphibians are endemic to the state which speaks volumes about the high level of biodiversity in this state. Thus Kerala is a land of amazing biodiversity, has a vast storehouse of natural supplements and medicine. In this account, around 48 species are exported in the form of raw drugs and extracts (Sudhakar *et al.,* 2007; Rajith and Ramachandran, 2010).

The practices of traditional medicine are based on hundreds of years of belief and observation which predate the development of the modern medicine. Traditional medicine plays an important role in the health care of India. More than 35,000 plant species are being used around the world for medicinal purposes (Van, 1997). More than 8,000 plants are used in our country especially for their medicinal values by the rural people (Tiwari, 2000). Around 25,000 formulations in modern allopathic systems of medicine are derived from those plant species which are being used as folk medicines through out the world since ages. For any management and sustainable utilization of plant resource, documentation and categorization of species are more essential. Thirunelli forests of Wayanad district, Kerala encourages rich variety plant species which including many medicinal plants and to favourable climatic and soil factors. However, there is no proper documentation of plant species in general and medicinal species in particular. Hence, in the present study, a detailed attempt has been made to enlist the medicinal plants in this region which will be used as a base line data for the development of management plant apart from number of other factors.

Materials and Methods

The study was carried out in Thirunelli forests (900 m above msl) (11°53'N latitude and 76°01' E longitude), located in the northern boundary of southern Western Ghats forests in Wayanad in the Nilgiri Biosphere Reserve region (5520.4 km^2).

The folklore medicinal plants were enumerated and the information about their uses were collected during the field trips made during the year 2009-10 monthly interval. Plant species were identified with the regional and local floras (Gamble and Fischer, 1957; Nair and Hentry, 1983). Local inhabitants were consulted to know the uses of these plants. Informations regarding medicinal properties were gathered from persons having familiarity and knowledge with herbal medicine. Routine herbarium

Table 21.1: List of medicinal plants in Tirunelli forest, Wayanad district, Kerala.

Sl.No.	Botanical Name	Family	Habit	Parts used	Medicinal Uses
1.	*Abrus fruticulosus* Wight and Arn.	Fabaceae	Climber	Root, leaves and seed	Used to treat tetanus and to prevent rabies
2.	*Abutilon indicum* Sweet.	Malvaceae	Shrub	Leaves	Cure dysentery and piles
3.	*Acacia pennata* Willd.	Mimosaceae	Tree	Bark	Odontalgia
4.	*Adhatoda zeylanica* Medik.	Acanthaceae	Shrub	Leaves	Cough, Heart pain
5.	*Adiantum caudatum* L.,	Adiantaceae	Herb	Leaves and rhizome	Healing wound
6.	*Aerva lanata* Juss.	Amaranthaceae	Herb	Whole plant	Kidney stone
7.	*Ageratum conyzoides* L.	Rutaceae	Herb	Leaves	Diarrhoea, skin diseases
8.	*Allium sativum* L.	Liliaceae	Herb	Seed and Bulbous	Seed oil is applied over the affected area till cure
9.	*Aloe vera* Burm.f.	Liliaceae	Herb	Succulent leaves	Reduce the body temperature
10.	*Alstonia scholaris* R.Br.	Apocynaceae	Tree	Bark	Nervous disorders
11.	*Amaranthus spinosus* L.	Amaranthaceae	Herb	Root	Temporary insanity
12.	*Andrographis paniculata* Wall. ex. Nees.	Acanthaceae	Herb	Root, leaves	Fever, snake bite, and skin diseases
13.	*Asparagus racemosus* Willd	Liliaceae	Armed vine	Root	Root paste is applied on the affected parts till cure
14.	*Biophytum sensitivum* Dc.	Oxalidaceae	Herb	Whole plant	Gonorrhoea, Asthma
15.	*Boerhavia diffusa* L.	Nyctaginaceae	Herb	Root	Diarrhoea, kidney problems
16.	*Calophyllum inophyllum* L.	Clusiaceae	Tree	Bark	Wound
17.	*Calotropis gigantean* R.Br.	Asclepidaceae	Shrub	Leaves, latex	Reduce the body pain
18.	*Cardiospermum halicacabum* L.	Sapindaceae	Climber	Whole plant	Gout, diabetes
19.	*Cassia alata* L.	Caesalpinaceae	Shrub	Leaves	Fungal diseases and worm infection
20.	*Cassia fistula* L.	Caesalpinaceae	Tree	Leaves	Kill the worms in the intestine
21.	*Cathranthus roseus* Don.	Apocynaceae	Herb	Root, Leaves	Fever, Dysentry and Piles

Contd...

Table 21.1–*Contd...*

Sl.No.	Botanical Name	Family	Habit	Parts used	Medicinal Uses
22.	*Coleus amboinicus* Lour	Lamiaceae	Herb	Root and Leaves	Plant used in the treatment of diabetes
23.	*Costus speciosus* Smith.	Cornaceae	Herb	Stem	Plant used as greens
24.	*Curculigo orchioides* Gaertn.	Amarylidaceae	Herb	Tuber	Snake poison
25.	*Curcuma domestica* Valeton.	Zingiberaceae	Herb	Rhizome	Digestion, antiseptic properties
26.	*Cyclea peltata* Diels.	Mensipermaceae	Twiner	Tuber	Stomach pain, Insect bite
27.	*Drosera indica* L.,	Droseraceae	Herb	Whole plant	Anti-cancerous
28.	*Eclipta alba* Hassk.	Asteraceae	Herb	Leaves	Used for the wound of dog bite
29.	*Elephantopus scaber* L.	Asteraceae	Herb	Leaf	Leaf extract is administered internally till cure
30.	*Emilia sonchifolia* Dc.	Asteraceae	Herb	Leaves	Wounds
31.	*Exacum bicolor* Roxb.	Gentianaceae	Herb	Whole plant	Diabetics, Fever and eye diseases.
32.	*Glycosmis pentaphylla* Dc.	Rutaceae	Shrub	Root	Fever
33.	*Glycyrrhiza glabra* L.	Fabaceae	Climber	Whole plant	Tuberculosis
34.	*Helicteres isora* L.	Sterculiaceae	Shrub	Fruits, Root	Skin diseases, stomachache
35.	*Hemidesmus indicus* R.Br.	Asclepidaceae	Shrub	Whole plant	Skin diseases
36.	*Indigofera tintoria* L.	Fabaceae	Shrub	Root	Snake poison
37.	*Jatropha curcas* L.	Euphorbiaceae	Shrub	Stem bark	Injuries
38.	*Lantana camara* Moldenk.	Verbinaceae	Shrub	Root, leaves	Wounds healing
39.	*Lawsonia inermis* L.	Lythraceae	Shrub	Leaves	Jaundice
40.	*Leucas aspera* Spreng.	Lamiaceae	Herb	Whole plant	Gas trouble
41.	*Mimosa pudica* L.	Mimosaceae	Herb	Leaves	Sexual diseases, cold
42.	*Murraya koenigi* Spreng.	Rutaceae	Tree	Root, Bark and Leaves	Skin diseases, dysentery and Ulcer
43.	*Naravelia zeylanica* Dc.	Ranunculaceae	Climber	Leaves	Worm infection

Contd...

Table 21.1–*Contd...*

Sl.No.	Botanical Name	Family	Habit	Parts used	Medicinal Uses
44.	*Passiflora calcarata* Mast.	Passifloraceae	Climber	Fruit	Plant used as greens
45.	*Phyllanthus niruri* L.	Euphorbiaceae	Herb	Whole plant	Plant used as greens
46.	*Plumbago zeylanica* L.	Plumbaginaceae	Shrub	Leaves, root	Snake bite
47.	*Portulaca oleracaea* L.	Portulaceae	Herb	Seed	Used in the treatment of diabetes
48.	*Rauvolfia serpentine* Benth.ex.kurtz.	Apocynaceae	Shrub	Root	Snake bite
49.	*Ricinus communis* L.	Euphorbiaceae	Shrub	Leaves, latex	Wound healing
50.	*Santalum album* L.	Santalaceae	Tree	Wood	Black heeds on the face
51.	*Scoparia dulcis* L.	Scrophulariaceae	Herb	Leaf	Used in the treatment of fever
52.	*Scleichera trijuga* Lour.	Sapindaceae	Tree	Bark, Seed oil	Seed oil is massaged over the affected area till cure
53.	*Sida acuta* Burm.	Malvaceae	Herbs	Root and Leaves	Apply fresh leaves as poultices on boils
54.	*Sida rhombifolia* L.	Malvaceae	Shrub	Leaves	Headache
55.	*Tectona grandis* L.	Verbinaceae	Tree	Bark	Curing blood related diseases
56.	*Tinospora cordifolia* Willd.	Menispermaceae	Climber	Stem, leaves	Diabetics
57.	*Tridax procumbens* L.	Asteraceae	Herb	Flowers, leaves	Wound, reduce toothache
58.	*Vitex negundo* L.	Verbenaceae	Shrub	Leaves and root	Rheumatism

methods were followed in pre-serving the plant specimens (Jain, 1965; Jain and Rao, 1977). The medicinal uses of species were cross checked through the literature available (Sukumaran and Raj, 2010; Rahman *et al.,* 2008).

Results and Discussion

Biogeographically, Wayanad is a transition area between the moist and dry deciduous forests in southern Western Ghats. It harbours habitat restricted endemic species as well as disjunct populations of species that are found in both regions (Pascal, 1988; Rodgers and Panwar 1988; WWF, 2001).The present attempt, the documentation of 58 medicinally important plant species belonging to 35 families has been made in Thirunelli forest, Wayanad district, Kerala and these medicinal plants were used for the treatment of various ailments like ulcer, poisonous bites, cuts, rheumatics, jaundice, nervous disorders, asthma, fever and diabetes etc. (Table 21.1). Asteraceae was the dominant family which contributes four species followed by Fabaceae, Malvaceae, Liliaceae, Apocynaceae, Rutaceae, Euphorbiaceae, Verbinaceae (three species each), Acanthaceae, Amaranthaceae, Asclepidaceae, Caesalpinaceae, Lamiaceae, Mensipermaceae (two species each) and 21 families represented by single species to the list of medicinal plants in Thirunelli forest.

One important sources of locating potential bioresources is indegeous knowledge of the folk (Jain and Sikarwar, 1997). The inventory findings reinforce that indigenous knowledge is dynamic and botanical knowledge diminishing (Posey, 1983; Boom, 1987; Lee *et al.,* 2001; Prance, 1991; Anand *et al.,* 2006,; Sukumaran and Raj, 2010). However, people of the modern generation learn from their ancestors on the basis of keen observation only. The people have been using plant remedies against various ailments from time immemorial without knowing their effective constituents.

The study revealed that pharmacologists, pharmacogosists and phytochemists have screened only a few plants for their active principles. Clinical investigation take along time and are highly expensive to analysis large numbers of plants. Hence, traditional folklore medicinal knowledge is the best source of information for preliminary screening in such instances. Even through their medicinal value and economic importance are elaborated, it is important to go for in-depth investigations particularly experimental and clinical studies for findings uses for future generations. Conservation of medicinal plant diversity of these grooves through therefore most important for management and sustainable development in managing the fragile ecological processes and life-support system.

References

Anand A.M., Nandakumar, N., Karunakaran, L., Ragnathan, M. and Murugan, V., 2006. A survey of medicinal plants in Kollimalai Hills tracts, Tamil Nadu, 5(2): 139–143.

Boom, B.M., 1987. Ethnobotany of the Chacabo Indians, Beni, Bolivia. *Adv. Econ. Bot.*, 4: 1–68.

Gamble, J.S and Fischer, C.E.C., 1957. *Flora of the Presidency of Madras, Vols. 1–3,* Botanical Survey of India, Calcutta.

Jain, S.K and Rao, R.R., 1977. *A Handbook of Field and Herbarium Methods,* Todays and Tomorrows Printers and Publishers, New Delhi.

Jain, S.K. and Sikarwar, R.L., 1997. Prospective under utilized bioresources: Clues from indigenous knowledge in Latin America. *J. Indian Bot. Soc.,* 76: 253–260.

Jain, S.K., 1965. Medicinal plant lore of the tribals of Bastar. *Econ. Bot.,* 19: 236–250.

Lee, R.A., Balick, M.J., Link, D.L. Sohl, F., Brosi, B.J. and Roynor, W., 2001. Cultural dynamism and change: An example from the Federated states of Micronosia, *Econ. Bot.,* 55(1): 9–15.

Nair, N.C. and Henry, A.N., 1983, 1987 and 1989. *Flora of Tamil Nadu, India, Vols. 1–3,* Botanical Survey of India, Coimbatore.

Pascal, J.P., 1988. *Wet Evergreen Forests of the Western Ghats of India: Ecology, Structure, Floristic Composition and Succession.* French Institution, Pondicherry.

Posey, D.A., 1983. Indigenous knowledge and development: An ideological bridge to the future. *Cieniae Cultura,* 35(7): 887–894.

Possingham, H.P. and Wilson, K.A., 2005. Biodiversity: Turning up the heat on hotspots. *Nature,* 436: 919–920.

Prance, G.T., 1991. What is ethnobotany today? *J. Ethanopharmacol.,* 32: 209–216.

Rahman, A.H.M.M., Anisuzzaman, M., Haider, S.A., Ahmed, F., Rafiul Islam, A.K.M. and Naderuzzaman, A.T.M., 2008. Study of medicinal plants in the graveyards of Rajshahi city. *Res. J. Agric. and Biol. Sci.,* 4(1): 70–74.

Rajith, N.P. and Ramachandran, V.S., 2010. Ethnomedicines of *Kurichyas,* Kannur district, Western Ghats, Kerala. *Indian Journal of Natural Products and Resources,* 1(2): 249–253.

Rodgers, W.A. and Panwar, H.S., 1988. Planning a wildlife protected areas network in India Department of Environment, Forests, and Wildlife/Wildlife Institution of India report, Wildlife Institution of India. pp. 1–2.

Sudhakar Reddy, C., Chiranjibi, P., Reddy, K.N. and Raju, V.S., 2007. Census of endemic flowering plants of Kerala, India. *Journal of Plant Sciences,* 2(5): 489–503.

Sukumaran, S. and Raj, A.O.S., 2010. Medicinal plants of sacred grooves in Kanyakumari district. Southern Western Ghats. *Indian Journal of Traditional Knowledge,* 9(2): 294–299.

Tiwari, D.N., 2000. Report of the task force on conservation and sustainable uses of medicinal plants. (Bull planning commission, Govt. of India, New Delhi).

Van Seters, A.P., 1997. *Non-wood Forest Products,* FAO, p.116.

WWF, 2001. Wild world, WWF full report, Southern Western Ghats montane rain forests (IMO151).

2013, Perspectives in Plant Biodiversity   *Pages* **169–180**
Editor: **Dr. K. Muthuchelian,** *Vice Chancellor, Periyar University, Salem*
Published by: **Daya Publishing House, NEW DELHI**

Chapter 22

Screening and Characterization of Taxol: An Anticancer Drug from Endophytic Fungi

M. Pandi[1] *, P. Rajapriya[2], Yogeswari[3]
and J. Muthumary[3]

[1]*Department of Molecular Microbiology, School of Biotechnology,*
Madurai Kamaraj University, Madurai.
[2]*Srinivasan College of Arts and Science, Perambalur, Tamil Nadu*
[3]*Centre for Advanced Studies in Botany, University of Madras,*
Guindy Campus, Chennai

ABSTRACT

Paclitaxel (Taxol) is the most effective antitumor agent developed in the past three decades. Taxol has been hailed by many in the cancer community as a major breakthrough in the treatment of cancer. Taxol was originally isolated from bark of the pacific yew *Taxus brevifolia* in 1971. The pacific yew is a slow-growing under story tree found mainly in climax forests throughout the pacific North West. While it is not currently threatened or endangered, continuous harvesting for commercial preparation of Taxol is untenable. Due to the shortage and natural exploitation of yew trees, the taxol is produced by a semi-synthetic and Plant cell culture method has been suggested as an attractive alternative technique that could overcome the limitation of extracting useful metabolites from natural resources. In spite of many reports from various groups, commercial

* Corresponding Author: E-mail: an_pandi@rediffmail.com

paclitaxel production by semi-synthetic and plant cell culture has not yet been successful. The major obstacle to commercialization has been the low yield of paclitaxel from plant cell culture. Since, scientists all over the world have been searching for new routes and sources to improve taxol production to meet the demand, the most significant finding during the last decade might be the discovery of taxol production by endophytic fungi. The purpose of the present study is to isolate and identify the endophytic fungi and screened for the taxol production. The production of taxol by using endophytic fungi has given rise to the possibility for its cheaper and wide availability, which may eventually be available *via* large scale industrial fermentation. The fungus can serve as a potential material for genetic engineering to improve the taxol production.

Keywords: *Endophytic fungi, Taxol.*

Introduction

Taxol is a chemotherapic drug specifically effective against prostate, ovarian, breast and lung cancer. Its primary mechanism of action is related to the ability to stabilize the microtubules and to disrupt their dynamic equilibrium (Horwitz, 1993). Taxol, though isolated more than 28 years ago, has been the subject of extensive investigation aimed at overcoming its toxicity and low solubility. The search of an alternative source of taxol other than the bark of the yew trees (*Taxus* sp)., has been carried out by scientists all over the world to meet the demand in clinics. The most significant finding in the last decade might be the discovery of endophytic Taxol producing fungus in Gymnosperm, particularly, in yew trees. It is remarkable that Taxol produced by the endophytes is identical to that produced by *Taxus* spp., chemically and biologically (Stierle *et al.*, 1993; Wang *et al.*, 2000).

Recently, taxol producing endophytic fungi have been successfully isolated from different medicinal plants in our group (Gangadevi and Muthumary, 2007 and 2008, SenthilKumaran *et al.*, 2008a, 2008b and 2009a and b and Pandi *et al.*, 2008). Thus, fermentation processes using taxol producing microorganisms may be an alternative way to produce taxol. The fungus can serve as a potential material for genetic engineering to improve the taxol production. The present study deals with the isolation of endophytic fungi from medicinal plant and extraction and characterization of taxol from an endophytic fungus.

Materials and Methods

Isolation and Identification of Endophytic Fungi

The leaves of *Morinda citrifolia* was collected at University of Madras, Guindy Campus, Chennai. The plant materials were subjected to endophytic isolation within 3 hours after harvest. The endophytic fungal cultures were separated from the healthy leaves according to the general mycological procedure (Gangadevi and Muthumary, 2007). Specifically, the leaves were washed with running tap water, sterilized with ethanol (75 per cent V/V) for 1 min and sodium hypochloride (2.5 per cent V/V) for 15 min, then rinsed in sterile water for three times and cut into 1cm long segments. Plant segments were then transferred to potato dextrose agar plates amended with ampicillin (200µg/ml) and streptomycin (200µg/ml) to inhibit bacterial growth. After two days, fungi were observed growing from the inner leaf segments in the plates. Individual hyphal tips of the various fungi were removed from the agar plates, placed on new PDA medium and incubated at 25°C for at least 2 weeks. Each fungal culture was checked for purity and sub cultured to another agar plate by the hyphal tip method. Fungal identification methods were based on the morphology of the fungal culture, the mechanism of spore production and the characteristics of the spores.

Cultivation and Extraction of Taxol from Selected Fungal Isolates

The selected endophytic isolates were inoculated into 2000 ml Hopkins flask of MID medium supplemented with 1 g soy tone L^{-1}, incubated for 12 hours under light and dark cycle at temperature between 22°C to 25°C for 21 days (Strobel *et al.,* 1996 a and b). After 21 days, the cultures were passed through four layers of cheese cloth and 0.25g of Na_2Co_3 (0.025 per cent W/V) was added to the culture filtrate to avoid fatty acid contamination. The culture filtrate was further extracted with twice the volume of dichloromethane and the organic phase was taken to dryness under reduced pressure at 50°C. The dry solid residue was re-dissolved in methanol for the subsequent separation. The crude extracts were analyzed by chromatographic separation and spectroscopic methods.

Thin Layer Chromatography (TLC)

TLC analysis was carried out on Merk 1mm (20 x 20 cm) silica gel precoated plates. The plates were developed by the solvent system reported by Strobel *et al.* (1996 a and b). The taxol was detected with 1 per cent vanillin sulfuric acid (w/v) and heating. It appears as a bluish spot that faded to dark grey after 24 hours. Then the area of the plate containing putative taxol was carefully removed by scraping off the silica at the appropriate R_f value and eluted with methanol.

Ultra Violet (UV) Spectroscopic Analysis

After chromatography, the area of the TLC plate containing putative taxol was carefully removed by scrapping off the silica at the appropriate R_f and exhaustively eluting it with methanol. The purified sample of taxol was dissolved in 100 per cent methanol and analyzed by Beckman DU-40 UV -Spectrophotometer and compared with authentic taxol.

Infra-Red Spectroscopic Analysis

The IR spectra of the compound were recorded on Shimadzu FTIR 8000 series instrument. The purified taxol was ground with IR grade potassium bromide (KBr) (1:10) pressed into discs under vacuum using spectra lab Pelletiser and compared with authentic taxol. The IR spectrum was recorded in the region between 4000–5000 cm^{-1}.

High Performance Liquid Chromatography (HPLC) Analysis

To confirm the presence of taxol, the fungal extract was subjected to high performance liquid chromatography. Taxol was analyzed by HPLC (Shimatzu 9A model) using a reverse phase C_{18} column with a UV detector. Twenty µl of the sample was injected each time and detected at 232 nm. The mobile phase was methanol/ acetonitrile/water (25:35:40, by vol). at a flow rate of 1.0 ml min⁻¹. The sample and the mobile phase were filtered through 0.2 µm PVDF filter before injecting into the column. Taxol was confirmed by comparing the peak area of the samples and authentic taxol.

Results

Description of Fungus

Botryodiplodia theobromae Pat.

E. Punithalingam, Journal of Agriculture of the University of Puerto Rico **52**: 260, 1968, CMI, Descriptions of Pathogenic fungi and Bacteria No. 519.

Bull. Soc. Mycol. Fr. **8**: 132, 1892.

Sohi, H.S. and Maholay. 1974. Indian Journal of Mycology and Plant Pathology 4: 26.

Sreemali, J.L. and K. S. Bilgrami. 1968. Indian Phytopath. **21**: 357.

Rajak, R.C. and Pandey, A. K. 1985. Indian journal of mycology and plant pathology **15**: 186-194.

Bhatnagar, M.K. and Mali, B.L. 1986. Indian Journal of mycology and plant pathology **16**(2): 169.

Rao, V.G. 1986. Biovigyanam. **12**(1): 14-16.

The test fungus was sub-cultured and maintained on potato dextrose agar (PDA) medium and was grown in MID medium supplemented with 1g soytone 1⁻¹ for taxol production (Pinkerton and Strobel, 1976). The culture was incubated for 21 days, at 23–24°C. After 10 days the colony was gray to dark gray, cottony, reverse of the colony turning dark grayish green with abundant fruit bodies (Figure 22.1A). The conidiomata were pycnidial, erumpent, separate, simple, 250-350 µm in diameter, globose, unilocular to multilocular with ostiole circular. The conidia were initially unicellular, hyaline, subovoid to ellipsoidal, thick walled with grainy inclusions with truncate base. The mature conidia were dark brown, single septate, often longitudinally striate, 19-20 x 12-13 µm, paraphyses up to 50µm long (Figure 22.1B). Based on these results the selected test fungus was identified as *Botryodiplodia theobromae* Pat. (BTMUBL-5).

Extraction of Taxol

The fungus was grown in three litres Hopkins flasks containing 1500 ml of MID medium supplemented with 1g soytone 1⁻¹. The cultures were incubated for 21 days. After the incubation period, the cultures were filtered through four layers of cheesecloth to remove mycelia and the fruit bodies. To the culture filtrate 0.25 g of Na₂CO₃ was added with frequent shaking in order to reduce the amount of fatty acids that may contaminate taxol in the culture. Then the culture filtrate was extracted with two equal volumes of solvent dichloromethane. The organic phase was collected and the solvent was then removed by evaporation under reduced pressure at 35 °C using rotary vacuum evaporator. The dry solid residue was re-dissolved in methanol for the subsequent

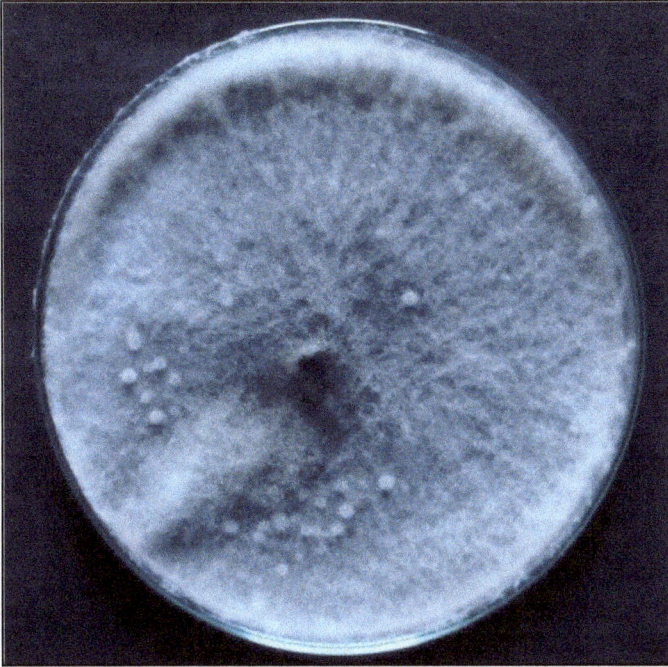

Figure 22.1A: Morphology of *Botryodiplodia theobromae* Pat.

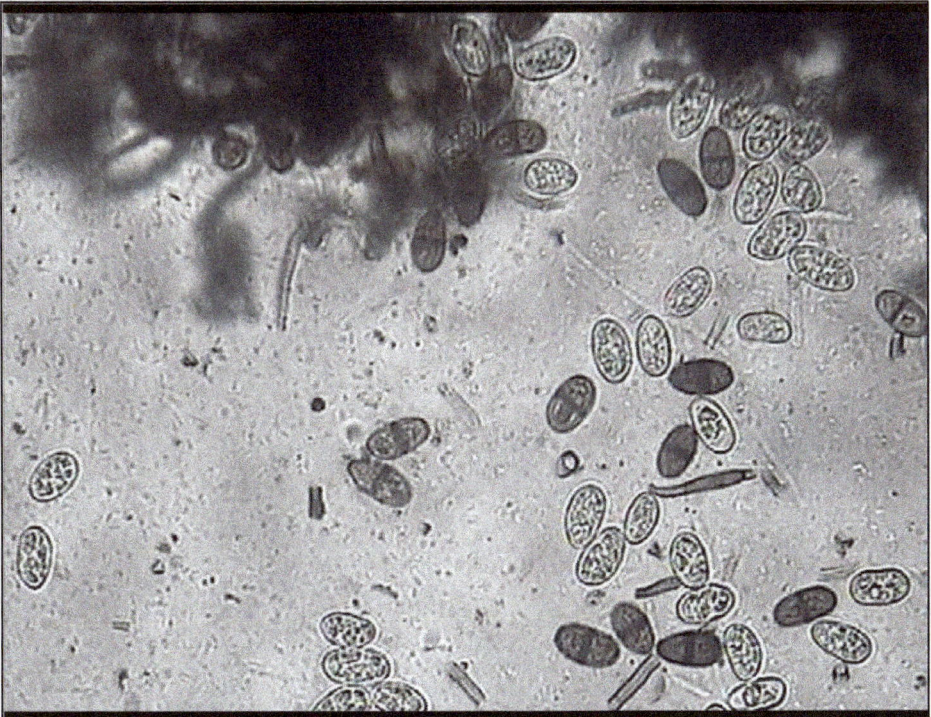

Figure 22.1B: Conidia of *Botryodiplodia theobromae* Pat.

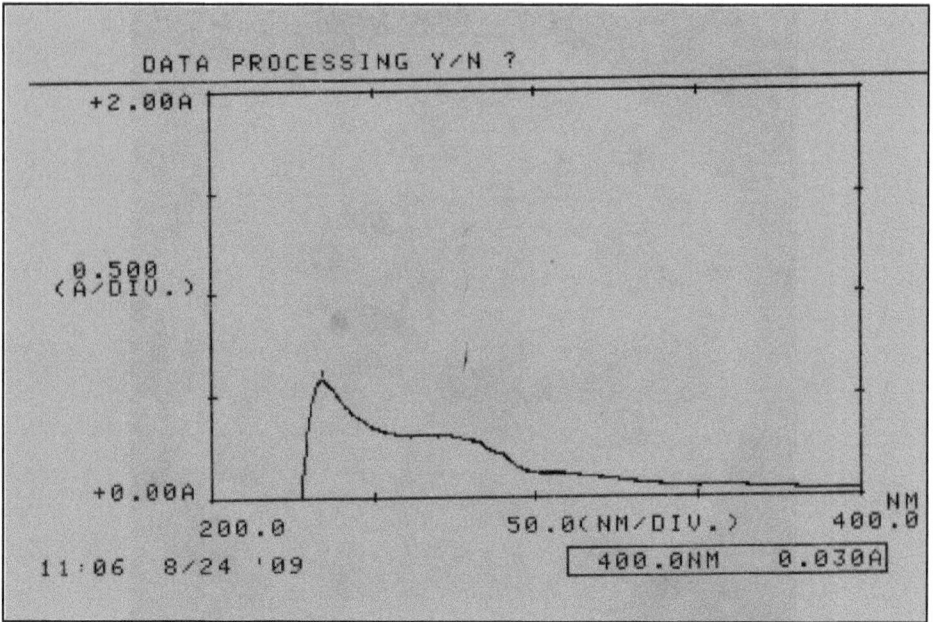

Figure 22.2A: Authentic taxol.

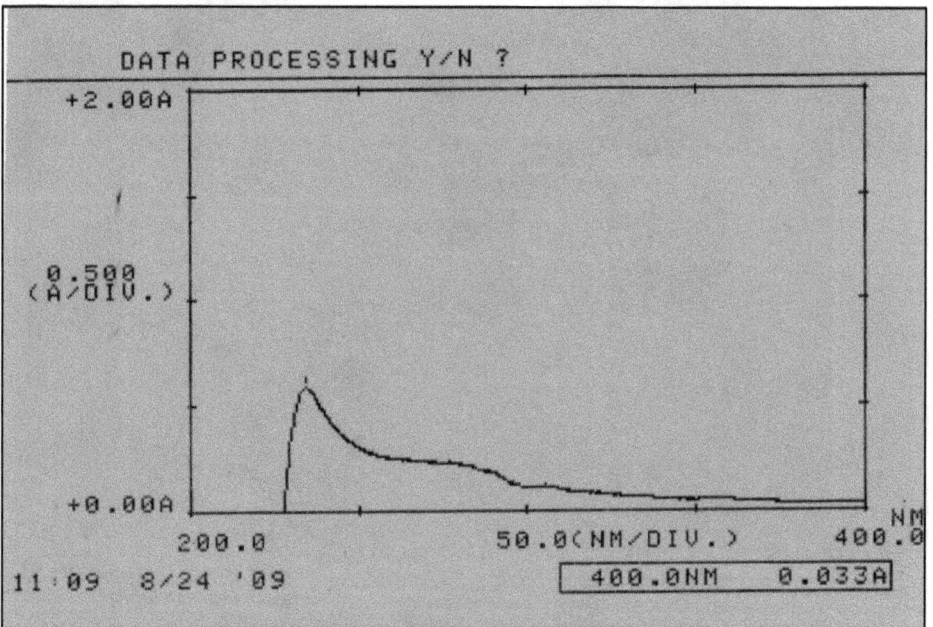

Figure 22.2B: Fungal taxol.

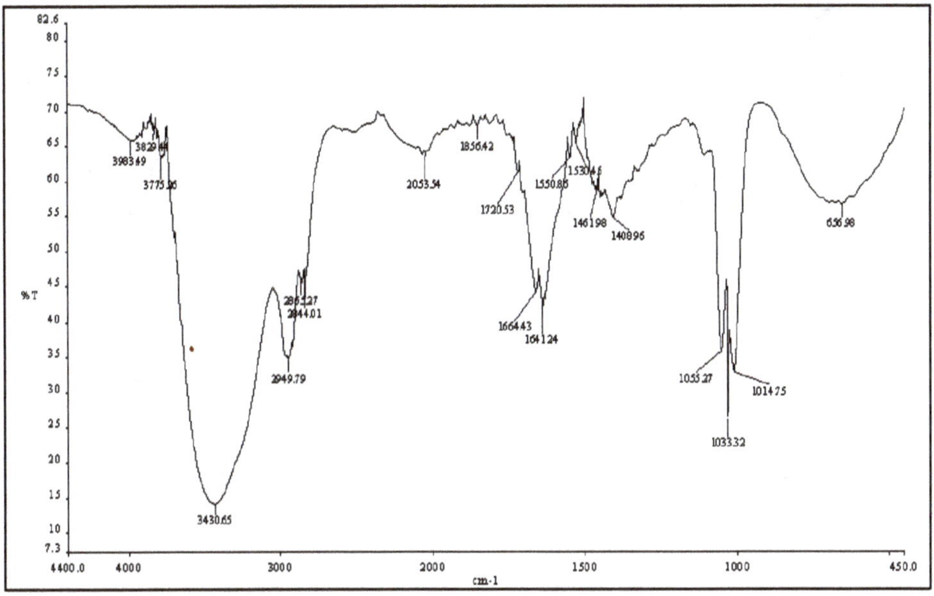

Figure 22.3A: Authentic taxol.

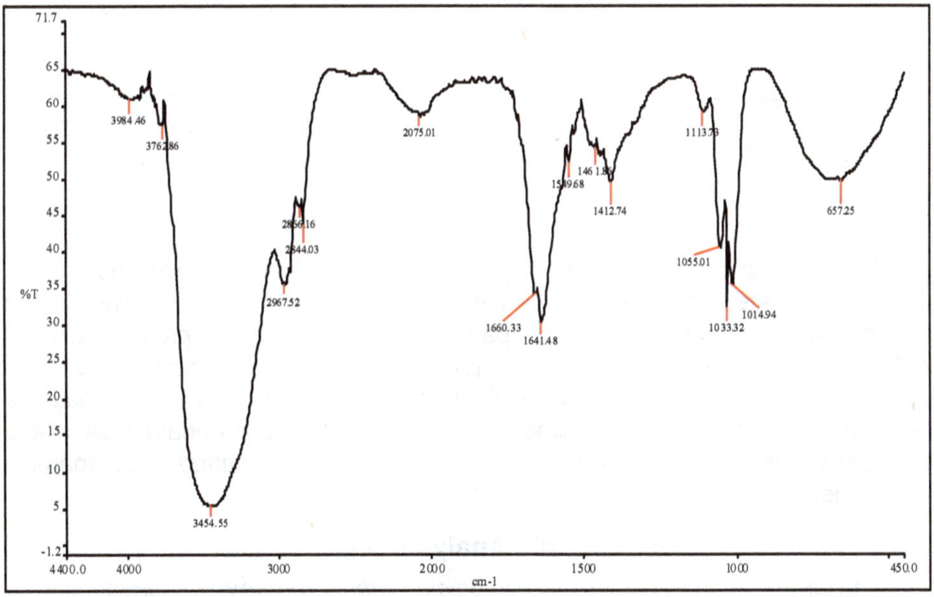

Figure 22.3B: Fungal taxol.

separation. The crude extracts were analyzed by chromatographic separation and spectroscopic analysis.

Thin Layer Chromatography (TLC)

TLC analysis was carried out on Merck 1mm (20x20cm) silica gel precoated plates. The plates were developed by the solvent system reported by Strobel *et al.*

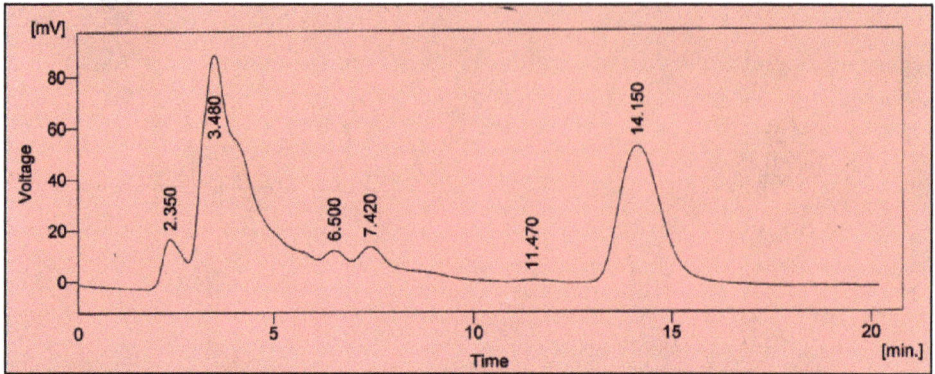

Figure 22.4A: Authentic taxol.

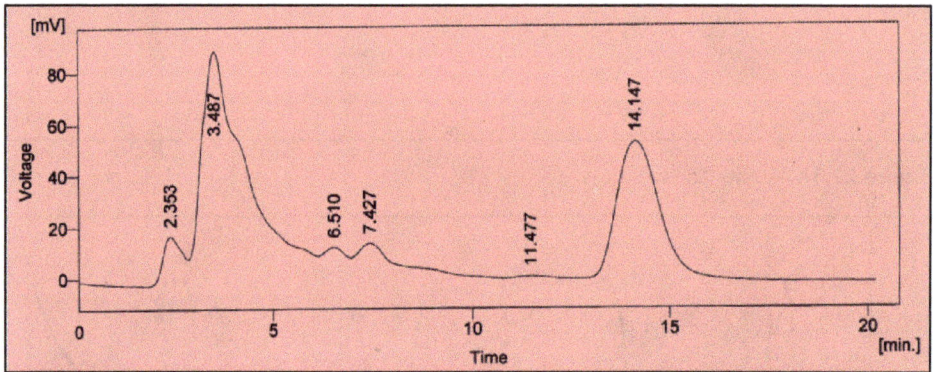

Figure 22.4B: Fungal taxol.

(1996a). The taxol was detected with 1 per cent vanillin sulfuric acid (w/v) and heating. The taxol isolated from the fungus *Botryodiplodia theobromae* showed similar chromatographic properties when compared to authentic taxol. The presence of taxol in *Botryodiplodia theobromae* extract was confirmed by the appearance of a bluish spot fading to dark gray after 24 hours. Then the area of the plate containing putative taxol was carefully removed by scraping off the silica at the appropriate R_f value and eluted with methanol. Further, the presence of taxol was identified by comparison with authentic taxol using UV, IR and HPLC.

Ultra Violet (UV) Spectroscopic Analysis

The fungal taxol isolated from *Botryodiplodia theobromae* was characterized by UV spectrum. UV absorption spectrum of fungal taxol is similar to that of authentic taxol with a maximum absorption at 235nm (Figures 22.2A and 22.2B).

Infra-Red Spectroscopic Analysis

The IR spectrum of the compound was recorded on Shimadzu FTIR 8000 series instrument. The fungal taxol was ground with IR grade potassium bromide (KBr) (1:10) pressed into discs under vacuum using spectra lab Pelletiser and compared with authentic taxol. The IR spectrum was recorded in the region between 4000–500 cm⁻¹. The appearance of bands in IR spectra convincingly illustrates the identical chemical nature of the extracted taxol from the fungus with that of authentic taxol. A broad peak in the range of 3336-3454 cm⁻¹, which was due to hydroxyl (-OH) groups stretch. The aliphatic CH stretch is observed in the range of 2920-2964 cm⁻¹. The aromatic ring (C=C) stretching frequency was observed in the range of 1412-1549 cm⁻¹. The registration of peak observed in the range of 1014–1068 cm⁻¹ is due to the presence of aromatic C, H bends. The IR spectra of the fungal samples were superimposed on the spectrum of authentic taxol (Figures 22.3A and 22.B).

High Performance Liquid Chromatography (HPLC) Analysis

To confirm the presence of taxol, the fungal extract was subjected to high performance liquid chromatography. Taxol was analyzed by HPLC (Shimadzu 9A model) using a reverse phase C_{18} column with a UV detector. 20 μl of the sample was injected each time and detected at 232 nm. The mobile phase was methanol/acetonitrile/water (25:35:40, by vol). at a flow rate of 1.0 ml min⁻¹. The sample and the mobile phase were filtered through 0.2 μm PVDF filter before injecting into the column. The fungal sample gave a peak with similar retention time as that of authentic taxol (Figure 22.4A and 22.B). The quantity of taxol produced by the fungus was calculated and it was estimated to be 160μg/L.

Discussion

Taxol is an important and expensive anticancer drug. Taxol is produced by all yew species (Georg *et al.,* 1994) and is found in extremely low amounts in the needles, bark and roots (Vidensek *et al.,* 1990). However, this plant is not suitable for cultivation because of its slow growth, and there is a limitation on gathering this plant which grows naturally. After many years of research total synthesis of paclitaxel has been achieved (Nicolaou *et al.,* 1994). The process, however, is multi-stepped, and the overall yield has made this approach economically unfeasible. Nevertheless, semisynthesis of paclitaxel from its natural precursor, 10–deacetylbaccatin III, is now possible. This precursor can be extracted from the needles and twigs of the European yew *Taxus baccata.* Production by the semisynthetic route from the 10-deacetylbaccatin III precursor replaces the bark of *T. brevifolia* as the major source of paclitaxel (Cragg *et al.,* 1993). Plant cell culture of *Taxus* spp. is another approach (Roberts and Schuler, 1997). But all these processes are too expensive for commercialization and give low yield (Guenard *et al.,* 1993; Frense 2007). Therefore, in order to lower the price of paclitaxel and make it more available, a fermentation process involving a microorganism would be the most desirable means of supply.

Recently several taxol producing endophytic fungi have been identified such as *Bartalinia robillardoides* Tassi, *Pestalotiopsis terminaliae, Collectotrichum gleosporioides, Fusarium* sp., *Periconia* sp. and *Nodulisporium sylviforme.* The spectroscopic and chromatographic analyses of these fungal taxol found to be are identical to that of authentic taxol. (Wang *et al.,* 2000; Zhou 2001a and b; Guo *et al.,* 2006; Yuan *et al.,* 2006; Gangadevi and Muthumary 2007; Gangadevi and Muthumary

2008a; Gangadevi and Muthumary 2008b). The production of taxol from a microbial source has many advantages over other sources. Industrial production of bioactive compounds like taxol requires reproducible, dependable productivity. If a fungus is the source organism, it can be grown in tank fermenters to produce an inexhaustible supply of taxol. The added advantage is that the fungi usually respond favorably to routine cultural techniques.

The present study was taken up in order to investigate taxol production from an endophytic fungus *Botryodiplodia theobromae* Pat. isolated from a medicinal plant *Morinda citrifolia*. Microbes which live inside the plants (endophytes), may offer tremendous potential source of therapeutic compounds. Strobel *et al.* (1996 b) isolated number of fungal endophytes including *Pestalotia, Pestalotiopsis, Fusarium, Alternaria, Pithomyces* and *Monocheatia* from stem and phloem of *Taxus* sp. One of the most commonly isolated endophytic species is *Pestalotiopsis microspora*. Furthermore, several other *P. microspora* isolates were obtained from bald cypress in South Carolina. In the case of one strain of *P. microspora* (CP-4), Taxol was isolated from culture medium and was also shown to be identical to authentic Taxol by chromatographic and spectroscopic studies (Li *et al.,* 1996). The discovery of *Taxomyces andreanae* was the first demonstration that any organism other than *Taxus* spp. could produce taxol. However, the yields of taxol and taxanes have been low. Taxol is positively identified via its chromatographic mobilities with authentic taxol in a multitude of thin layer chromatographic systems (Stierle *et al.,* 1993).

In the present study, the taxol from the fungus, *Botryodiplodia theobromae* was extracted and the presence of taxol was confirmed using TLC, UV, IR and HPLC. The UV absorption spectrum of fungal compound isolated from *Botryodiplodia theobromae* yielded similar absorption to authentic Taxol with a maximum absorption at 235nm (Figures 22.2A and 22.2B). The appearance of bands in IR spectra convincingly illustrates the identical chemical nature of the extracted Taxol from the fungus with that of authentic taxol (Figures 22.3A and 22.2B). The fungal sample was analyzed by HPLC to confirm the presence of Taxol and gave a peak when eluting from a reverse phase C_{18} column, with about the similar retention time as authentic Taxol (Figures 22.4A and 22.2B). The quantity of Taxol produced by the fungus was calculated and it was estimated to be 160µg/L.

The techniques like TLC, UV,IR, HPLC, the high resolution ^{1}H–and ^{13}C–NMR, MS are the tools applied in the confirmation test for the antitumor compound taxol isolated from fungi and are supported by many workers (Stierle *et al.,* 1993, Strobel *et al.,* 1996a, Strobel *et al.,* 1996b). The investigations and the significance in the discovery of fungi that produce taxol indicate that there are abundant resources of fungi that produce taxol. Most of them are endophytes of *Taxus* spp. However, the present study has reported the isolation of taxol-producing endophytic fungus from tropical medicinal plants in India. Hence, from both the ecological and economic view point, a microbial source will be the replacement for the yew. Such a great number of fungi that produce taxol strongly suggest that the kingdom of the fungi is a gold mine of taxol. The endophytic fungi are able to produce taxol has created a possibility for a cheaper and more widely available product to be eventually produced via industrial fermentation.

Acknowledgements

One of the authors (Mohan Pandi) is thankful to the Indian Council of Medical Research; Government of India for the Senior Research Fellowship during the investigation was carried out.

References

Cragg, G.M., Schepartz, S.A., Suffness, M. and Grever, M.R., 1993. The Taxol supply crisis, New NCI policies for handling the large-scale production of novel natural product anticancer and anti-HIV agents. *Journal of Natural Products,* 56: 1657–1668.

Frense, D., 2007. Taxanes: perspectives for biotechnological production. *Applied Microbiology and Biotechnology,* 73: 1233–1240.

Gangadevi, V. and Muthumary, J., 2007. Taxol, an anticancer drug produced by an endophytic fungus *Bartalinia robillardoides* Tassi, isolated from a medicinal plant, *Aegle marmelos* Correa ex Roxb. *World Journal of Microbiology and Biotechnology,* Published online: DOI 10.100/3 11274–007–9530–4.

Gangadevi, V. and Muthumary, J., 2008a. Taxol production by *Pestalotiopsis terminaliae,* an endophytic fungus of *Terminalia arjuna. Biotechnology and Applied Biochemistry,* DOI: 10.1042/DA20070243.

Gangadevi, V. and Muthumary, J., 2008b. Isolation of *Colletotrichum gleosporioides,* a novel endophytic Taxol–producing fungus from the leaves of a medicinal plant, *Justicia jendarussa. Mycologia Balcanica.,* Vol. 5: 1–4.

Georg, G.I., Chen, T.T., Ojima, I. and Vyas, D.M. (editors) 1994. Taxane Anticancer Agents (Basic Science Current Status, American Chemical Society. Symposium Series no. 583). American Chemical Society: Washington, D.C.

Guenard, D., Gueritte–Voegelein, F. and Potier, P., 1993. Taxol and taxotere: discovery, chemistry and structure–activity relationships: *Acc. Chen.Res.,* 26: 160–167.

Guo, B.H., Wang, Y.C., Zhou, X.W., Hu, K., Tan, F., Miao, Z.Q. and Tang, K.X., 2006. An endophytic taxol–producing fungus BT2 isolated from *Taxus chininsis* var. *mairei. African Journal of biotechnology,* 5: 875–877.

Horwitz, S.B., Cohen, D., Rao, S., Ringel, I., Shen, H.J. and Yang, C.P., 1993. Taxol: Mechanisms of action and resistance. *J Natl Cancer Inst.,* 15:55–61.

Li, J.Y., Strobel, G.A., Sidhu, R., Hess, W.M. and Ford, E., 1996. Endophytic taxol producing fungi from bald cypress *Taxodium distichum. Microbiology,* 142: 2223–2226.

Nicolau, K.C., Yang, Z., Liu, J.J., Ueno, H., Nantermet, P.G., Guy, R.K., Clairborne, C.F., Renaud, J., Couladouros, E. A., Paulvannan, K. and Sorenson, E.J., 1994. Total synthesis of taxol. *Nature,* 367: 630–634.

Pandi, M., Radhai, R. and Muthumary, J., 2008. Characterization and Optimization of Taxol production from an endophytic fungus *Botryodiplodia theobromae* Pat. *Indian Journal of Applied Microbiology,* 8(1): 44–53.

Roberts, S.C. and Schuler, M.L., 1997. Large–scale plant cell culture. _Curr. Opin. Biotechnol.,_ 8: 154–159.

Senthil Kumaran, R., Muthumary, J. and Hur, B.K., 2009b. Isolation and identification of an Anticancer drug, Taxol from _Phyllosticta tabernaemontanae,_ a leaf spot fungus of an Angiosperm, _Wrightia tinctoria. The Journal of Microbiology,_ Vol. 47(1): 40–49.

SenthilKumaran, R., Hur, B.K. and Muthumary, J., 2008a. Production of Taxol from _Phyllosticta spinarum,_ an Endophytic Fungus of _Cupressus_ sp. _Engineering in Life Science_s, 8(4): 1–10.

SenthilKumaran, R., Muthumary, J. and Hur, B.K., 2008b. Taxol from _Phyllosticta citricarpa,_ a leaf spot fungus of the angiosperm _Citrus medica. Journal of Bioscience and Bioengineering._ Vol. 106. No.1: 103–106.

Senthilkumaran, R., Muthumary, J. Eun–Ki Kim and Byung–ki Hur., 2009a. Production of Taxol from _Phyllosticta dioscoreae,_ a leaf spot fungus isolated from _Hibiscus rosa–sinensis, Biotechnology and Bioprocess engineering,_ 14: 76–83.

Stierle, A., Strobel, G. and Stierle, D., 1993. Taxol and taxane production by _Taxomyces andreanae,_ an endophytic fungus of Pacific Yew. _Science,_ 260: 214–216.

Strobel, G., X, Yang, J. Sears, R. Kramer, R.S. Sidhu and W.M. Hess., 1996a. Taxol from _Pestalotiopsis microspora,_ an endophytic fungus of _Taxus wallichiana. Microbiology,_ 142: 435–440.

Strobel, G.A., W.M. Hess, E. Ford, R.S. Sidhu and X. Yang., 1996b. Taxol from fungal endophytes and the issue of biodiversity. _Journal of Industrial Microbiology,_ 17: 417–423.

Vidensek, N., Lim, P., Campbell, A. and Carlson, C., 1990. Taxol content in bark, wood, root, leaf, twig and seedling from several _Taxus_ species. _Journal of Natural Products,_ 53: 1609–1610.

Wang J, Li, G., Lu H, Zheng Z, Huang, Y. and Su, W., 2000. Taxol from _Tuberfularia_ sp, strain TF5, an endophutic fungus of _Taxus mairei, FEMS. Microbiol. Lett._ 193: 249–253.

Yuan J I, Jian–Nan B I, Bing Yan and Xu–Dong Zhu., 2006. Taxol producing fungi: a new approach to industrial production of taxol, _Chinese Journal of Biotechnology,_ 22: 1–6.

Zhou, D. P., Ping, W. X. and Sun, J. Q., 2001a. Study on isolation of taxol producing fungi. _Journal of Microbiology,_ 21: 18–20.

Zhou, D.P., Sun, J.Q. and Yu, H.Y. _et al.,_ 2001b. _Nodulisporium,_ a genus new to China. _Mycosystema,_ 20(2):277–278.

2013, Perspectives in Plant Biodiversity   *Pages* **181–185**
Editor: **Dr. K. Muthuchelian,** *Vice Chancellor, Periyar University, Salem*
Published by: **Daya Publishing House, NEW DELHI**

Chapter 23

In vitro Antagonistic Activity of Diverse Isolates of *Pseudomonas fluorescens* Against *Pythium aphanidermatum*: A Plant Pathogenic Fungus

K. Manonmani[1] and K. Packialakshmi[2]
[1]Agricultural Research Station (TNAU), Vaigai Dam – 625 562
[2]Department of Biotechnology, Lady Doak College, Madurai – 625 002

ABSTRACT

Pseudomonas fluorescens is a common bacterial antagonist present in soil and widely used for the biocontrol of plant diseases. Seven isolates of *P. fluorescens* were isolated from various locations of Ramnad District and their identity was characterised both morphologically and biochemically. All the isolates were found to produce the characteristic fluorescent green colour pigment in the culture medium. *Pythium aphanidermatum* is a most serious plant pathogenic fungus causing damping off disease in major horticultural crops. *In vitro* antagonistic activity of the seven isolates of *P. fluorescens* against *P. aphanidermatum* was assessed by dual culture technique. The results revealed that the strain *P. fluorescens* LDC 1 exerted maximum reduction in mycelial growth of *P. aphanidermatum*. It recorded the diameter of mycelial growth of 3.0cm as against 9.0 cm in control. It exhibited 67 per cent reduction over control. *P. fluorescens* LDC 3 ranked next by recording diameter of mycelial growth of 3.5cm with 61 per cent reduction over control. Among the seven isolates *P. fluorescens* LDC 1 was found to be the best in arresting the growth of *P. aphanidermatum* and further studies could be taken up to use *P. fluorescens* LDC 1 as a biocontrol agent for the management of damping off disease affecting important crops.

Keywords: *Pseudomonas fluorescens, Pythium aphanidermatum, In vitro antagonistic activity.*

Introduction

Pythium aphanidermatum is an Oomycetous fungus belongs to the division oomycetes of chytridiomycotina. The genus *Pythium* includes a number of readily recognized species with wide distributions and host ranges. In the early 1900s pathologists found *Pythium* spp consistently associated with root diseases and it soon became apparent that these fungi were important plant pathogens. They cause severe yield loss due to pre emergence and post emergence damping off disease.

Current control measures of *P. aphanidermatum* include the use of fungicides (carbendazim, captan, thiram), soil fumigants (methyl bromide, chloropicrin, meta sodium), crop rotation with non susceptible crops and cultivation of disease tolerant or resistant varieties. The exorbitant use of chemical fungicides has also led to development of fungicide tolerant species that again difficult to control. But long term crop rotation was found to be effective in eliminating *Pythium* spores from soil only after several years and it will also influence the cropping pattern of the specific area. Likewise the development of disease tolerant/resistant plant varieties also consumes considerable time.

So an alternate method with antagonists possessing antifungal activity can be successfully employed for the effective management of the disease (Alam *et al.*, 1992). There has been large body of literature describing the potential use of *Pseudomonas fluorescens* as an antagonist for plant disease management. It makes up a diverse group of bacteria that can be successfully used as a biocontrol agent. This fast growing rhizobacteria out competes the fungal pathogens by expressing various antagonistic mechanisms *viz.,* antibiosis, plant growth promotion, lysis of hyphae and rhizosphere competency. With a view to isolate newer and efficient strains in terms of antagonistic activity, the present study has been planned.

Materials and Methods

Isolation of Antagonist

Soil samples were collected from seven different places in Ramnad District, Tamil Nadu. One gram of rhizosphere soil was serially diluted and the antagonist was isolated by dilution plate technique on the selective King's B (KB) medium. The strains were characterized biochemically.

Isolation of *P. aphanidermatum*

Tomato seedlings infected with damping off disease were collected from various areas and were cut into 1mm size bits with sterile scalpel and surface sterilized in 0.1 per cent mercuric chloride. After repeated washings in sterile distilled water, the 2-3 bits were placed equidistantly on sterile Petri plates containing Potato Dextrose Agar (PDA) medium supplemented with streptomycin sulphate. Then the plates were incubated at 25–27°c in an inverted position for 5-7 days. The mycelial strand was purified on slants of PDA medium supplemented with streptomycin sulphate. The morphological character of the fungal culture was observed under the microscope and its identity and pathogenicity was established by proving Koch's postulates on tomato seedlings (Seetharaman, 1989).

Screening of an Effective Strain of *P. fluorescens* for Antagonistic Activity

The effective strain of *P.fluorescens* was screened by assessing the *in vitro* antibiotic activity against the plant pathogenic fungus, *P.aphanidermatum* by dual culture technique (Dennis and Webster, 1971).

An actively growing culture disc of *P. ahanidermatum* was placed at 1.5cm away from the rim of the Petri plate containing sterilized PDA medium. An overnight culture of *P.fluorescens* was streaked on the opposite side, leaving 1.5cm space from the rim. For each isolate three replicates were maintained in completely randomized design (CRD). Petri plate with the *P.aphanidermatum* alone served as a control. The plates were incubated for 78 hrs and the inhibition zone was measured.

Growth inhibition was calculated in per cent using the following formula:

Per cent inhibition = $[(A_{1-}A_{2})/A_{1}] \times 100$

A_1: Mycelial growth in control

A_2: Mycelial growth in treatment

Results and Discussion

Isolation and Characterization of Antagonist

Seven isolates of the antagonist, *P.fluorescens* was isolated from rhizosphere soil samples and the colonies were found to be circular, convex with entire margin and produced characteristic fluorescent green colour pigment after two days of incubation at 37-°C. The isolates were confirmed to be *Pseudomonas* by using varius biochemical tests and the results were represented in Table 23.1.

Isolation and Characterization of *P. aphanidermatum*

P. aphanidermatum was isolated from the tomato plants infected with damping off disease and its pathogenicity was confirmed by establishing the damping off disease on young tomato plants. Dull white cottony mycelial growth was observed on PDA medium. Phase contrast microscopic view revealed the characteristic features of *P. aphanidermatum*. Hyphae were found to be hyaline to dull white colour. Mycelium was aseptate and oogonium was aplelurotic.

Table 23.1: Biochemical characters of the *Pseudomonas* isolates.

Sl.No.	Biochemical Tests	Observation	Results
1.	Gram's staining	Safranin stain on cell wall	Negative rod
2.	Methyl red test	Yellow colour development	Negative
3.	Indole test	No red colour development	Negative
4.	Starch hydrolysis test	Absence of lytic zone	Negative
5.	Nitrate test	Absence of red coloration	Positive
6.	Citrate test	Colour changed from green to blue	Positive
7.	H_2S production test	No black colour development	Negative
8.	Catalase test	Presence of bubble formation	Positive

Screening of Effective Isolate of an Antagonist

In dual culture technique, all the seven isolates of *P.fluorescens* ($Pf_{LDC1, 2,3,4,5,6 \text{ and } 7}$) were found to display various degree of antagonistic activity against the *in vitro* growth of *P. aphanidermatum*. The maximum growth inhibition was observed with the isolate Pf_{LDC1}, which recorded the mycelial growth of 3.0cm diameter (Table 23.2 and Figure 23.1). It was found to exhibit greater antagonistic activity of 67 per cent reduction over control (PROC) against *P. aphanidermatum* compared to other strains.

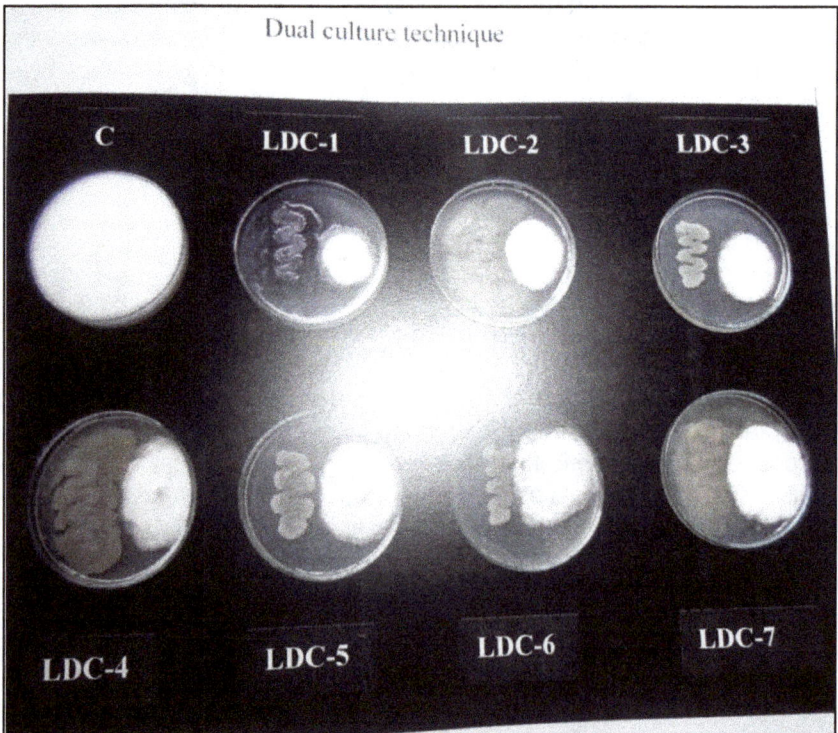

Figure 23.1: *In vitro* antagonistic activity of *Pseudomonas fluorescens* against *Pythium aphanidermatum*.

Table 23.2: *In vitro* antagonistic activity of *P. fluorescens* against *P. aphanidermatum.*

Sl.No.	Antagonistic Strains	Diameter of Mycelial Growth (cm)*	Per cent Reduction Over Control
1.	Pf_{LDC1}	3.0^a	67.00
2.	Pf_{LDC2}	4.0^c	55.50
3.	Pf_{LDC3}	3.5^b	61.00
4.	Pf_{LDC4}	4.3^c	53.00
5.	Pf_{LDC5}	4.5^d	50.00
6.	Pf_{LDC6}	4.5^d	50.00
7.	Pf_{LDC7}	5.0^e	44.40
8.	Pf_{LDC8}	9.0^f	–

*: Mean of three replications

Cartwright *et al.* (1995); Rosales *et al.* (1995) supported the *in vitro* antagonistic activity of *P.fluorescens* against the growth of *P. aphanidermatum.* Ramamoorthy *et al.* (2007) reported that the antibiosis was the major mechanism behind the antagonistic activity of *P.fluorescens* against *P. aphanidermatum.* The antibiotics may be pyoluteorin, pyrrolnitrin, phenazine 1 carboxylic acid and 2,4 diacetyl phloroglucinol.

References

Cartwright, D.K., Chilton, W.S and Benson D.M., 1995. Pyrrolnitrin and phenazine production by *P.cepecia* strain 5.5 B a biological agent of *Rhizoctonia solani.Appl. Microbial.Biotechnol.,*43:121–127.

Dennis, C and Webster, J., 1971. Antagonistic properties of species groups of *Trichoderma*II. Hyphal interaction. *Trans. British Mycological Society,* 57: 363–369.

Ramamoorthy, V., Raguchander,T and Samiyappan, R., 2007. Enhancing resistance of tomato and hot pepper to *Pythium* disease by seed treatment with fluorescent pseudomonas. *Eur.J.Plant Pathol.,*108: 429 –441.

Rosales, A.M. ThomashowL.,Cook, R.J and Mew, T.W., 1995. Isolation and identification of antifungal metabolites produced by rice associated antagonistic *Pseudomonas* spp. *Phytopathology,* 85:1028–1032.

Seetharaman, K., Whitehead, E, Keller, N.P.,Waniska, R and Rooney, L.W., 1989. *In vitro* activity of sorghum seed antifungal proteins against grain mold phytopathogens. *J.Agric. Food Chem.,*45: 3666–3671.

2013, Perspectives in Plant Biodiversity   *Pages* **186–192**
Editor: **Dr. K. Muthuchelian,** *Vice Chancellor, Periyar University, Salem*
Published by: **Daya Publishing House, NEW DELHI**

Chapter 24

Seasonal Diversity of AM Fungal Spores in Leguminous Plants of Dharward in Karnataka, India

M. Mirdhe Romana and H.C. Lakshman,*
P.G. Department of Studies in Botany,
Microbiology Lab, Karnatak University, Dharwad

Introduction

Healthy and fertile soils are characterized by the presence of a diverse population of microorganisms an important component of which is arbuscular mycorrhiza. AM Fungi are most common type of mycorrhiza which constitute major group of soil microbiota. Associations with AM Fungi increase plants access to scare and immobile nutrients particularly phosphorous. The presence of these fungi has been shown to be essential for sustained growth and competitive ability of plants (Koide *et al*, 1994). AM Fungi are considered to be host specific. Evidence shows that fungi spore density differs seasonally, with some fungi sporulating in late winter and others sporulate at end of summer. As the spores represent the dormant stage of the fungus, the physiological activity is similar to that seasonal spore count (Mohammadi *et al.,* 2008). However the competitive balance between AM Fungal species in terms of their ability to colonize roots may be affected by environmental conditions.

Legumes play an important role in natural ecosystems and they are useful in the revegetation if deficient ecosystems that have low availability of Nitrogen and

---
* Corresponding Author: E-mail: romeebot@gmail.com

phosphorous and other nutrients. The scarcity of available phosphorous and the imbalance of trace elements actually limit legume establishment (Barea *et al.,* 1992). Moreover legumes exhibit a considerable degree of dependence on mycorrhizae to thrive in stressed situations and the legumes which are colonized by AM Fungi represent a significant carbon sink. (Kucey and Paul, 1982). Legumes have a higher phosphorous requirement for nodule formation, nitrogen fixation and optimum growth and mycorrhizal condition of legume crops found to increase its vegetation in addition to improve nodulation.

Materials and Methods

Rhizosperic soil sample were collected from different localities of Dharwad district. For each plant species three different soil samples were collected and they were mixed to get a homogenized mixture of a particular soil sample and this mixture was taken for the study of AM Fungal spores. The procedure was followed for three different seasons namely summer, rainy and winter. At the same time the roots of all experimental plants were collected and stored in FAA.

Isolation and Assessment of Root Colonization

The spores were isolated from the rhizosperic soil by wet sieving and decanting method (Gedermann and Nicholson, 1963). The percent of root colonization was assessed by root slide technique (Phillips and Hayman, 1970). The percentage of root colonization was determined by formula:

$$\text{Per cent of root colonization} = \frac{\text{No of root bits showing colonization}}{\text{Total number of root bits assessed}} \times 100$$

Results and Discussion

Legumes are most important plants which contribute rich protein from pods. In the present investigation all the fifteen leguminous plants studied exhibited AM Fungal colonization in roots and spore population in the rhizosphere soil samples. All the fifteen plants, showed a wide range of variation, under different seasons (Table 24.1). The level of AM fungal association depends on root morphology, metabolism, and rate of plant growth (Warner and Mosse, 1980).

It was found that rhizospheric soil samples from different locations of Dharwad district had AM Fungal spores. It was observed that on these plants the percentage of root colonization was found to be more in the winter season compared to that of summer and rainy season (Figure 24.1). Similar observation was recorded by Lakshman (1994). The spore population was minimum in rainy season and gradually increased by winter highest spore population was found in summer season (Figure 24.2).

This is because fungi are believed to require well aerated soils and have thought to have problems adapting to water–logged conditions. This finding is consistent with Mosse, 1981., Ingam and Wilson, 1999. The highest number of root colonization was recorded in *Albizzia lebbeck* while lowest was in *Cassia auriculata* during winter season. In other two seasons there was not much increase in percentage of root colonization in all experimental plants. Spore population was found to be more on

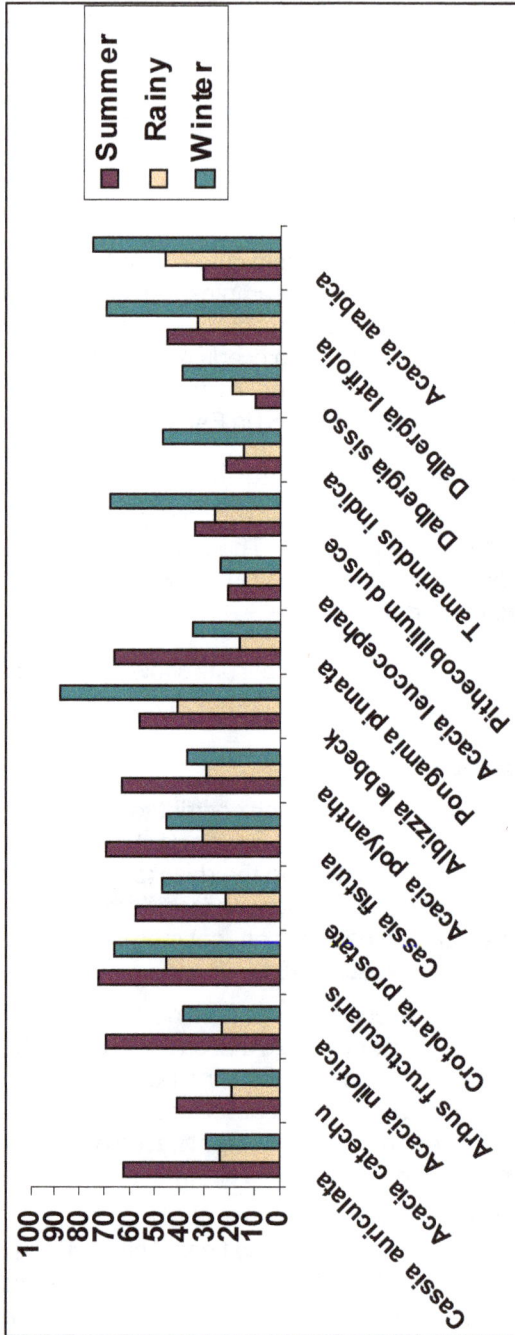

Figure 24.1: Showing the percentage of root colonization in fifteen leguminous plants in three seasons.

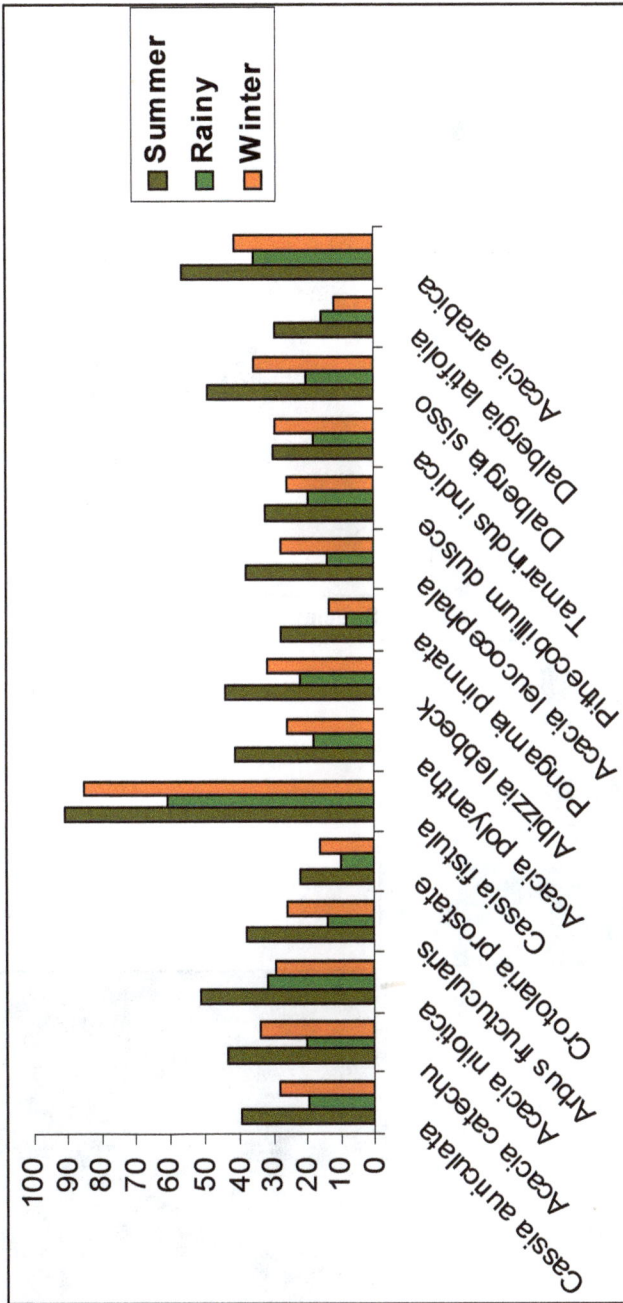

Figure 24.2: Showing the number of spores/50 gm of soil in fifteen leguminous plants in three seasons.

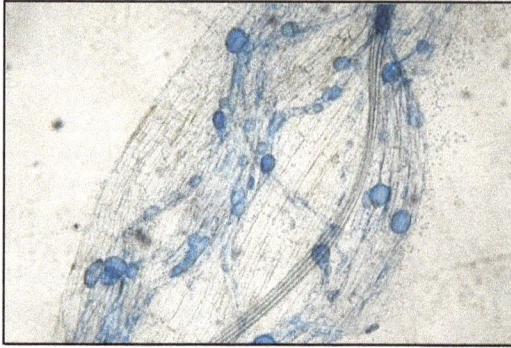

***Glomus* Species**

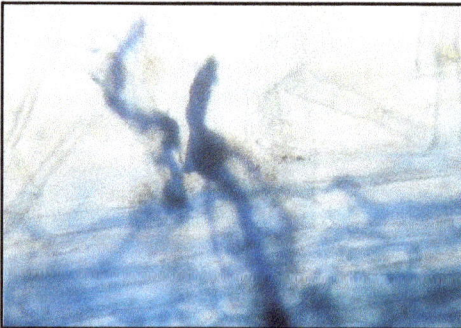

***Acaulospora* Species**

Figure 24.3: Macerated root section of *Albizzia lebbeck* showing Vesicles and Hyphae of AM Fungi and spores of AM Fungi.

Cassia fistula and lowest was found in *Pongamia pinnata* during summer when compared to all other plants in different seasons. This may be due to the presence of secondary metabolites.

Table 24.1: Showing the Per cent of root colonization and Spore number in Fifteen Leguminous plants in three different seasons.

Sl.No.	Name of the Plant	Per cent Root Colonization			No. of Spores/50 gm of Soil		
		Summer	Rainy	Winter	Summer	Rainy	Winter
1.	*Cassia auriculata*	63	24	29	39	19	28
2.	*Acacia catechu*	41	19	25	43	20	34
3.	*Acacia nilotica*	69	23	35	51	32	29
4.	*Arbus fructucularis*	73	45	66	38	14	26
5.	*Crotolaria prostate*	58	22	47	22	10	16
6.	*Cassia fistula*	69	31	45	91	61	85
7.	*Acacia polyantha*	64	29	37	41	18	26
8.	*Albizzia lebbeck*	56	41	88	44	22	31
9.	*Pongamia pinnata*	66	16	35	27	08	13
10.	*Acacia leucocephala*	21	14	24	38	14	27
11.	*Pithecobillium dulsce*	34	26	68	32	19	26
12.	*Tamarindus indica*	22	15	47	30	18	29
13.	*Dalbergia sisso*	10	19	39	49	20	35
14.	*Dalbergia latifolia*	45	33	69	29	15	11
15.	*Acacia arabica*	31	46	75	56	35	41

The temporal variation in the spore density is a consequence of many interacting factors such as plant communities, soil characteristics and climate.(Escurdero and Medoza, 2005, and Lakshman *et al*, 2001). AM fungal spore isolated from the Rhizosphere soil samples of plants of family leguminosae under study belonging to five genera *viz Glomus, Acaulospora, Scutellopspora, Gigaspora* and *Sclerocystis*. Altogether spores belong to 38 AM fugal species were recorded, among which *Glomus* predominated over other genera. Experimental results strongly support the work of Lakshman *et al*, 2001 and Gautam and Roy, 2009.

Conclusion

It can be concluded that many legumes are growing in this part of Karnataka. In the present study brought out the seasonal screening of AM Fungi on important fifteen legumes is quite interesting. Though nodulation, physico-chemical characters of soil, growth response studies would be warranted. The seasonal variation and diversity of AM fungi may be a preliminary work.

References

Barea,J.M., Azcon, R. and Azcon-Aguilar, C., 1992. Vesicular-arbuscular mycorhizal fungi in nitrogen fixation systems. *Methods Microbiol.*, 24:391–416.

Escurdero, V. and Medoza, R., 2005. Seasonal variation of Arbuscular mycorrhizal fungi in temperate grasslands. *Mycorrhiza*, 15: 291–299.

Gautam, N.K. and Roy, A.K., 2009. Seasonal Diversity in arbuscular mycorrhizal fungi associated with some medicinal plants of Jharkhand. *J. Indian Bot. Soc.*, 88(1 and 2): 56–59.

Gerdemann, J.W. and Nicolson, T.H., 1963. Spores of mycorrhizal Endogone extracted from soil by wet-sievings and decznting. *Trans. Br. Mycol. Soc.*, 46: 235–244.

Ingam, E. and Wilson, M., 1999. The mycorrhizal colonization in six wetland species at sites differing in land use history. *Mycorrhiza*, 9: 233–235.

Koide, R.T., Shumway, D.L. and Mabon, S.A., 1994. Mycorrhizal fungi and reproductions of field population of Abutilon theophrasti Medic (Malvaceae). *New Phytolgist*, 126: 123–130.

Kucy, R.M.N. and Paul, E.A., 1982. Carbon flow, photosynthesis and N_2 Fixation in mycorrhizal and nodulated faba beans (*Vicia faba* L). *Soil Biol. Biochem.*, 14: 407–412.

Lakshman, H.C., Mulla, F.I. and Inchal, R.F., 2001. Seasonal fluctuations of VAM fungi in some commonly cultivated crops of Dharwad District. *Asian J. of Microbiology and Biotechnology*, 3(1–2): 59–64.

Lakshman, H.C., 1994. Arbuscular mycorrhiza in hydrophytic and xerophytic plants. *Nat. Con.*, 73: 114–118.

Mohammadi, E., Danesh, Y.R., Prasad, R. and Varma, A., 2008. *Mycorrhizal Fungi: What we Know and What should we Know?* Springer-Verlag, 3–27 pp.

Mosse, B., 1981. *Vesicular-Arbuscular Mycorrhiza Research for Tropical Agriculture*. Reaearch Bulletien 194. University of Hawaii, pp. 82.

Philips, J.M. and Hayman, D.S., 1970. Improved procedure for clearing roots and staining parasite and VAM fungi for rapid assessment of infection. *Trans. Brit. Mycol. Soc.*, 55: 158–161.

2013, Perspectives in Plant Biodiversity   *Pages* **193–198**
Editor: **Dr. K. Muthuchelian,** *Vice Chancellor, Periyar University, Salem*
Published by: Daya Publishing House, NEW DELHI

Chapter 25

Effect of Arbuscular Mycorrhizal Fungi in Improvement of Alkaline Phosphatase Activity on *Gloriosa superba* L. Under Water Stress Condition

S.B. Gadi[1], H.C. Lakshman[2] and A. Channabasava[2]
[1]Department of Botany, J.S.S. College, Dharwad – 580 003
[2]Department of Botany (Microbiology Lab), Karnatak University,
Dharwad – 580 003

Introduction

Studies on the relation between arbuscular mycorrhizal fungi (AMF) and host tree water and on the drought-resistance increase of host plants by AMF showed that AMF could increase the absorption and utilization of nu-trient and water by plants, improve the water metabolism, and enhance the drought-resistance of host plants Ellis *et al.,* 1985; Bethlenfavay *et al.,* 1987; Ianson *et al.,* 1988; *Auge et al.,* 1992; Zhang *et al.,* 1994). Research proved that mycorrhizal fungi could improve water absorption by citrus under dry conditions (Levy *et al.,* 1983). Inoculation of AMF to *Phaseolus aureus* resulted in one-half water required for dry material accumulation of I g compared with the control (Wang *et al.,* 1989). The inoculation of AMF to poplar seedlings increased the water absorption of the seedlings, enlarged the expansion degree of bark, and postponed the water potential compensation point of net photosynthesis rate by 2-3 days compared with the control under water stress conditions (Chen and Tang, 1997). While the mechanism of the host water condition improvement and plant drought resistance increase by mycorrhizae still remained unclear, the effects of AMF alkaline phosphatase (ALP) activities on *Gloriosa superba* L. drought-resistance under water stress were studied with histochemical techniques, and the mechanism was worked out.

Materials and Methods

Experimental Soil

Sandy soil collected from a field near Naremdra was mixed with river sand (v/v 1:1). The sand and soil mixture, which had the following properties: available phosphorus 8.36 ppm, available potassium 11.5 ppm, total nitrogen 0.21 per cent, organic matter 1.0 per cent and pH 7.2, was prepared for use after autoclaving.

AMF and Hosts

Glomus mosseae (Nicol. and Gerd). Gerd. and Trappe (NX 96078), *G. mosseae* (ShB 96(06), *G. geosporum* (Nicol. and Gerd). Walker, and *G. constrictum* Trappe were isolated from rhizosphere soil of *Hippophae rhamnoides* L. in dry and desert areas of Shaanxi and Ningxia as experimental strains, and *Gloriosa superba L.* was used as host.

Preparation and Inoculation Culture

The culture of the experimental AMF used *Zea mays* L. as the host plant; the above-ground and surface soil (2 cm) were re-moved after infection for 4-6 weeks, and the mixture (including spore, hyphae and mycorrhiza), was used as inoculation agent. Aseptic inoculation of AMF (50 g/pot) was completed 50 g of aseptic sand was used as the control, and 3-5 germinating *Gloriosa superba* L. seeds were shown per pot. The plants were grown in the culture room with a natural light duration of 12 h/day and room temperature of 16-22°C. 40 pots in each treatment. Forty days after emergence, the soil water content was controlled at -2.54 Mpa so as to enable water stress to be measured sixty days after emergence, the mycorrhiza Infection rate, height, fresh weight and wilting coefficient index were determined.

At the 30th, 60th and 90th days, the mycorrhiza samples were collected from three points on the lateral root system of the seedling (40 plants, I plant/pot); the roots were removed from the soil then cleared and stained. The staining of trypen blue (TB), succinate dehydrogenase (SOH) and alkaline phosphatase (ALP) were conducted with histochemical techniques (Gianinazzi and Gianinazzi-Pearson 1992; Tisserant *et al.,* 1993), mycorrhiza infection degrees of TB, SOH and ALP were determined with the line intersect method, and the mycorrhizal root length and its percentage were calculated with the following formula (Giovannetti and Mosse, 1980; Newman, 1966).

$$R = \frac{pAN}{2AH}$$

where,

     R: Mycorrhizal root length

     A: Petri dish area

     N: Line intersect mycorrhiza number

     H: Totalline length

$$\text{Mycorrhizal root length percentage (per cent)} = \frac{\text{Mycorrhizal length}}{\text{Total roots length}} \times 100$$

Results and Discussion

The amount total hyphae, functional hyphae and active hyphae. In AMF hyphae, the total hyphae (TB), functional hyphae (SDH) and active hyphae (ALP) amounts of various AMF were different with the trend TB>SDH>ALP. TB could reach 72 per cent at the highest and 38 per cent at the lowest, SDH 61 per cent and 36 per cent, and ALP 49 per cent and 27 per cent respectively. Generally, when TB was high, SDH and ALP were high too, while when TB lowered, the other two lowered. However, when TB of *G. constrictum* was relatively lower, the SDH and ALP were higher; when TB of *G. geosporum* was relatively higher, the SDH and ALP were lower (Figure 25.1).

Total hyphae and functional hyphae are the base of active hyphae. The early infection of mycorrhizal fungi affected the later functions of its SDH and ALP (Gianinazzi and Gianinazzi-Pearson, 1992). The mycorrhizal hyphae included the live hyphae and dead hyphae, the former participated in the change and metabolism of phosphoric nutrition, which was the main source if the P absorption capacity increase of the host.

The effect of total hyphae, functional hyphae and active hyphae on plant growth After 30,60 and 90 days' culture of the inoculated AMF under water stress, the relationship between the three kinds of hyphae (TB, SDH and ALP) of *Glomus mosseae* and the host growth were determined through samples (Figure 25.2). The results showed that over 30-60 days, the three kinds of hyphae rose slowly and so did the plant growth; over 60-90 days, the TB and SDH continuously rose slowly, but the ALP rose faster and the fresh weight of the plants increased sharply, which indicated that ALP played an important role in plant growth. The more the ALP, the faster the symbiote of AMF and hosts formed, and the SDH had a very close relationship with the nutrient accumulation and metabolism of the hosts (Tisserant *et al.,* 1993).

The function of ALP activity to plant drought-resistance. All four strains of AMF collected from the rhizosphere of *Gloriosa superba* L. could form arbuscular mycorrhizae with the original hosts, and the plants inoculated with AMF under water stress grew stronger, the fresh weight and seedling height were higher, and the wilting coefficient was lower than those of the control (Table 25.1). The plants grew slowly at an early stage; the plant fresh weights and wilting coefficients determined on the 60th day had no obvious differences except the strain of *G. mosseae*. When determined on the 90th day, the fresh weights and wilting coefficients of plants inoculated with AMF showed marked differences, while the plant height did not. However, the plants in the control looked weak, and few changes happened in the fresh weights and wilting coefficients of the control on the 60th and 80th day determinations. With the rise of ALP activity, plant fresh weight increased and the wilting coefficient lowered, which indicated that the mycorrhizae formed by AMF and *Gloriosa superba* L. could obviously raise the drought-resistance of the hosts. However, different kinds of fungi had different drought-resistances. Of the four strains, *G. mosseae* (NX96078) was the strongest in accelerating the host drought-resistance, followed by G. *mosseae* (ShB96006);

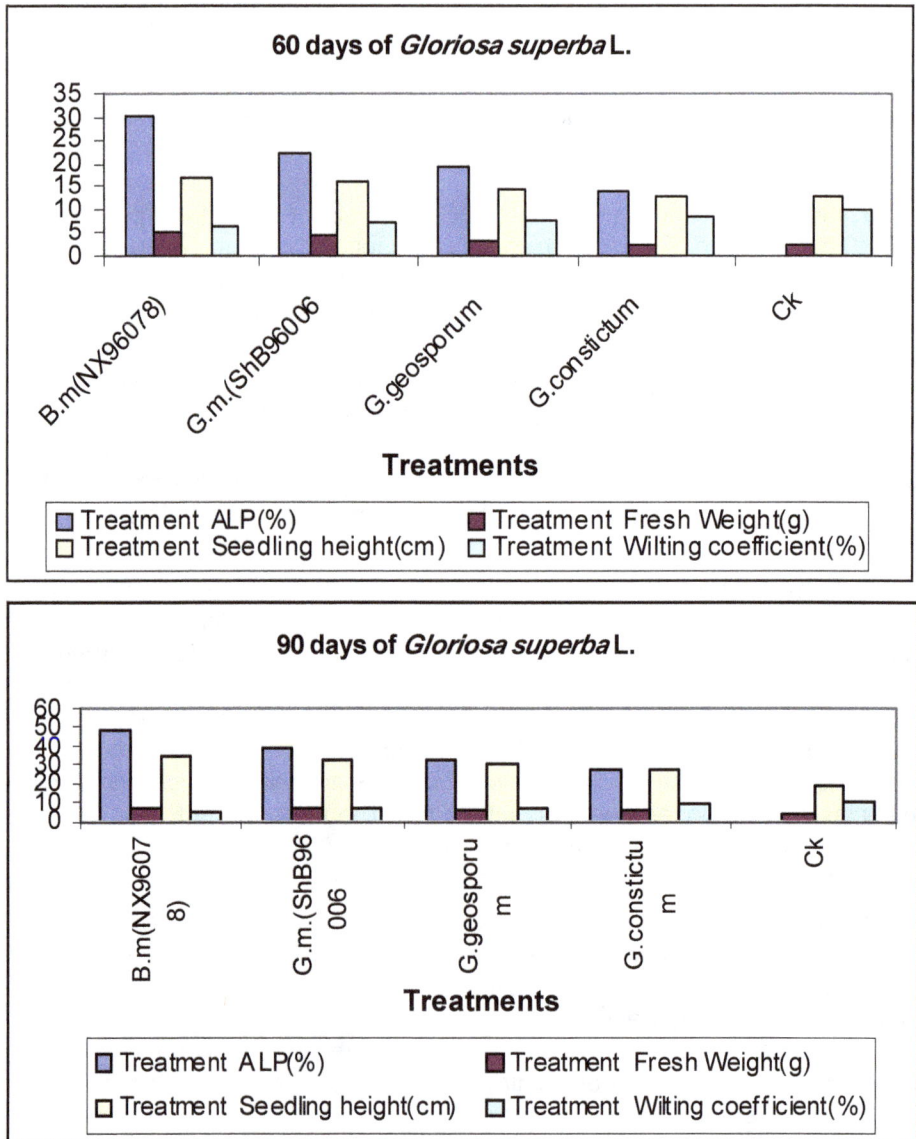

Figure 25.1: Effect of ALP activity on host water stress condition in *Gloriosa superba* L. for 60 and 90 days.

G. geosporum and *G. constrictum* were the weakest. ALP participated directly in the nutrient material ex-change of host trees. Large proportions of ALP and SDH were the key factors for AMF to increase the biomass accumulation and drought-resistance of the host plants. In conclusion, in recent years, most conducted research used total hyphae infection rate to show the improvement function of AMF to plant growth. In this study, the total hyphae, functional hyphae and active hyphae of AMF, the plant growth and drought-resistance were comprehensively analyzed, which proved that ALP of the hyphae greatly affected biomass accumulation of the hosts. The hyphae with ALP activity had the strongest impact on plant growth and drought resistance. The research results not only show that AMF could improve plant P nutrition and increase the plant drought-resistance, but also could improve the absorption and utilization of P in soil by host under water deficiency and raise the infertility-resistance capacity of host trees while increasing the drought-resistance of the host. Different AMF affected differently the growth improvement and drought-resistance of host trees. Therefore, it is important to use mycorrhizae biotechniques in dry and desert areas to overcome the two restricting factors of drought and infertility.

Table 25.1: Effect of ALP activity on host water stress condition in *Gloriosa superba* L. for 60 and 90 days. (P=0.05).

Treatment	60 Days				90 Days			
	ALP (%)	Fresh Weight (g)	Seedling Height (cm)	Wilting Coefficient (per cent)	ALP (%)	Fresh Weight (g)	Seedling Height (cm)	Wilting Coefficient (%)
B.m	30.1a	5.4a	17.1b	6.62a	48.8a	8.1a	34.7b	5.22a
G.m.	22.3b	4.4b	16.2b	7.21b	39.1a	7.0a	32.5b	7.03a
G.geosporum	19.6b	3.1c	14.5c	7.53b	33.0a	6.5a	29.6b	7.91a
G.constictum	14.4c	2.2c	12.8c	8.37c	27.2a	5.9a	27.8c	8.71a
Ck	0.0c	2.2c	12.8c	9.84c	0.0c	3.2b	18.4c	9.76b

References

Allen, M.F., Smith, W.K., Moore, T.S. and Christensen, M., 1981 Comparative water relations and photosynthesis of mycorrhizal and nonmycorrhizal *Bouteloua gracilis* H. B. K. Lag ex stead. *New Phytol.*, 88: 683–693.

Auge, R.M., Stodola, A.J.W., Brown, M.S and Bethlenfalvay, G.J., 1992 Stomatal response of mycorrhizal cowpea and soybean to short-term osmotic stress. *New Phytol.*, 120: 117–125.

Bethlenfalvay, G., Brown, M.S., Mihara, K.L. and Stafford, A.E., 1987 *Glycine–Glomus–Rhizobium* symbiosis. V. Effects of mycorrhiza on nodule activity and transpiration in soybean under drought stress. *Plant Physiol.*, 85: 115–119.

Chen, H. and Tang, M., 1997. Advances in mycorrhizae research on poplar. *Sci. Silvae Sinicae,* 33: 183–188.

Ellis, J.R., Larsen, H.J. and Boosalis, M.G., 1985. Drought resistance of wheat plants inoculation with VA mycorrhizas. *Plant Soil*, 86: 369–378.

Gianinazzi, S. and Gianinazzi-Pearson, V., 1992. Cytology, histochemistry and immunocytochemistry as tools for studying structure and function of endomycorrhiza. *Methods. Microbiol.*, 24: 109–139.

Giovannetti, M. and Mosse, B., 1980. An evaluation of techniques for measuring vesicular-arbuscular mycorrhizal infection in roots. *New Phytol.*, 84: 489–500.

Ianson, D.C. and Linderman, R.C., 1988. Variation in VA mycorrhizal enhancement of N fixation and induction of drought tolerance in pigeon pea. *Proceedings of 7th NACOM*, p. 149.

Levy, Y., Syvertsen, J.P. and Nemec, S., 1983. Effect of drought stress and vesicular–arbuscular mycorrhiza on Citrus transpiration and hydraulic conductivity of roots. *New Phytol.*, 93: 61–66.

Liu, R. and Hao, W., 1994. Effects of VA mycorrhizal fungi on water metabolism of plants. Acta Pedol Sin 31 (Supplement), p. 46–53.

Newman, E.L., 1966. A method of estimating the total length of root in a sample. *J. Appl. Ecol.*, 3: 139–145.

Safir, G.R., Boyer, J.S. and Gerdemann, J.W., 1972. Nutrient status and mycorrhizal enhancement of water transport in soybean. *Plant Physiol.*, 49: 700–703.

Tang, M. and Chen, H., 1995. The effect of VA mycorrhizas on the resistance of poplar to canker fungus (*Dothiorella gregaria*). Mycorrhizas for Plantation Forestry in Asia. *ACIAR*, 62: 67–71.

Tisserant, B., Gianinazzi-Pearson, V., Gianinazzi, S. and Collotte, A., 1993. In plant histochemical staining of fungal alkaline phosphatase activity for analysis of efficient arbuscular mycorrhizal infections. *Mycol. Res.*, 97: 245–250.

Wang, H. and Wu, G. and Li, H., 1989. Effects of VA mycorrhiza on the growth of *Phaseolus aureus* and its water use. *Acta Pedol Sin.*, 26: 393–400.

Zhang, M., Wang, Y. and Xing, L., 1994. The strain (CX–91) of VA mycorrhizal fungi for drought–resistance. *Beijing Agric. Sci.*, 12: 25–26.

2013, Perspectives in Plant Biodiversity   *Pages* **199–204**
Editor: **Dr. K. Muthuchelian,** *Vice Chancellor, Periyar University, Salem*
Published by: **Daya Publishing House, NEW DELHI**

Chapter 26

Physiological Studies on Evolution of Soybean Genotypes for Drought Tolerance

C. Rajendran, R. Amutha, T. Sivakumar and K. Sivasubramaniam

Unit of Crop Physiology, Department of Seed Science and Technology, Agricultural College and Research Institute, TNAU, Madurai

ABSTRACT

Studies were undertaken during 1998-99 in the Department of Crop Physiology, Agricultural College and Research Institute, Madurai to screen soybean genotypes for drought tolerance. The extent of drought tolerance was tested by photosynthetic rate, leaf water potential, leaf temperature, transpiration rate, stomatal difference resistance. The genotypes culture 425, UGM 34 and EC 2541 showed tolerance to drought due to higher photosynthetic rate, higher pod number per plant, higher yield among the genotypes tested.

Keywords: *Drought, Genotypes, Soybean, Photosynthetic rate, Yield, Tolerance.*

Introduction

The area under soybean cultivation in India has grown from 300 hectares in 1967 to seven lakhs hectares as on today. But the average productivity per unit area is low due to several factors (Trikha, 1985). Raising soybean crop offers a number of advantages *viz.,* ability to fix nitrogen, low phosphorus requirement (Tadano and Tanaka,

1980), tolerance to low pH and high soil moisture content (Tadano *et al.,* 1979). Many scientists have reported the adverse effects of water stress on soybean (Wein *et al.,* 1979). Drought stress has been observed to be one of the factors causing low yields since soybean cultivation is confirmed mostly to drought prone areas.

The main objective of this study which was undertaken during the year 1998-99 was to evaluate different soybean genotypes and identify suitable cultivars for adoption in drought prone areas.

Materials and Methods

The experiment was conducted to evaluate different soybean genotypes under rainfed condition with a view to select the drought tolerant types. A total of 30 genotypes of soybean were taken up for the study. The experiment was carried out during dry season of 1998-99 at Agricultural College and Research Institute, Madurai. The experiment was laid out in a Randomized Block Design replicated thrice with a spacing of 30 x 10 cm. The plot size was 5.0 x 8.0 m. Observations were recorded at the reproductive stage of the crop for the following physiological parameters. The water potential was measured using Pressure bomb Apparatus.

Leaf temperature, transpiration rate and stomatal diffusive resistance were determined by means of "Steady State Prometer" of make Licor, Lincoln, Nebraska, U.S.A. Photosynthetic rate was recorded in the top fully expanded leaf with the help of "Infra Red Gas Analyser" of model 225-2B-SS. At harvest date, yield components such as pod number/plant and pod yield/plant were collected in randomly selected ten plants in each replication and in each genotype.

Results and Discussion

During summer, the period of drought is common under dry land conditions. Crop plants modify their physiological activities in response to the water deficit conditions. The degree of adaptation of the crops varies with the genotypes. Some types exhibit higher ability to survive the unfavourable dry environment by modifying the physiological mechanism. In the present study as well, it was noticed that wide variations existed among the thirty genotypes of soybean in respect of the various physiological attributes studied.

The observations recorded on leaf water potential, leaf temperature, transpiration rate and stomatal diffusive resistance are furnished in Table 26.2. The leaf water potential which ranged from -10.75 bars (culture 1018790 to -24 bars (EC 109548) showed wide variations in response to water deficit conditions. The fact that some of the genotypes maintained leaf water status at a higher level in spite of drought indicated their ability to survive under dry environment. Other parameters such as leaf temperature and transpiration rate also indicated wider genotypic variations. From the observations on these two parameters *viz.,* leaf temperature and transpiration rate, it could be observed that those genotypes which had higher rate of transpiration showed reduced leaf temperature. The stomatal diffusive resistance of leaf showed lower values in genotypes with higher transpiration rate. The existence of such an interrelationship among these physiological characters was already established in many field crops in previous studies (Roark and Quinseberry, 1977). In the study also, the negative relationship existing among these leaf parameters was brought out in the correlation studies (Transpiration rate Vs leaf temperature r = -0.69741**. Stomatal diffusive resistance Vs transpiration rate r = -0.6049*. Leaf water potential Vs leaf temperature r = -0.5617).

Table 26.1: Observations on photosynthesis and yield components.

Sl. No.	Cultivars	Photosynthetic Rate $(mg\ CO_2\ dm^{-2}\ hr^{-1})$	Pod No. Plant^{-1}	Dry Pod Weight $(g\ plant^{-1})$
1.	Culture 70	19.45	45	9.5
2.	EC 100776	20.63	40	7.4
3.	AM 5549	18.31	34	6.8
4.	Culture 189	19.07	29	5.7
5.	Culture 101879	26.36	56	15.5
6.	EC 109548	22.41	37	13.0
7.	EC 50084	16.75	38	8.4
8.	EC 15086	19.43	36	10.0
9.	MACS 125	23.61	28	8.8
10.	ACC No.2004	19.01	36	12.9
11.	EC 18733	19.09	54	11.4
12.	Culture 15590	18.63	43	9.0
13	SDDC	23.01	33	12.0
14.	PLSO-1	20.70	22	4.8
15.	**UGM 34**	**26.41**	**57**	**16.8**
16.	EC 62384	28.01	41	9.6
17.	DS 295	19.36	81	14.9
18.	EC109545	22.61	38	10.5
19.	PLSO-1	20.63	49	11.0
20.	**EC 2541**	**24.10**	**58**	**15.8**
21.	**Culture 425**	**20.71**	**52**	**18.3**
22.	UGM 30	28.69	43	13.6
23.	EC 95287	26.61	16	2.8
24.	EC 50082	24.31	23	5.0
25.	EC 109548	21.02	42	9.5
26.	EC 18226	19.41	37	13.5
27.	EC 141390	18.61	33	9.5
28.	EC 95258	22.01	41	13.0
29.	DC 13007	24.63	31	8.8
30.	EC 24058	23.07	41	13.3
	SE(d)	1.410	1.112	1.323
	CD (5 per cent)	3.969	3.762	3.996

Table 26.2: Observations on physiological parameters in soybean genotypes.

Sl.No.	Cultivars	Leaf Water Potential (Bars)	Leaf Temperature (°C)	Transpiration Rate (m.moles m^{-2} s^{-1})	Stomatal diffusive Resistance (dcm^{-1})
1.	Culture 70	−16.25	33.00	8.28	2.13
2.	EC 100776	−15.75	32.45	7.72	2.01
3.	AM 5549	−17.25	32.70	7.57	3.66
4.	Culture 189	−17.25	33.40	6.36	2.71
5.	Culture 101879	−10.75	32.45	10.03	2.19
6.	EC 109548	−11.25	31.85	11.47	1.24
7.	EC 50084	−13.25	31.60	12.03	2.45
8.	EC 15086	−14.75	31.65	11.11	1.27
9.	MACS 125	14.25	31.90	14.03	1.25
10.	ACC No.2004	−14.75	32.55	10.57	3.45
11.	EC 18733	−14.00	32.20	10.56	3.33
12.	Culture 15590	−21.75	32.55	15.20	3.19
13	SDDC	−20.75	32.10	11.43	1.11
14.	PLSO-1	−12.75	32.35	7.83	5.91
15.	UGM 34	−11.25	32.05	6.64	2.51
16.	EC 62384	−15.50	32.10	5.23	2.77
17.	DS 295	−19.75	38.10	8.24	2.07
18.	EC109545	−16.75	31.55	15.02	1.89
19.	PLSO-1	−17.75	32.90	8.71	3.01
20.	EC 2541	−14.75	33.10	3.84	2.58
21.	Culture 425	−12.25	33.50	4.59	3.95
22.	UGM 30	−15.75	32.35	7.31	2.46
23.	EC 95287	−16.25	32.20	5.29	2.30
24.	EC 50082	−21.40	33.25	6.41	3.07
25.	EC 109548	−24.00	33.20	7.16	3.14
26.	EC 18226	−15.25	32.55	7.75	3.32
27.	EC 141390	−20.25	32.35	10.20	1.54
28.	EC 95258	−15.75	32.50	8.61	2.85
29.	DC 13007	−15.75	32.80	8.36	2.63
30.	EC 24058	−11.75	31.60	14.23	1.95
	SE(d)	1.430	1.730	1.267	0.281
	CD (5 per cent)	3.792	3.962	3.393	1.064

It has been stated that because of lower stomatal resistance which means higher leaf conductance, transpiration rate increases with concomitant reduction in leaf temperature brought about by evapotranspiration process.

Rate of photosynthesis varied significantly among the soybean genotypes. Among the cultivars studied cultures *viz.*, UGM 30 (28), EC 62384 (28.01, EC 95287) (26.61) and UGM 34 (26.4) recorded higher photosynthetic rate. Photosynthesis is the major physiological process that determines the yield in crop plants.

In soybean, photosynthesis may influence yield through N_2 fixation because the root nodules consume a considerable amount of carbohydrates. It was further observed that the genotypes which possessed higher stomatal resistance exhibited decreased photosynthesis, the reason being lower diffusion rate of CO_2.

It was seen that the genotypes exhibited wide variability in respect of pod number per plant. The high yielder identified in this study *viz.*, culture 425, UGM 34 and EC 254 possessed greater number of pods. Previous studies in soybean also indicated that yield variation was mainly due to variation in pod number. Thus, it could be seen that the pod number was an important factor contributing for increased yield in soybean.

Even though some of the genotypes such as UGM 30, EC 62384 and EC 95287 had shown higher rate of photosynthesis, their yield was not greater. Although photosynthetic efficiency is the primary component of drymatter productivity, it has however, been found to be inconsistently related to economic yield (Gaskel and Pearce, 1981) because of several factors like photorespiration, dark respiration, translocation of assimilated and partitioning efficiency, sink size, etc. It was evident that both higher photosynthetic rate and pod number should be essential for achieving higher pod yield in soybean. It could be said that the high yielders, culture 425, UGM 34 and EC 2541 produced greater pod yields by virtue of their high rate of photosynthesis and pod number per plant.

References

Gaskel, M.L. and Pearce, R.B., 1981. Growth analysis of maize hybrids differing in photosynthetic capacity. *Agron. J.*, 73: 817–821.

Roark, B. and Quinseberry, J.E., 1977. Environmental and genetic components of stomatal behaviour in two genotypes of upland cotton. *Plant Physiol.*, p. 354–359.

Rode, I.A., Spurway, R.A. and Mewhirter, K.S., 1983. Breeding dryland soybeans for northern inland. New South Wales. In: *Proc. Aust. Plant Breed. Conf. Cont.* pp. 91–93 Addaide. Screening soybeans for tolerance to moisture stress. A field procedure. *Field Crop Research*, 3: 321–335.

Tadano, T., Kirimoto, K., Aoyama, I. and Tanaka, A., 1979. Comparison of tolerance to high moisture conditions of the oil among crop plants. *J. Sci. Soil and Manure*, 50: 261–263.

Tadano, T. and Tanaka, A., 1980. Comparison among crop species in response to low phosphorus concentration in culture solution during early growth stages. *J. Sci. Soil and Manure*, 51: 359–404.

Tanaka, A. and Hayakawa, Y., 1975. Comparison of tolerance to soil acidity among crop plants. Part 3. Tolerance to soil acidity. *J. Sci. and Manure*, 46: 26–32.

Trikha, R.N., 1985. The potential of soybean in Indian cropping systems. In: *Soybean in Tropical and Subtropical Cropping Systems,* (Eds.) S. Shanmugasundaram and E.W. Sulzberger. The Asian Vegetable Research and Development Centre, Taiwan, pp. 77–80.

Wein, H.C., Littleton, E.J. and Ayanaba, A., 1979. Drought stress of cowpea and soybean under tropical conditions. In: *Stress Physiology in Crop Plants*, (Eds.) H. Mussell and R. Staples. John Wiley and Sons, New York, pp. 284–301.

2013, Perspectives in Plant Biodiversity   *Pages* **205–223**
Editor: **Dr. K. Muthuchelian,** *Vice Chancellor, Periyar University, Salem*
Published by: **Daya Publishing House, NEW DELHI**

Chapter 27

Allometric Variation and Reproduction in *Panax wangianus* S.C. Sun (Araliaceae): An Endangered Medicinal Plant in the Sacred Grove Forest of Meghalaya, North-East India

N. Venugopal and Preeti Ahuja*
Plant Anatomy Lab, Department of Botany,
North-Eastern Hill University, Shillong – 793 022, Meghalaya

ABSTRACT

Panax L. (Araliacea) consists of approximately 18 species, out of which 16 are from eastern Asia and two from eastern North America. *Panax wangianus* S. C. Sun (Syn. *Panax pseudoginseng* Wall). is a critically endangered, medicinal plant of north-east India. The objective of this study was to determine the allometric relationship with reproductive efficiency and population size of *Panax wangianus*. Data on vegetative and reproductive characters were collected from 2016 individuals of the population in Law Lyngdoh, Smit sacred grove in Nongkrem, Shillong, India during 2007–2009. Larger plants (in terms of age

---

* Corresponding Author: E-mail: nvenugopal3@gmail.com

class 35–50 years and above 50 years) are reproductively more successful because of its vegetative growth and flowering phenology that influences the more production of fruits and seed set in Nongkrem sacred grove. Morphological variations were observed in natural conditions based on prong numbers and the number of carpellate conditions. Age class was significant for to predict the size of the plant and its reproductive capacity. Therefore, from conservation and management point of view, the age class 35–50 years and above 50 years is the most important for population sustainability. Plants of age class 35–50 years may facilitate the floral rewards to the pollinators. Therefore, in *Panax wangianus,* there is a clear-cut correlation between the age of the plant and pentacarpellate condition, formation of axillary inflorescences with more productivity.

Keywords*: Allometric relationship with reproductive efficiency, Panax wangianus, neoteny, Nongkrem sacred grove.*

Introduction

The family Araliaceae consists of approximately 55 genera and 1500 species which include a number of important medicinal and ornamental plants (Wen *et al.,* 2001b). The members of this family are mostly distributed in the tropics and subtropics especially in southern and Southeast Asia and the Pacific islands, but there are several well-known genera from the temperate zones as well (*e.g. Aralia, Hedera, Oplopanax* and *Panax*). In India, Araliaceae is represented by 16 genera distributed mostly in the northern and north-eastern regions of Himalayas. *Panax* consists of approximately 18 species out of which 16 are from eastern Asia and two from eastern North America (Lee and Wen, 2004; Reunov *et al.,* 2008). The generic name *Panax* derived from Greek term meaning "cure all" for its reputed medicinal use in China (Anderson *et al.,* 2002). The Chinese have been using ginseng for over 2000 years as a tonic, a stimulant and a fatigue-resistance medicine (Wen and Nowicke, 1999).

The two American species, *Panax trifolius* and *Panax quinquefolius,* were defined on the basis of their morphology, molecular and pollen ultrastructure. Numerous studies of *P. quinquefolium* and *Panax ginseng* have examined various ecological relationships (Lewis and Zenger, 1982; Lewis, 1984; Anderson *et al.,* 1984; Anderson, 1996; Hackney, 1999; Shahi, 2007; Van der Voort 2005 and Stathers and Bailey, 1986), reproductive biology (Carpenter and Cottam, 1982; Lewis and Zenger, 1982; Schlessman, 1985,1987), chemical composition (Shim *et al.,* 1983; Tomoda *et al.,* 1985; Hansen and Boll, 1986; Hikino *et al.,* 1986; Oshima *et al.,* 1987), medicinal properties (Hu, 1976; Carlson, 1986; Dubrick, 1983) and growth, yield and rates of photosynthesis, respiration and transpiration under varied solar radiation and/or temperature conditions (Hu, 1976; Lee *et al.,* 1980; Konsler, 1986; Proctor and Tsujita, 1986; Stoltz, 1982; Strick and Proctor, 1985).

Yoo *et al.* (2001) reported that several Himalayan species of *Panax* have been taxonomically problematic due to sympatry of morphologically distinct taxa and the existence of intermediates. With the advent of molecular techniques efforts have been made to resolve various taxonomic disputes by using nuclear ribosomal DNA,

ITS sequences, AFLP and cpDNA restriction site variations (Wen and Zimmer, 1996; Zhou *et al.,* 2005; Choi and Wen, 2000).

Panax wangianus (syn. *Panax pseudoginseng*) S. C. Sun is a perennial, critically endangered, herb native to sub-tropical wet forests of North-East Himalayan regions especially in Meghalaya (Pushpangadan and Nair, 2005). Its rhizome has been used as a blood-regulating medicine and as a tonic (Wen, 2007). *P. wangianus* was abundantly distributed about a century ago, but now their population is decreasing alarmingly because of various human impacts such as urbanization, over exploitation of natural resources, pollution of soil, water and atmosphere due to coalmine activities which contribute to the global climatic change (Pushpangadan and Nair, 2005). Relationship between age, size, reproduction and demographic studies of *P. quinquefolius* has extensively been studied in North America (Mooney and McGraw, 2009; Charron and Gagnon, 1991). However, *P. wangianus* growing in sacred groves of Meghalaya in wild conditions has not been thoroughly investigated on its growth, behaviour and reproductive biology. Therefore, the present investigation reveals that the morphological and reproductive changes in relation to the age of the plant. This study will be helpful to understand the reproductive biology for its conservation and sustainable utilization.

Materials and Methods

Study site: Populations of *P. wangianus* were located in the Law Lyngdoh, Smit sacred grove in Nongkrem (east khasi hills) 25°31'N, 91°52'E at an altitude 1833 m amsl, Law Lyngdoh (Mawphlong) 25°26'N, 91°44' E at an altitude 1796 m amsl and Shillong peak 25°32' N, 91°49' E at an altitude 1965 m amsl in Shillong, Meghalaya. There are no anthropogenic activities like cutting, grazing etc. in Nongkrem sacred grove because of various religious beliefs of tribal communities of Meghalaya in northeast India. Based on these beliefs, certain patches of forests are designated as sacred grove under customary law and are protected from any product extraction by the community (Tiwari *et al.,* 2008). Therefore, *P. wangianus* grew well and 4.16 per cent plants reached up to the age of 50 years and above. It occurs in colonies of a few plants in rich, shady, deciduous forests, in deep leaf litter.

Species description: *P. wangianus* has a whorl of digitately compound leaf at the summit of aerial shoot; the leaf (prong) consists of three to eight palmately compound leaflets with a petiole, of which the terminal three to five leaflets are larger than the lateral two to three leaflets. The flowers were arranged in an umbel inflorescence on a long peduncle arising from the centre of the leaf attachment at the top of the aerial stem. The flowers were bisexual, pentamerous with an inferior ovary. The fruits were berry. The population was monitored for three consecutive years (2007, 2008 and 2009), with an interval of two to three weeks duration during the growing season April to September of *P. wangianus*. In each visit both the vegetative and reproductive status was recorded (Tables 27.1 and Table 27.2). The vegetative and reproductive characters were taken from the juvenile stage to fruit formation. After full expansion the length and width of leaflets were measured and their leaflet area was measured by a Planimeter (Anderson *et al.,* 1993). Age was recorded by counting the number of bud scale scars on the rhizome that form as a result of the annual abscission of the

Table 27.1: Morphological features of *Panax wangianus* growing in Nongkrem sacred grove.

Age Class	Number of Prongs	Number of Individuals (n) and per cent of Population	Height of Aerial Shoot (cm)	Aerial Shoot Diameter (cm)	Range of Rhizome Diameter (cm)	Total Number of Leaflets	Breadth of Leaflet (cm)	Average Leaflet Area (cm²)	Range of Petiole length (cm)
1-10 years	1	187= 9.27 per cent	20.00 ± 6.20	0.3±0.02	0.5-0.7	3-5	0.4-0.8	2.00 ± 0.12	3-4
10-15 years	2	243=12.05 per cent	20.00 ± 6.20	0.4±0.03	0.7-1.5	6-10	0.4-1.0	4.00 ±0.02	4-7
16-24 years	3	482=23.90 per cent	30.00 ± 6.20	0.5±0.02	0.8-1.5	9-15	1.0-2.0	8.00 ± 0.58	6-10
25-34 years	4	602=29.86 per cent	80.00 ± 6.20	0.7±0.06	1.0-2.0	12-20	2.0-3.0	16.00±1.73	10-15
35-50 years	5	418=20.73 per cent	95.00 ± 9.09	1.0±0.01	1.0-3.0	22-36	3.0-4.0	32.5 ± 3.76	26
50 years and above	6	84=4.16 per cent	110.00 ±6.20	1.5±0.12	2.0-3.5	30-42	3.5-4.5	37.5 ± 3.25	28-32

± indicates standard deviation.

Table 27.2: Reproductive features of *Panax wangianus*.

Age Class	Number of Prongs	Length of Peduncle of Terminal Inflorescence (cm)	Length of Peduncle of Axillary Inflorescence (cm)	Number of Carpels/Plant	Range of Number of Stigmatic Lobes/Plant	Range of number of Flowers/Plant	Range of Number of Axillary Inflorescence/Plant	Range of Number of Fruits/Plant
1-10 years	1	–	–	–	–	–	–	–
10-15 years	2	–	–	–	–	–	–	–
16-24 years	3	15	"	3	1-2	13-56	–	1-3
25-34 years	4	27	"	3	1-3	25-60	–	4-9
35-50 years	5	38	7	3-5	1-5	25-73	2-3	6-17
50 years and above	6	40	9	3-5	1-5	45-75	5-8	7-30

– indicates no flowering.

aerial stem (Figure 27.1). The aerial stem is produced only once in a year per rhizome. The relationship between age of the plant and morphological characters (height of the plant, number of prongs and leaflets and average leaflet area) and reproductive characters (length of peduncle, initiation of flower buds, mature flowers, anthesis of flowers, numbers of axillary inflorescence, young and mature fruit formation and seed dispersal) were calculated statistically by using Pearson's correlation coefficient. The mean value and standard deviation were calculated from 2016 individuals of the population. Multiple regression analysis was calculated to determine the influence of age on morphological and reproductive characters by using Statistica version 5.0 software (Zar, 1974).

For scanning electron microscope (SEM), glutaraldehyde fixed (post-fixed with osmium tetroxide) and dehydrated flowers were critical point dried in a Jeol JCPD-5 critical point dryer, 3-methyl butyl acetate solution as the exchange liquid. Dried flowers were dissected with razor blades, sputter coated with gold in an Eiko ion coater, and examined with a Jeol JSM 6360 at 30KV.

Results

The height of aerial shoot ranges from 30 cm to 112 cm depending upon the age of the plant. Similarly the aerial shoot diameter as well as rhizome diameter also varies from 0.3 cm to 1.5 cm and 0.5 cm to 3.5 cm respectively in different age classes (Table 27.1). Morphological variations were observed in natural conditions based on prong number. Leaflet area increased concurrently with age. Plants of age class 1–10 years had one prong with three to five leaflets and 10–15 years had two prongs with six to ten leaflets respectively. The average leaflet area of one-pronged and two-pronged plants was 2 cm^2 and 4 cm^2 respectively. Plants of age class 16–24 years had three prongs with 9–15 leaflets and 25–34 years age class contained four prongs with 12–20 leaflets. The average leaflet area of three-pronged and four-pronged plants was 8 cm^2 and 16 cm^2 respectively. While plants of age class 35–50 years and above 50 years had five prongs with 22–36 leaflets and six prongs with 30–42 leaflets respectively. The average leaflet area of five-pronged and six-pronged plants was 32.5 cm^2 and 37.5 cm^2 respectively. Petiole length also increases with age of the plant. It ranges from 3cm to 32cm depending upon age and number of prongs of the plant (Table 27.1). Occasionally only one to two percent one-pronged and two-pronged plants were of more than 40 years old. It indicated that neoteny do exist in *P. wangianus* (Figure 27.1).

In Nongkrem sacred grove, plants of age class 1–10 years and 10–15 years were either juvenile or vegetative stages only *i.e.* they were non-fecundity. During 2007–2009, the plants did not flower until they were grown and attained the minimum age of 16 years. Plants older than 16 years with more than three prongs were producing flowers from April to early May. The length of peduncle of terminal inflorescence also varies from 15 cm to 40 cm in different age classes. The number of flowers in an umbel was 13–56 and 25–60 flowers in the age class 16–24 and 25–34 years respectively (Table 27.2). Plants older than 35 years had two to three axillary inflorescences in addition to the terminal inflorescence (Figure 27.3). The number of flowers in an umbel varied from 25–73 flowers. The numbers of axillary inflorescence were five to eight and the number of flowers ranged from 45–75 in the 50 years old plant (Figure 27.2). The length of peduncle of axillary inflorescence was 7 cm in age class 35–50 years and 9 cm in age class 50 years and above (Table 27.2).

The number of style and stigmatic lobes varies in each inflorescence (Table 27.2). Generally each ovary consists of three carpels. In the age class 35–50 years and above 50 years, the number of carpels varied from three to five which would be readily observed from the exterior by counting the number of lobes in fruits (Figure 27.3). Each carpel had single pendulous, anatropous and unitegmic ovule. The stigma was dry with unicellular papillate outgrowths. In the age class of 16–24 years, the flowers had one to two stigmatic lobes; the flowers of age class 25–34 years had one to three numbers of stigmatic lobes. While plants of age class of 35–50 years and above, the number of stigmatic lobes in each inflorescence ranged from one to five (Figure 27.3). The terminal inflorescence matured first followed by axillary inflorescence (Figure 27.3).

Young fruits were green in color which appeared in July and matured during October. As fruits mature, it became bright red on the bottom and black on the top (Figure 27.2). The color of pedicel also changed from green to crimson red. Fruits contained one to five seeds. The number of fruits produced by the age class 16–24 years and age class 25–34 years was one to three and four to nine respectively. Age class 35–50 years and above produced 6–17 and 7–30 number of fruits respectively. In *P. wangianus* the percentage of one-seeded fruit is 26 per cent, two-seeded fruit 31 per cent, three seeded fruit 22 per cent, four-seeded fruit 12 per cent and five seeded fruits are 9 per cent. But the ability to produce flowers that mature into fruit increased with the age of the plant. Phenological and reproductive characters revealed that most of the individuals of age class 35–50 and 50 years above plants contributed most of the flowering, fruiting and seed production.

Size Relationship between the Age, Morphological and Reproductive Features of the Plant

In *P. wangianus*, during the study period from 2007–2009, a strong positive correlation was observed between the age versus height of the plant ($r^2 = +0.93$), aerial shoot diameter ($r^2 = +0.92$), rhizome diameter ($r^2 = +0.94$), total number of leaflets ($r^2 = +0.97$), breadth of leaflet ($r^2 = +0.98$), average leaflet area ($r^2 = +0.96$) and petiole length ($r^2 = +0.94$) (Table 27.3).

Similarly, the reproductive parameters also showed positive correlation with the age of the plant. The value of correlation coefficient of length of peduncle ($r^2 = +0.95$), length of peduncle of axillary inflorescence ($r^2 = +0.78$), number of carpels ($r^2 = +0.81$), number of stigmatic lobes ($r^2 = +0.92$), number of flowers ($r^2 = +0.88$), number of axillary inflorescence ($r^2 = +0.69$) and number of fruits ($r^2 = +0.93$) were also very high (Table 27.4).

Discussion

The vegetative as well as reproductive, phenological relationships in *P. wangianus* showed that nearly 90 per cent of the features correlated with age. Wen (2001a) and Shu (2007) demonstrated that the high level of morphological and reproductive plasticity

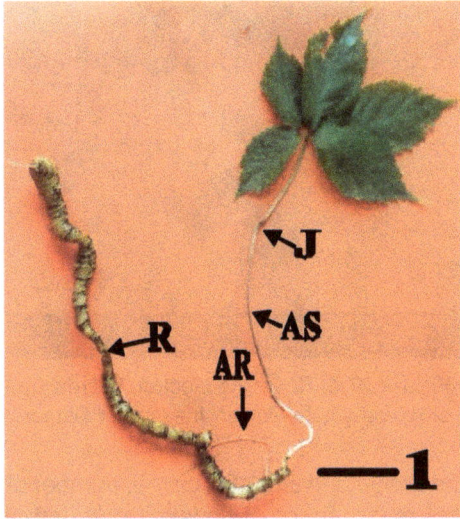

Figure 27.1: An entire 48 years old neoteny plant of *Panax wangianus* with one-prong consists of five leaflets. Aerial shoot (as); joint (j) between petiole and aerial shoot; Rhizome (r) with adventitious root (ar). Bar = 4.6 cm.

Figure 27.2: Above 50 years old plant with six prongs (38 leaflets), seven axillary inflorescence and one terminal inflorescence with fruits. Mature fruits are bright red on the bottom and black on the top. Bar = 11.40 cm. (Inset An enlarged view of mature fruits. Bar = 5.5cm).

Figure 27.3: An enlarge view of terminal (ti) and axillary inflorescences (ai) showing unistigmatic to pentastigmatic as well as tri-pentacarpellary condition. Terminal inflorescence bearing green young fruits. Unistigmatic (us); bistigmatic (bs); terminal inflorescence (ti); axillary inflorescence (ai) and pentacarpellate (pc). Note the terminal inflorescence mature earlier than the axillary one. Bar =10 cm. (Inset uni–pentacarpellate fruits; Bar =5cm).

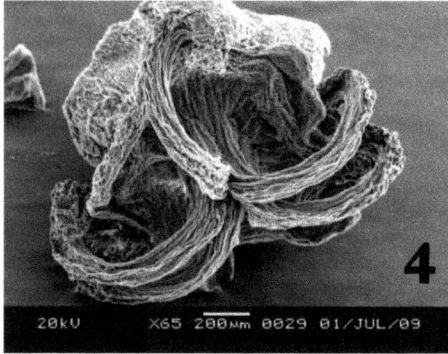

Figure 27.4: Scanning electron microscope of pentastigmatic flower.

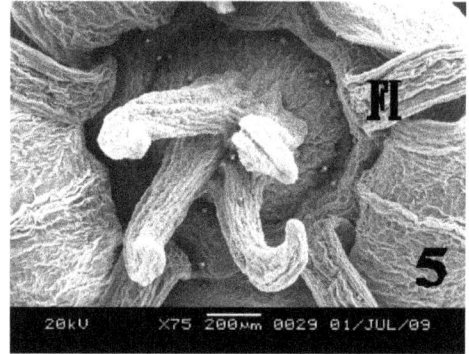

Figure 27.5: Scanning electron microscope of tetrastigmatic flower. fi = anther filament.

in eleven species of *Panax* with respect to the shape and size of rhizome, number of leaves, leaflet shape, pubescence of leaves and number of flowers in an umbel inflorescence (Tables 27.5 and 27.6). Ghou *et al.* (2010) also reported that the morphological variations in stem, leaf, root, flower and fruit in the cultivated populations of *Panax notoginseng*. However, in population of *P. wangianus* growing in Nongkrem sacred grove, there is a clear-cut correlation between the different age classes with respect to the height of the plant, aerial shoot diameter, rhizome diameter, total number of leaflets, breadth of leaflet, average leaflet area, petiole length, length of peduncle, length of peduncle of axillary inflorescence, number of carpels, number of

Table 27.3: r^2 value of different morphological features versus age of the plant.

	Vegetative Parameters						
Dependent Variable	Height of the Plant	Aerial Shoot Diameter	Rhizome Diameter	Total Number of Leaflets	Breadth of Leaflet	Average Leaflet Area	Petiole Length
Age of the plant	r^2	r^2	r^2	r^2	r^2	r^2	r^2
	+0.93	+0.92	+0.94	+0.97	+0.98	+0.96	+0.94

Table 27.4: r^2 value of different reproductive features versus age of the plant.

	Reproductive Parameters						
Dependent Variable	Length of Peduncle of Terminal Inflorescence	Length of Peduncle of Axillary Inflorescence	Number of Carpels	Number of Stigmatic Lobes	Number of Flowers	Number of Axillary Inflorescence	Number of Fruits
Age of the plant	r^2	r^2	r^2	r^2	r^2	r^2	r^2
	+0.95	+0.78	+0.81	+0.92	+0.88	+0.69	+0.93

Table 27.5: Morphological variations in different species of _Panax_ reported by Anderson _et al._ (1993), Grushvitskii (1961), Khrolenko _et al._ (2007), Shu (2007) and Wen (2001a).

Characters	Panax wangianus	Panax ginseng	Panax japonicus	Panax zingiberensis	Panax notoginseng	Panax pseudoginseng	Panax vietnamensis	Panax stipuleanatus	Panax bipinnatifidus	Panax quinquefolius	Panax trifolius
Age (years)	1-50	140-150	–	–	–	–	–	–	–	25-30	–
Number of prongs	1-6	3-6	3-5	3-7	3-6	4	3-5	3-4	3-5	1-5	3
Height of aerial shoot (cm)	20-110	30-60	50-100	20-60	20-60	50	40-100	–	–	20-50	5-20
Aerial shoot diameter (cm)	0.3-0.12	–	–	–	–	–	–	–	–	–	–
Range of rhizome diameter (cm)	0.5-3.5	–	–	–	–	–	–	–	–	–	–
Total number of leaflets	3-42	9-30	15-25	9-32	9-42	12-16	–	–	–	3-20	–
Breadth of leaflet (cm)	0.4-4.5	–	–	–	–	–	–	–	–	–	–
Average leaflet area (cm²)	2-37.5	–	–	–	–	–	–	–	–	–	–
Range of petiole length (cm)	3-32	–	–	–	–	–	–	–	–	–	–

– indicates data not available.

Table 27.6: Reproductive variations in different species of *Panax* reported by Anderson *et al.* (1993), Fiebig *et al.* (2001), Shu (2007) and Wen (2001a).

Characters	Panax wangianus	Panax ginseng	Panax japonicus	Panax zingiberensis	Panax notoginseng	Panax pseudoginseng	Panax vietnamensis	Panax stipuleanatus	Panax bipinnatifidus	Panax quinquefolius	Panax trifolius
Length of peduncle of terminal inflorescence (cm)	15-40	15-30	12-21	24-26	7-25	12	–	–	–	–	–
Length of peduncle of axillary inflorescence (cm)	7-9	–	–	–	–	–	–	–	–	1	–
Number of carpels/plant	3-5	2	2-5	2	2	2	–	–	–	2-3	–
Range of number of stigmatic lobes/plant	1-5	–	–	–	–	–	–	–	–	–	–
Range of number of flowers/plant	13-75	30-50	50-80	–	80-100	20-50	–	–	–	32-79	Male umbels = 15-30; female umbels = 3-14
Range of number of axillary inflorescences/plant	2-8	0	2-4	–	–	1-3	2-4	–	2-5	1	–
Range of number of fruits/plant	1-30	–	–	–	–	–	–	–	–	30-40	–

– indicates data not available.

stigmatic lobes, number of flowers, number of axillary inflorescences and number of fruits (Tables 27.1 and 27.2).

But with respect to the age classes in *P. wangianus* the height of the aerial shoot did not vary much in age class 1–10 years and 10–15 years with one and two-pronged plants respectively. However, in age class 16–24years, 25–34 years, 35–50 years and 50 years and above, there was a steady increase in the height of aerial shoot, aerial shoot diameter, rhizome diameter, petiole length and breadth of the leaflet (Table 27.1). The height of aerial shoot of *P. wangianus* showed similar growth trend when compared to *P. japonicus* and *P. vietnamensis*. While in the remaining six species *P. ginseng*, *P.zingiberensis*, *P. notoginseng*, *P. pseudoginseng*, *P. quinquefolius* and *P. trifolius,* the height of aerial shoot ranged from 5–60 cm (Wen, 2001a and Shu, 2007).

The number of prongs in *P. wangianus* was 1–6 while Wen (2001a) and Shu (2007) reported 1–5 in *P.quinquefolius* and in *P. ginseng*, *P. japonicus*, *P.zingiberensis*, *P. notoginseng*, *P. pseudoginseng*, *P. vietnamensis*, *P.stipuleanatus*, *P. bipinnatifidus*, *P. quinquefolius* and *P. trifolius* it ranged from 3–7 (Table 27.5).

Anderson *et al.* (1993) mentioned that in *P. quinquefolium* one-prong plant of two years of age possessed three to five leaflets; two-prong plants (with 10 leaflets) ranged from three to six years of age, three-prong plants (with 15 leaflets) ranged from seven to nine years of age, and four-prong plants (with 20 leaflets) ranged from 10 to 11 years of age. However, in *P. wangianus* plants of age class 1-10 years had one prong with three to five leaflets and 10–15 years had two prongs with six to ten leaflets respectively. Plants of age class 16–24 years had three prongs with 9–15 leaflets and 25–34 years age class contained four prongs with 12–20 leaflets. While plants of age class 35–50 years had five prongs with 22–36 leaflets. Lewis and Zenger (1983) observed that four–and five-prong plants represented the oldest individual of a population of *P. quinquefolium*. But in *P. wangianus*, six-prong plants with 30–42 leaflets fell under the category of 50 years and above age class. Similar observations were reported in *P. quinquefolius,* using the combination of leaf area and leaf number which are the good indicators of age (Furedi, 2004 and McGraw and Furedi, 2005). In *P. quinquefolius* the maximum number of leaflets was 20 (Anderson *et al.,* 1993). While in *P.ginseng*, *P. japonicus*, *P. zingiberensis*, *P. notoginseng* and *P. pseudoginseng*, it ranged from 9–42 (Shu, 2007 and Wen, 2001a).

In Nongkrem sacred grove, only one to two percent was one-pronged and two-pronged plants of *P. wangianus* exhibited neoteny and their age was above 40 years. The retention of ancestral juvenile characters by mature plants was reported in few members of family Araliaceae *e.g. Schefflera racemifera* and *S. longipetiolata* by Fiasch and Frodin (2006).

Lewis and Zenger (1982); Carpenter and Cottam (1982); Lewis (1984); Anderson *et al.* (1993) and (Gagnon, 1999) reported that the height of aerial shoot, number of leaves and leaflet and leaflet area increased with which were the good predictors of age in *P. quinquefolium* as estimated by the number of bud scars on the rhizome. Similar observations have been encountered in *P. wangianus* (Table 27.1). The maximum age of *P. quinquefolium* estimated by Anderson *et al.* (1993) was 25–30 years and in *P. ginseng* it was 140–150 years (Grushvitskii, 1961 and Khrolenko *et al.*, 2007); while in *P. wangianus* it was above 50 years. Rhizome damage could make the plant appear younger than its actual age, but growth probably would continue with adventitious roots assuming the storage and absorption functions (Anderson *et al.*, 1993).

Similarly the reproductive features were also positively correlated with length of peduncle, length of peduncle of axillary inflorescence, number of carpels, number of stigmatic lobes, number of flowers, number of axillary inflorescences, number of fruits and flowering season, age versus fecundity of the plant and ripening of fruits etc. In *P. wangianus*, the length of peduncle of terminal inflorescence ranged from 15–40 cm while in *P. ginseng, P. japonicus, P. zingiberensis, P. notoginseng* and *P. pseudoginseng* it ranged from 7–30 cm (Shu, 2007 and Wen, 2001a).

Farrington (2006) reported two pronged plants also showed fecundity in *P. quinquefolium.* However, in *P. wangianus* the fecundity has been observed from three-pronged plants and onwards.

Schlessman (1985) reported three and four-pronged plants tended to flower earlier to one and two-pronged plants. In *P. quinquefolius*, 20 per cent –50 per cent of two-year-old plants have inflorescences (Proctor *et al.*, 2003). Fiebig *et al.* (2001) reported that there were 32 to 79 flowers in an inflorescence in a three-year-old plant of *P. quinquefolius.* Shu (2007) and Wen (2001a) reported that the numbers of flowers per plant ranged from 20–100 in *P. ginseng, P. japonicus, P. notoginseng* and *P. pseudoginseng.* Out of literature surveyed of 11 species of *Panax*, the only species *P. trifolius* is a dioecious plant in which male and female umbel consists of 15–30 flowers and 3–14 flowers respectively (Wen, 2001a). But in the case of *P. wangianus*, all the flowers in an umbel were hermaphrodite. Only in the three-pronged (six + leaflets) plants, the inflorescence consisted of 13–56 number of flowers. The earliest flowering plants were 16 years of age; they had few flowers and only rarely formed fruit, but after 24 years of age, all plants developed flowers and most had fruits. Five and six–pronged plants produced more flowers and fruits, whereas one and two-pronged plants remained vegetative or in juvenile stage (Table 27.2).

Anderson *et al.* (1993) reported only one axillary inflorescence *in P. quinquefolium* and the length of peduncle of axillary inflorescence was one cm and rarely does it produce fruits. Shu (2007) and Wen (2001a) reported 1–5 numbers of axillary inflorescences in *P. japonicus, P. pseudoginseng, P. vietnamensis* and *P. bipinnatifidus.* While in *P. wangianus* the numbers of axillary inflorescence is two to eight and the length of peduncle of axillary inflorescence has been measured from seven cm to nine cm depending upon the age of the plant and all the axillary inflorescence also possessed fruits. The young fruits of *P. wangianus* collected in July remained greenish and turned into red bottom and black on the top after six to eight weeks (Figure 27.2. Inset). But in *P. quinquefolium* all the ripened fruits are reddish in color (Proctor *et al.*, 2003).

In addition to the present investigation, as well as that of others (Anderson *et al.*, 1993; Carpenter and Cottam, 1982; Lewis and Zenger, 1982), has shown that fruit production is correlated with plant size. The initiation of flower and fruit production varies in different population of *P. quinquefolium*. Anderson *et al.* (1993) reported that the fruit production started in four year old plants of *P. quinquefolium* in Illinois and produced 30–40 numbers of fruits. While Carpenter and Cottam (1982) reported fruit production began after eight years in *P. quinquefolium* in Wisconsin. Farrington (2006) mentioned that all the plants irrespective to age produce berries containing one to three seeds. In *P. wangianus* fruit production started after 16 years and it increased with age of the plant. Plants of 50 years and above produced 7–30 numbers of fruits. Schlessman (1985) reported that flowers containing one, two and three ovaries produced one, two and three-seeded fruit, respectively, and the ratio of two-to one-ovuled flowers increases with plant size (age). In *P. wangianus* the percentage of one-seeded fruit is 26 per cent, two-seeded fruit 31 per cent, three seeded fruit 22 per cent, four-seeded fruit 12 per cent and five seeded fruits are 9 per cent. Stoltz and Garland (1980) also reported about one-seeded (16.3 per cent), two-seeded (77.0 per cent), three-seeded (6.5 per cent) and four-seeded (0.2 per cent) fruits in *P. quinquefolium*.

Shu (2007) reported all the species of *Panax* in China are two carpellate and bistigmatic except *P. japonicas* which have two to five carpels and two to five styles growing in China. Wen (2001a) reported one to three carpellate conditions except in *P. trifolius*. Wen *et al.* (2001b) also mentioned pentacarpellate condition was an evolutionarily liable and developmental abnormality in Araliaceae. Philipson (1970) and Eyde and Tseng (1971) also reported two to five carpellate condition is common in Araliaceae. The pluricarpellate condition (upto 20 carpels) is a primitive condition in Araliaceae. The American ginseng with one styled flowers represent an intermediate step in the evolution of functional maleness. But Asiatic ginseng is functional hermaphrodite because all the individuals of *P. wangianus* in the population growing in Nongkrem sacred grove showed one to five stigmatic with three to five carpellate conditions.

Larger plants (in terms of age class 35–50 years and above 50 years) are reproductively more successful because of its vegetative growth and flowering phenology that influences the more production of fruits and seed set in Nongkrem sacred grove. Morphological variations were observed in natural conditions based on prong numbers and the number of carpellate conditions. Age class was significant for to predict the size of the plant and its reproductive capacity. Therefore, from conservation and management point of view, the age class 35–50 years and above 50 years is the most important for population sustainability. Plants of age class 35–50 years may facilitate the floral rewards to the pollinators. Therefore, in *Panax wangianus,* there is a clear-cut correlation between the age of the plant and pentacarpellate condition, formation of axillary inflorescences with more productivity.

Acknowledgements

We would like to thank Prof. M.S. Dkhar, Head, Department of Botany for providing facilities. Our sincere thanks to Headman, Nonkgrem sacred grove, for collection of materials for research purpose. One of us (PA) expressed gratitude to the University Grant Commission, New Delhi under University with Potential of excellence (UPE) programme, NEHU, Shillong.

References

Anderson, R.C., Fralish, F.S., Armstrong, J. and Benjamin, P., 1984. *Biology of Ginseng (Panax quinquefolium) in Illinois.* Department of Conservation, Division of Forest Resource and Natural Heritage, Springfield, Illinois.

Anderson, R.C., Fralish, J.S., Armstrong, J.E. and Benjamin, P.K., 1993. The ecology and biology of *Panax quinquefolium* L. (Araliaceae) in Illinois. *The American Midland Naturalist*, 129: 357–372.

Anderson, J.M. and Ingram, J.S.I., 1993. *Tropical Soil Biology and Fertility: A Handbook of Methods*, 2nd Edn. CAB International, Wallingford.

Anderson, M.P., 1996. American ginseng (*Panax quinquefolium* L). in southwestern Wisconsin: patterns of demography and habitat association. *M.Sc. Thesis*, University of Wisconsin, Madison.

Anderson, R.C., Anderson, M.B. and Houseman, G., 2002. Wild American ginseng. *Native Plants Journal*, 3: 93–105.

Ashton, P.S., Givnish, T.J. and Appanah, S., 1988. Staggered flowering in the Dipterocarpaceae: New insights into floral induction and the evolution of mast fruiting in the aseasonal tropics. *American Naturalist*, 132: 44–66.

Bawa, K.S., 1983. Patterns of flowering in tropical plants. In: *Handbook of Experimental Pollination Biology*, (Eds.) C.E. Jones, R.J. Little. Van Nostrand Reinhold, New York, p. 394–410.

Borchert, R., Meyer, S.A., Felger, R.S. and Porter-Bolland, L., 2004. Environmental control of flowering periodicity in Costa Rican and Mexican tropical dry forest. *Global Ecology and Biogeograph*, 13: 409–425.

Brady, C.N. and Well, R.R., 2002. *The Nature and Properties of Soils*. Pearson Education Pte Ltd., Singapore, p. 86–89.

Carlson, A.W., 1986. Ginseng: America's botanical drug connection to the orient. *Economic Botany*, 40: 233–249.

Carpenter, S.G. and Cottam, G., 1982. Growth and reproduction of American ginseng (*Panax quinquefolius*) in Wisconsin, U.S.A. *Canadian Journal of Botany*, 60: 2692–2696.

Champion, H.G. and Seth, S.K., 1968. *Revised Survey of the Forest Types of India.* Government of India, New Delhi.

Charron, D. and Gagnon, D., 1991. The demography of northern populations of *Panax quinquefolium* (American ginseng). *Journal of Ecology*, 79: 431–445.

Choi, H.K. and Wen, J., 2000. A phylogenetic analysis of *Panax* (Araliaceae): Integrating cpDNA restriction site and nuclear rDNA ITS sequence data. *Plant Systematics and Evolution*, 224: 109–120.

Dahlgren, J.P., VON Zeipel, H. and Ehrlen, J., 2007. Variation in vegetative and flowering phenology in a forest herb caused by environmental heterogeneity. *American Journal of Botany*, 94: 1570–1576.

Dhirendra, Singh N. and Venugopal, N., 2010. Cambial activity and annual rhythm of xylem production of *Pinus kesiya* Royle ex. Gordon (Pinaceae) in relation to phenology and climatic factors growing in sub-tropical wet forest of North-East India. *Flora* doi: 10.1016/j. flora.2010.04.021

Dominguez, C.A. and Dirzo, R., 1995. Rainfall and flowering synchrony in a tropical shrub: variable selection on the flowering time of *Erythroxylon havanense*. *Evolutionary Ecology*, 9: 204–216.

Dubrick, M.A., 1983. Dietary supplements and health aids: a critical evaluation. *Journal of Nutrition Education*, 15: 123–129.

Eyde, R.H. and Tseng, C.C., 1971. What is the primitive floral structure of Araliaceae? *Journal of the Arnold Arboreum*, 52: 205–239.

Farrington, S.J., 2006. An ecological study of American ginseng (*Panax quinquefolius* L). in the Missouri Ozark Highlands: effects of herbivory and harvest, ecological characterization and wild simulated cultivation. *M.Sc. Thesis*, University of Missouri, Columbia.

Fiasch, P. and Frodin, D.G., 2006. A new species of *Schefflera* (Araliaceae) from Espirito Santo State, Brazil. *Kew Bulletin*, 61: 187–191.

Fiebig, A.E., Proctor, J.T.A., Posluszny, U. and Murr, D.P., 2001. The North American ginseng inflorescence: development, floret abscission zone, and the effects of ethylene. *Canadian Journal of Botany*, 79: 1048–1056.

Furedi, M.A., 2004. Effects of herbivory by white-tailed deer (*Odocoileus virginianus* Zimm). on the demography and conservation biology of American ginseng (*Panax quinquefolius* L). *Ph.D. Dissertation*, Department of Biology, West Virginia University, Morgantown, West Virginia.

Gagnon, D., 1999. An analysis of the sustainability of American ginseng harvesting from the wild: The problem and possible solutions. First draft of a report to the Office of Scientific Authority of the U.S. Fish and Wildlife Service.

Grushvitskii, I.V., 1961. Zhen 'shen': voprosy biologii (Ginseng: Biological Problems), Leningrad: Akad. Nauk SSSR.

Guo, H.B., Cui, X.M., An, N. and Cai, G.P., 2010. Sanchi ginseng (*Panax notoginseng* (Burkill) F. H. Chen) in China: distribution, cultivation and variations. *Genetic Resources Crop Evolution*, 57: 453–460.

Hackney, E.E., 1999. The effects of small population size, breeding system, and gene flow on fruit and seed production in American ginseng (*Panax quinquefolius* L., Araliaceae). *M.Sc. Thesis*, West Virginia University.

Hansen, L. and Boll, P.M., 1986. Polyacetylenes in Araliaceae: Their chemistry, biosynthesis and biological significance. *Phytochemistry*, 25: 285–293.

Hikino, H.M., Takahashi, M., Otake, K. and Konno, C., 1986. Isolation and hypoglycaemic activity of eleutherans A, B, C, D, E, F and G: glycans of *Eleuthercoccus serticosus* roots. *Journal of Natural Products*, 49: 293–297.

Hu, S.Y., 1976. The genus *Panax* (ginseng) in Chinese medicine. *Economic Botany*, 30: 11–28.

Inouye, D.W., Saavedra, F. and Lee-Yang, W., 2003. Environmental influences on the phenology and abundance of flowering by *Androsace septentrionalis* (Primulaceae). *American Journal of Botany*, 90: 905–910.

Khrolenko, Y.A., Burundukova, O.L., Bezdeleva, T.A., Muzarok, T.I. and Zhuravlev, Y.N., 2007. Age stages in the ontogeny of cultivated *Panax ginseng* C. A. Mey. *Biology Bulletin*, 34: 120–125.

Konsler, T.R., 1986. Effect of stratification temperature and time on rest fulfilment and growth in American ginseng. *Journal of the American Society for Horticultural Science*, 111: 651–654.

Lee, J.C., Cheon, S.K., Kim, Y.T. and Jo, J.S., 1980. Studies on the effect of shading materials on the temperature light intensity, photosynthesis and root growth of Korean ginseng (*Panax ginseng* C. A. Meyer). *Korean Society of Crop Science*, 25: 91–98.

Lee, C. and Wen, J., 2004. Phylogeny of *Panax* using chloroplast trnC–trnD intergenic region and the utility of trnC–trnD in interspecific studies of plants. *Molecular Phylogenetics and Evolution*, 31: 894–903.

Lewis, W.H. and Zenger, V.E., 1982. Population dynamics of the American ginseng *Panax quinquefolium* (Araliaceae). *American Journal of Botany*, 69: 1483–1490.

Lewis, W.H. and Zenger, V.E., 1983. Breeding systems and fecundity in the American ginseng, *Panax quinquefolium* (Araliaceae). *American Journal of Botany*, 70: 466–468.

Lewis, W.H., 1984. Population structure and environmental corollaries of *Panax quinquefolium* (Araliaceae) in Delaware country, New York. *Rhodora*, 86: 431–438.

Liu, D., Cui, X.M., Wang, C.L., Lu, F. and He, C.F., 1992. Preliminary study on photosynthetic characteristics of *Panax notoginseng*. *Southwest Journal of Agricultural Science*, 2: 41–43.

McGraw, J.B. and Furedi, M.A., 2005. Deer browsing and population viability of a forest understory plant. *Science*, 307: 920–922.

Mooney, E.H and McGraw, J.B., 2009. Relationship between age, size and reproduction in populations of American ginseng *Panax quinquefolius* (Araliaceae) across a range of harvest pressures. *Ecoscience*, 16(1):84–94.

Opler, P.A., Frankie, G.W. and Baker, H.G., 1976. Rainfall as a factor in the release, timing, and synchronization of anthesis by tropical trees and shrubs. *Journal of Biogeography*, 3: 231–236.

Oshima, Y., Sato, K. and Hikino, H., 1987. Isolation and hypoglycaemic activity of quinquefolans A, B, and C, glycans of *Panax quinquefolium* roots. *Journal of Natural Products*, 50: 188–190.

Pfeifer, M., Heinrich, W. and Jetschke, G., 2006. Climate, size and flowering history determine flowering pattern of an orchid. *Botanical Journal of the Linnean Society,* 151: 511–526.

Philipson, W.R., 1970. Constant and variable features of the Araliaceae. *Journal of Linnaean Society Botany,* 63: 87–100.

Porwal, M.C., Talukdar, G., Singh, H., Triparthi, O.P., Tripathi, R.S. and Roy, P.S., 2000. Biodiversity characterization at landscape level using remote sensing and geospatial modeling in Meghalaya (India). In: *Biodiversity and Environment,* (Eds.) P.S. Roy, S. Singh, A.G. Toxopeus. Indian Institute of Remote Sensing, Dehradun, p. 206–219.

Proctor, J.T.A. and Tsujita, M.J., 1986. Air and root-zone temperature effects on the growth and yield of American ginseng. *Journal of Horticulture Science,* 61: 129–134.

Proctor, J.T.A., Dorais, M., Bleiholder, H., Willis, A., Hack, H. and Meier, V., 2003. Phenological growth stages of North American ginseng (*Panax quinquefolius*). *Annals of Applied Biology,* 143: 311–317.

Pushpangadan, P. and Nair, K.N., 2005. Medicinal plant wealth of Meghalaya: Its conservation, sustainable use and IPR issues. In: *Biodiversity Status and Prospects,* (Eds.) P. Tondon, M. Sharma and R. Swarup, p. 16–23.

Reunov, A.A., Reunova, G.D., Alexandrova, Y.N., Muzarok, T.I. and Zhuravlev, Y.N., 2008. The pollen metamorphosis phenomenon in *Panax ginseng, Aralia elata* and *Oplopanax elatus.* An addition to discussion concerning the *Panax* affinity in Araliaceae. *Zygote,* 17: 1–17.

Schlessman, M.A., 1985. Floral biology of American ginseng (*Panax quinquefolium*). *Bulletin of the Torrey Botanical Club,* 112: 129–133.

Schlessman, M.A., 1987. Gender modification in North American ginsengs: dichotomous choice versus adjustment. *Bioscience,* 37: 469–475.

Shahi, D.P., 2007. Effects of density on reproduction and demographic structures of American ginseng (*Panax quinquefolius* L). population in Ohio. *M.Sc. Thesis.* Graduate College of Bowling Green State University.

Shim, S.C., Chang, S.K., Hur, C.W. and Kim, C.K., 1983. Polyacetylenes from *Panax ginseng* roots. *Phytochemistry,* 22: 1817–1818.

Shu, R.S., 2007. *Panax* Linnaeus, Sp. Pl. 2: 1058. 1753. *Flora of China,* 13: 489–491.

Stathers, R.J. and Bailey, W.G., 1986. Energy receipt and partitioning in ginseng shade canopy and mulch environment. *Agricultural and Forest Meteorology,* 37: 1–14.

Stevenson, P.R., Castellanos, M.C., Cortes, A.I. and Link, A., 2008. Flowering patterns in a seasonal tropical lowland forest in Western Amazonia. *Biotropica,* 40: 559–567.

Stoltz, L.P. and Garland, P., 1980. Embryo development of ginseng seed at various stratification temperatures. In: *Proceedings of Second National Ginseng Conference*, Jefferson City, Mo., p. 43–51.

Stoltz, L.P., 1982. Leaf symptoms, yield, and composition of mineral-deficient American ginseng. *HortScience*, 17: 740–741.

Strick, B.C. and Proctor, J.T.A., 1985. Dormancy and growth of American ginseng as influenced by temperature. *Journal of the American Society for Horticultural Science*, 110: 319–321.

Tiwari, B.K., Barik, S.K. and Tripathi, R.S., 2008. Biodiversity value, status, and strategies for conservation of sacred groves of Meghalaya, India. *Ecosystem Health*, 4: 20–32.

Tomoda, M., Shimada, K., Konno, C. and Hikino, H., 1985. Structure of panaxan B, a hypoglycaemic glycan of *Panax ginseng* roots. *Phytochemistry,* 24: 2431–2433.

Tripathi, O.P., Pandey, H.N. and Tripathi, R.S., 2004. Distribution, community characteristic and tree population structure of subtropical pine forest of Meghalaya, northeast India. *International Journal of Ecology and Environmental Sciences,* 29: 207–213.

Tyler, G., 2001. Relationships between climate and flowering of eight herbs in a Swedish deciduous forest. *Annals of Botany*, 87: 623–630.

Van der Voort, M.E., 2005. An ecological study of *Panax quinquefolius* in central Appalachia: seedling growth, harvest impacts and geographic variation in demography. *Ph.D. Thesis*, West Virginia University.

Venugopal, N. and Liangkuwang, M., 2007. Cambial activity and annual rhythm of xylem production of elephant apple tree (*Dillenia indica* Linn.) in relation to phenology and climatic factor growing in sub-tropical wet forest of Northeast India. *Trees*, 21: 101–110.

Wen, J. and Zimmer, E.A., 1996. Phylogeny and biogeography of *Panax* L. (the ginseng genus, Araliaceae). Inferences from ITS sequences of nuclear ribosomal DNA. *Molecular Phylogenetics and Evolution*, 6(2): 166–177.

Wen, J. and Nowicke, J.W., 1999. Pollen ultrastructure of *Panax* (the ginseng genus, Araliaceae), an eastern Asian and eastern North American disjunct genus. *American Journal of Botany*, 86: 1624–1636.

Wen, J., 2001a. Species diversity, nomenclature, phylogeny, biogeography, and classification of the Ginseng genus (*Panax* L., Araliaceae). In: Utilization of biotechnological, genetic and cultural approaches for North American and Asian ginseng improvement, (Ed.) Z.K. Punja. *Proceedings of the International Ginseng Workshop*, Simon Fraser University Press, Vancouver, p. 67–88.

Wen, J., Plunkett, G.M. Mitchell, A.D. and Wagstaff, S.J., 2001b. The evolution of Araliaceae: A phylogenetic analysis based on ITS sequences of nuclear ribosomal DNA. *Systematic Botany,* 26(1): 144–167.

Wen J., 2007. *Mt. Sanqingshan: A Botanical Treasure.* The Plant Press, Smithsonian, National Museum of Natural History 10(4).

Yoo, K.O., Malla, K.J. and Wen, J., 2001. Chloroplast DNA variation of *Panax* (Araliaceae) in Nepal and its taxonomic implications. *Brittonia*, 53: 447–453.

Zar, J.H., 1974. *Biostatistical Analysis*, 2nd Edn. Prentice Hall, Englewood Cliffs, New Jersey.

Zhou, S.L., Xiong, G.M., Yi, L.Z. and Wen, J., 2005. Loss of genetic diversity of domesticated *Panax notoginseng* F H Chen as evidenced by ITS sequence and AFLP polymorphism: A comparative study with *P. stipuleanatus* HT Tsai *et* KM Feng. *Journal of Integrative Plant Biology*, 47(1): 107–115.

2013, Perspectives in Plant Biodiversity   *Pages* **224–229**
Editor: **Dr. K. Muthuchelian,** *Vice Chancellor, Periyar University, Salem*
Published by: **Daya Publishing House, NEW DELHI**

Chapter 28

Diversity of Arbuscular Mycorrhizal Fungi on Four Important Tree Species in Dharwad District of Karnataka

Krishna H. Waddar, H.C. Lakashman
and K. Sandeepkumar
P.G. Department of Studies in Botany (Microbiology Lab),
Karnatak University Dharwad – 580 003

Introduction

The mycorrhizoosphere micro biota differ qualitatively as well as quantitatively from the rhizahosphere of nonmycorrhizae plants. The microfauna influence the mycorrhiza formation as well as the host growth response (Fitter and Gabage, 1994). Many kinds of interactions occur between these microbial communities in the mycorhizosphere and mycorrhizae. The interaction between the mycorrhizae and soil microorganisms may be mutualistic or competitive and the effect the establishment and functions of mycorrhizal symbionts as well as modified the interactions of plant with other symbionts or pathogens in soil. The arbuscular mycorrhizal fungi found to be in world wide established symbiotic association with majority of plants (Smith and Read, 1997). Therefore in the present data on AM fungal diversity in four important tree species which are commonly growing in Dharwad district of Karnataka.

They are more prevalent in soils deficient in Phosphorus. Arbuscular mycorrhizal fungi are capable of scavenging Phosphorus from the soil in to the root system mainly through exploration of the soil by extraradial hyphae beyond the root hairs and the Phosphorus depletion zone.(Kunwar *et al.,* 1999).

Materials and Methods

Survey was conducted to collect the roots and rhizospheric soil samples of four tree plants form 10 different sites of Dharwad district in Karnataka. The AM fungal spores were quantified following the wet sieving and decanting technique (Gerdemann and Niocolson, 1963). AM fungal spores were identified by using Mycorrhizal identification manual (Schenck and Perez, 1990).

Roots samples were from four tree species washed in water then cut into 1cm bits and mixed in 10 per cent KOH solution, autoclaved for 30 minutes and acidified with 1N HCl. Then Stained with 0.05 per cent tryphan blue in lacto phenol (Philips and Hayman, 1970). 50-100 such root fragments were examined under a microscope, the mycorrhizal colonization was determined by Nicholson's formula (1955) as follows.

$$\text{Per cent of root colonization} = \frac{\text{Number of root Segments colonized}}{\text{Total number of root Segments examined}} \times 100$$

Results and Discussion

Dharwad is an adjacent to western ghat forests of Karnataka many valuable trees growing in this region. Presently four commonly growing trees were selected. Via, *Azadirachta indica* L., *Tamarindus indica* L., *Bauhinia varigata* L., *Pongamia pinnata* Vent. Survey work was carried out on the different places. The physicochemical characteristic feature of rhizospheric soil shown in (Table 28.1). The percent of root colonization and spore number in some places are correlated but most of the place do not correlated. This may be due to edaphic factors between different localities. Higher percent of root colonization was observed in *Azadirachta indica* L. and *Tamarindus indica* L., *Bauhinia varigata* L. respectively. Lower percent of root colonization was noted down in *Pongamia pinnata* Vent. Similarly, spore number per 50 gm of soil was determined, highest number of spores was recorded in *Azadirachta indica* L. rhizospheric soil samples. Similarly lower was recorded in the rhizospheric

Table 28.1: Physio-chemical properties of soils of Dharwad district in Karnataka.

Property	Mean Value of Soil
pH	6.7 ± 0.0115
Temperature (°C)	27± 0.000
Moisture	4.25 ± 0.144
Electrical conductivity (mm ohms/cm)	0.10 ± 0.00
P (mg/Kg)	0.96 ± 0.011
N (mg/Kg)	6.120 ± 0.0173
K (mg/Kg)	31.39 ± 0.785
Sand (per cent)	46.39 ± 0.624
Silt (per cent)	25.43 ±0.176
Clay (per cent)	28.18 ± 0.138
Organic matter	2.083 ± 0.044

soil of *Pongamia pinnata* Vent. These two plants per cent of root colonization and spore number were more or less correlated compared to *Tamarindus indica* L. and *Bauhinia varigata* L. Per cent of root colonization and spore number in the rhizosphere shown in (Table 28.2). Each plant had predominated arbuscular mycorrhizal fungal spore genera. Among these 5-6 most commonly occurring fungal spores were recorded.

Table 28.2: Per cent of root colonization and spore number of arbuscular mycorrhizal fungi in four tree species of Dharwad district in Karnataka.

Host Plant Species	Per cent Root Colonization	Spore Number 50 gm/Soil	AMF Species
Azadirachta indica L.	54.2	447	Gf, Gm, Al, Gm, Se, As, Sc, Gg, Gma, At
Tamarindus indica L.	50.8	346	Gm, Al, Gg, Sc, Ga, Sm, Glg
Bauhinia varigata L.	48.5	361	Gm,Gc, Sm,Glg, Al
Pongamia pinnata Vent.	44.7	298	Gma, Gf, Sc, Sm, As, Gc, Gf

Gf: *Glomus fasciculatum*; Gm: *Glomus mossae*; Gg: *Gegaspora gignata*; As: *Acaulospora scrobiulata*; Sc: *Sclerocystis calorpora*; Gma: *Gigaspora margrata*; Se: *Scutellospors erythropa*; Al: *Acaulospora levis*; Sm: *S. Microcarpa*; Gc: *Gigaspora claroidues*; Glg: *Glomus aggregatum*; At: *Acaulospora teberculata*.

In most of the soil samples genus *Glomus* considered to be most predominant spore genera and *Acaulospora* occupies second position compare to *Gigaspora*, *Scutellospora, Sclerocystis* species were least recorded spores in present study (Table 28.3). Higher numbers of arbuscular mycorrhizal fungal spores were recorded, where as these plants growing in waste land or forest soil, lower numbers of spores were recorded in the soil samples collected form agriculture field.

The results of the present study on four tree species of Dharwad district with regards to per cent of root colonization and spore number are in consistent with the earlier workers (Raman *et al.*,1992; Mosse, 1997 and Brundrett *et al.*, 1999).The mycorrhizal colonization consist of hyphae, vesicles and arbuscules. The percentage of colonization varied with plant species, this is mainly due to multiplication of AM fungi depending on its association with roots. The number of spores in soil at different depths is likely to differ. Similar observation was documented by (Mukerji *et al*,. 1984; Kunwer *et al.*, 1999 and Bhattacharya *et al.*, 2000). The lower percentage of root colonization and spore number in *Pongamia pinnata* Vent. may be due to presence of secondary metabolites, which exudates from the roots. This kind of observation also made by (Lakshman, 1996). There were five AM fungal spores such as *Acaulospora, Glomus, Scutellospora, Gigaspora* and *Sclrocystis* was recorded in the present work.

Conclusion

The diversity of AM fungal species on four tree species in Dharwad district is in it's of first kind. The twenty five Am species were recorded in the present study. Similar findings with other plants associated this four tree species may be warranted. To understand in better way can be achieved only after screening different isolates of AM fungi whether they actually colonize or not. Because most of the plants colonized

Table 28.3: Distribution of AM fungi in ten different places of Dharwad district in Karnataka.

AM Fungi Species	Places									
	A	B	C	D	E	F	G	H	I	J
Acaulospora species										
Acaulospora scrobiculata Trappe.	+	–	+	+	+	–	–	+	–	–
A. delicate Walker, Pfeiffer and Bloss.	–	+	–	–	–	–	–	+	–	–
A. laevis Gerdemann and Trappe.	+	+	+	+	+	–	+	–	+	+
A. spinosa Walker and Trappe.	–	–	–	–	–	+	–	–	–	+
A. teberculata Jonson and Trappe	–	+	–	–	–	+	+	–	–	–
A.gerdemanni Walker, Reed and Sander.	–	–	+	–	+	–	+	–	–	–
Gigaspora species										
G. margarita Becker and Hall.	+	–	–	+	+	+	–	+	–	+
G. gigantea (Nicolson and Gerdemann) Gerdemann and Trappe.	+	+	–	+	–	–	–	+	–	+
G. decipiens Hall and Abbott.	–	+	–	–	–	–	–	–	+	–
Glomus species										
Glomus mossae (Nicolson and Gerdemann) Gerdmann and Trappe.	+	+	+	+	–	+	+	–	+	+
G. fasciculatum Gerdemann and Trappe emend Walker and Koske.	+	–	+	–	–	+	–	+	+	+
G. aggregatum Schenck and Smith emend Koske.	–	+	+	–	–	–	–	+	+	+
G. microcarpum Tulasne and Tulasne.	–	+	–	–	+	+	–	–	–	–
G. constrictum Trappe.	–	+	+	–	–	–	–	–	–	–
G. botryoides Rothwell and Victor.	+	–	–	–	–	+	–	–	–	–
G. geosporum (Nicolson and Gerdemann) Walker.	+	+	–	–	–	–	+	–	–	–
G. macrocarpum Tulasne and Tulasne.	+	–	+	–	–	–	+	–	–	–
G. claroidues Schenck and Smith.	–	–	–	–	–	+	–	–	–	–
Glomus reticulatum Bhattacharjee and Mukerji										

Contd...

Table 28.3–*Contd...*

AM Fungi Species	Places									
	A	B	C	D	E	F	G	H	I	J
Sclerocystis species										
S. dussii (Patouillard) Von Hohnel.	+	+	+	–	–	+	+	–	–	+
S. microcarpa Iqbal and Bushra.	–	–	+	–	–	–	+	–	+	–
Scutellospra species										
S. calospora (Nicolson and Gerdmann). Walkerand Sanders	+	–	–	–	+	–	–	–	+	–
S. erythopa (Koske and Walker) Walker and Sanders	–	–	+	–	–	+	+	–	–	–
Scutellospra sp.	+	–	–	–	+	+	–	–	–	+

+: Present; –: Absent.

A: Garag; B: Mummigatti; C: Nuggikeri; D: Amargol: E: Kundagola; F: Kyarakoppa; G: Navalur; H: Yattinagudda; I: Byahatti; J: Karadigudda.

by AM fungi develop better root system, roots will be shorter, adventitious than those of lack of AM fungal association.

Acknowledgements

First author is thankful to UGC, New Delhi, for deputing under FDP XI plan and second author is indebted to UGC, New Delhi, for sanctioning major research project on Investigation of AM fungal role in four rare millets of North Karnataka.

References

Bhattacharya, P.M., Misra, D., Saha, J. and Chaudhuri, 2000. Arbuscular mycorrhizal dependency of *Eucalyptus tereticorni* SM, how real is it? *Mycorrhiza News,* 12(3).

Brundrett, M.C., Jasper, D.A. and Ashwath, N., 1999. Glomelean mycorrhiza fungi from tropical Australia–II: The effect of nutrient levels and host species on the isolation of fungi. *Mycorrhizal,* 8: 315–321.

Gerdemann, J.W. and Nicolson, T.H., 1963. Spores of mycorrhizal endogene Species extracted from soil by wet sieving and decanting Techniques of the British Mycological society, 46: 235–244.

Kunwar, I.K., Reddy, P.J.M. and Manoharachary, C., 1999. Occurrence of AMF associated with garlic rhizosphere soil. *Mycorrhizal News*, 11(2): 4–6.

Lakshman, H.C., 2009. Selection of suitable AM fungus to *Atrocarpus heterophyllus* Lam. A fruit/timber for an ecofriendly nursery. M.D. Publisher, New Delhi, pp. 50–61.

Lakshman, H.C., 1996. VA-Mycorrhizal studies in some economically important Timber tree species. *Ph.D. Thesis*, pp. 247.

Mosse, B., 1997. Techniques for studying mycorrhiza in tropical countries. *Mycorrhizal News*, 9(2): 7–13.

Mukerji, K.G., Sabharwal, A., Kochar, B. and Ardey, J., 1984. Vesicular Arbuscular Mycorrhizae, concepts and advances. In: *Progress in Microbial Ecology*, (Eds.) K.G. Mukerji, V.P. Agnihotri, and R.P. Singh. Printed House (India), Lucknow, pp. 489–525.

Phillips, J.M. and Hayman, D.S., 1970. Improved procedures for clearing roots and staining parasitic and vesicular arbuscular mycorrhizal fungi for rapid assessment of infection. *Transaction of the British Mycological Society*, 55: 158–161.

Raman, N., Nagarajan, N. and Sambandan, K., 1992. Vesicular arbuscular mycorrhizae of tree species of Mamandur forest of Tamil Nadu. *J. Tree Sci.,* 11: 135–139.

Schenck, N.C. and Perez, Y., 1990. *Manual for the Identification of VA Mycorrhizal Fungi,* 3rd Edn. Gainesville, Florida, USA, Synergistic Publications.

Smith, S.E. and Read, D.J., 1997. *Mycorrhizal Symbiosis,* 2nd Edn. Academic Press, London, 605 pp.

Trappe, J.M., 1989. The meaning of mycorrhizal to plant ecology. In: *Mycorrhizae for Green Asia*, (Eds.) A. Mahadevan, N. Raman and K. Natrajan, pp. 347–349.

2013, Perspectives in Plant Biodiversity   *Pages* **230–235**
Editor: **Dr. K. Muthuchelian,** *Vice Chancellor, Periyar University, Salem*
Published by: **Daya Publishing House, NEW DELHI**

Chapter 29

Isolation and Characterisation of Soil Fungi in Dry Deciduous Forest of Bhadra Wildlife Sanctuary, Karnataka

Shivakumar P. Banakar and B. Thippeswamy*
Department of P.G. Studies and Research in Microbiology,
Bioscience Complex, Kuvempu University,
Jnanasahyadri, Shankaraghatta – 577 451, Shivamogga Dist.

ABSTRACT

Soil is the natural medium for the growth of various types of saprophytic, parasitic and antagonistic microorganisms. Forest soil is highly rich in nutrient composition by the process of nutrient recycling mediated by saprophytic microorganisms. Present study was carried out for the soil fungal diversity in the Dry Deciduous forest. Soil sample was collected in the winter and summer season by random mixed sampling method. Serial dilution method was followed for the collected soil Sample to isolate and characterize the fungal species on Czapek Dox Agar and Martin's Rose Bengal Agar media. Environmental factors like, pH, Temperature, Soil moisture and nutrients were recorded. Diversity of soil fungi in the tropical dry deciduous forest majorly belongs to Dueteromycetes followed by Ascomycetes, Zygomycetes and Oomycetes were recorded. Indices of diversity were 2.136 and 2.102 (Shannon H), 0.8294 and 0.8278 (Simpson 1-D) and 5.208 and 6.45 (Fisher alpha) respectively in terms of genera.

Keywords: Soil fungi, Diversity, Dry deciduous forest, Environmental factors.

---

* Corresponding Author: E-mail: thippeswamyb205@gmail.com

Introduction

Forest soil composed of various types of microorganisms like Bacteria, Fungi, Actinomycetes, Protozoa and Algae in the different proportion. Fungi include macro and microfungi, majority of them are saprophytes, actively involved in the nutrient cycling in the environment. Turnover of the plant residue fraction of soil organic matter plays a dominant role in global carbon cycling and is fundamental to soil structure and fertility. Saprophytic basidiomycetes are likely to be the dominant recyclers of plant wastes in soil since they are the main producers of lignin modifying enzymes. In total 1.5 million fungal species are there, of which 74,000 species are identified (Hawksworth, 2001 and Manoharachary *et al.,* 2005). The classical methods of extrapolation from small samples in ecosystem analysis involve studying the species composition in quadrates and transects. Such approaches are not easy for fungi, as the production of a full species inventory even of a very small area is time-consuming and requires meticulous observation and repeated sampling exercises (Cannon, 1997).

Materials and Methods

Study Area

Study area was Bhadra Wildlife Sanctuary which is located at 13° 52' N and 75° 37' E with elevation of 690 ft msl Altitude range. Bhadra Wildlife Sanctuary in Western Ghat's belt, it has rich biodiversity hot spot and have different type of tree vegetation, Dry and Moist deciduous, Semi-Evergreen and Evergreen forests, it covers an area of 492.46 sq km.

Collection of Soil Sample

Organic soil sample was collected in the winter and summer season (November 2008 and March 2009) from a depth of 5-10 cm by random mixed sampling method (6 subsamples), samples were brought to the laboratory in a sterile polythene covers, air dried and preserved in a cold temperature for further use. Environmental conditions like soil temperature, atmospheric temperature, relative humidity (RH), soil moisture content (per cent), soil electric conductivity, pH and nutrients were analyzed and recorded frequently during collection of samples.

Isolation and Characterization of Fungal Species

Serial dilution was done for the collected soil sample on Czapek Dox Agar (CZA) and Martin's Rose Bengal Agar (MRBA) in duplicates, spread plate method was followed (Nilima *et al.,* 2007). After incubation of 72 to 120 hours at 25±2°C, characterization of isolated fungal species were done by observing macroscopic cultural characteristics (color, shape, surface, margin and size of the colony along with pigmentation and exudates) and morphological characteristics under microscopic observation (arrangement and size of the mycelium, conidiophores and conidia) by comparing with standard fungal identification manuals (Nagamani *et al.,* 2006; Barnett, 1960; Domsch *et al.,* 1980; and Gilman, 2001).

Results

The isolated fungal species belongs to the genus *Absidia, Aspergillus, Bipolaris, Cladosporium, Cordana, Eupenicillium, Gliocladium, Mucor, Paecilomyces, Penicillium, Pythium, Scopulariopsis, Trichoderma* and *Verticillium* (Table 29.1) (Figure 29.1). Indices of diversity in winter and summer season were 2.136 and 2.102 (Shannon H), 0.8294 and 0.8278 (Simpson 1-D) and 5.208 and 6.45 (Fisher alpha) respectively in terms of genera (Table 29.2). Environmental Conditions like soil temperature, atmospheric temperature, relative humidity (RH), soil moisture content (per cent), soil electric conductivity, pH, micro and macro nutrients were analyzed and documented (Table 29.3).

Table 29.1: Fungal genera over two Seasons during 2008-09.

Fungal Genus	Winter Season	Summer Season
Absidia	3	2
Aspergillus	8	5
Bipolaris	1	0
Cladosporium	5	3
Cordana	1	0
Eupenicillium	4	2
Gliocladium	2	1
Mucor	2	1
Paecilomyces	3	2
Penicillium	20	12
Pythium	0	1
Scopulariopsis	1	1
Trichoderma	2	1
Verticillium	6	4

Table 29.2: Diversity Indices over two Seasons during 2008-09.

Diversity Indices	Winter Season	Summer Season
Shannon H	2.136	2.102
Simpson 1-D	0.8294	0.8278
Fisher alpha	5.208	6.45
Equitability J	0.8326	0.8459
Evenness e H/S	0.6509	0.6819
Dominance D	0.1706	0.1722

* Genus diversity is measured on the basis of estimated number of species.

Table 29.3: Physico-chemical parameters over two Seasons during 2008-09.

Parameters	Winter Season	Summer Season
Atm. Temperature	28 °C	36°C
Soil Temperatue	25 °C	33°C
Relative Humidity (RH)	68 per cent	65 per cent
Soil Moisture Content	28 per cent	15.63 per cent
Soil pH	5.5	6.8
Soil EC (milli mhos)	0.41	0.24
Micro Nutrients		
Organic Carbon (per cent)	1.61 per cent	1.80 per cent
Phosphorus (Kg/Acre)	4.50	4.25
Potassium (Kg/Acre)	56.0	83.0
Macro Nutrients		
Zinc (ppm)	0.15	0.32
Copper (ppm)	1.39	0.42
Manganese (ppm)	2.29	5.23
Iron (per cent)	28.36 per cent	9.80 per cent

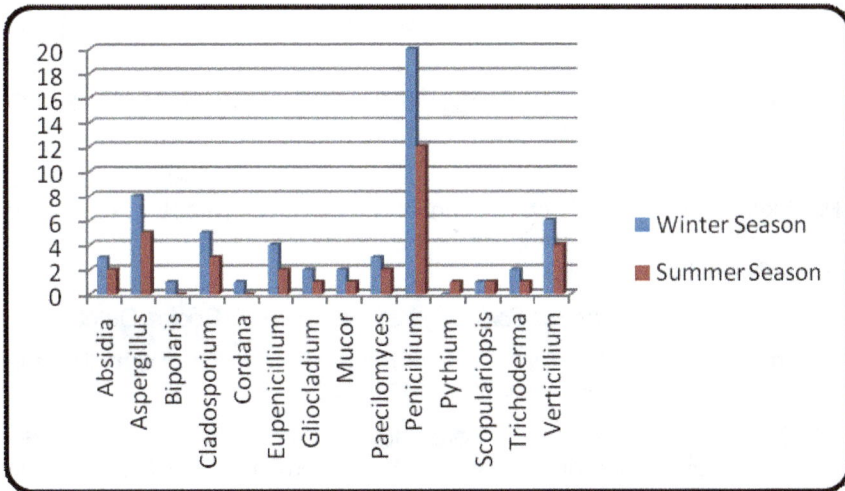

Figure 29.1. Graphical representation of occurrence of fungal genera over two Seasons in Dry deciduous forest during 2008-09.

Discussion

Diversity of microfungi in the tropical forest majorly belongs to *Dueteromycetes*, followed by *Ascomycetes* (Nilima *et al.*, 2007), highly organic layers supported the greatest number of fungal species and the numbers decreased with soil depth (Vardavakis, 1990). The classical methods of extrapolation from small samples in ecosystem analysis involve studying the species composition in quadrates and transects. Such approaches are not easy for fungi, as the production of a full species inventory even of a very small area is time-consuming and requires meticulous observation and repeated sampling exercises (Cannon, 1997). Saprophytic fungi are best adapted for decomposition of woody debris and leaf litter, as they can produce enzymes capable of breaking down recalcitrant substances (Hyde *et al.,* 2001). Seasonal changes in abundance or frequency of fungal species may be due to seasonal fluctuations of pedological factors (moisture, temperature, nutrient elements) (Vardavakis, 1990). Environmental factors like atmospheric temperature (25°C), soil temperature (28°C), relative humidity (68 per cent), soil moisture (28 per cent), soil pH (5.5), electric conductivity (0.41 mmhos) and nutrients like organic carbon (1.61 per cent), phosphorous (4.50/Kg). and Potassium (56.0/Kg). favors fungal diversity (Vardavakis, 1990). Environmental factors plays an important role on the fungal growth, some of the fungal species require specific environmental conditions in an optimum amount for the active growth, physiology and development.

References

Aneja, K.R., 2004. *Experiments on Microbiology, Plant Pathology and Biotechnology,* 4th Edn. New Age International Pvt. Ltd., p. 162.

Barnett, H.L., 1960. *Illustrated Genera of Imperfect Fungi,* 2nd Edn. Burgess Publishing Company, West Virginia, p. 1–225.

Cannon, P.F., 1997, Strategies for rapid assessment of fungal diversity. *Biodiversity and Conservation,* 6: 669–680.

Domsch, K.M., Gams, W. and Anderson, T.H., 1980. *Compendium of Soil Fungi,* Vol. 1. Academic Press Inc., New York, p. 1–391.

Fernaud, J.R.H., Marina, A., Gonzalez, K., Vazquez, J. and Falcon, M.A., 2006. Production, partial characterization and mass spectrometric studies of the extracellular laccase activity from *Fusarium proliferatum. Appl. Microbiol. Biotechnol.,* 70: 212–221.

Gilman, J.C., 2001. *A Manual of Soil Fungi,* 2nd edn. Biotech Books, Delhi, p. 1–365.

Hawksworth, D.L., 2001. The magnitude of Fungal Divers: The 1.5 million species estimate revisited. *Mycol. Res.,* 105: 1422–1432.

Hyde, K.D., Boonsom Bussaban, Barbara Paulus, Pedro W. Crous, Seonju Lee, Eric H.C. Mckenzie, Wipornpan Photita and Saisamorn Lumyong, 2001. Diversity of saprobic microfungi. *Biodivers. Conserv.,* 16: 7–35.

Manoharachary, C., Sridhar, K., Singh, R., Adholeya, A., Suryanarayanan, T.S., Rawat, S. and Johri, B.N., 2005. Fungal Biodiversity: Distribution, Conservation and Prospecting of Fungi from India. *Curr. Sci.,* 89: 58–71.

Nilima, S., Sadika, S. and Nanjundiah, V., 2007. Diversity of soil fungi in a tropical deciduous forest in Mudumalai, Southern India. *Curr. Sci.,* 93: 669–677.

Teather, R.M. and Wood, P.J., 1982. Use of Congo Red polysaccharide interaction in the enumeration and characterization of cellulolytic bacteria from rumen. *Appl. Environ. Microbiol.*, 43: 777–786.

Vardavakis, E., 1990. Seasonal fluctuations of soil microfungi in correlation with some soil enzyme activities and VA mycorrhizae associated with certain plants of a typical calcixeroll soil in Greece. *Mycologia,* 82: 715–726.

2013, Perspectives in Plant Biodiversity
Editor: Dr. K. Muthuchelian, *Vice Chancellor, Periyar University, Salem*
Published by: Daya Publishing House, NEW DELHI

Pages **236–240**

Chapter 30

Field Trail to Study the Effect of Fungicides in Controlling Brown Leaf Spot Disease of Rice and Altering the Microbial Diversity

P. Krishnan

Assistant Professor,
PG Department and Research Centre in Botany,
The Madura College, Madurai – 625 011, Tamil Nadu

ABSTRACT

Brown leaf spot disease caused by *Drechsleria oryzae* is a major disease of rice. Fungicide application is still the preferred control measures against this disease; nevertheless, frequent application of fungicides not only pollute the environment but also alters the biodiversity of the ecosystem. Hence, to minimize the fungicide use various fungicide treatments were adopted to control the disease.

Field trial was conducted in the rice field plots prepared in a randomized complete block design using the rice cv. IR 50. The seed treatment (6 g ai/Kg seed) and foliar spray (2 per cent aqueous solution), of bavistin were given individually as well as the combination of both the methods. The proportion of disease incidence and AUDPC was calculated for each treatment. Soil microbial population was also enumerated in all the experimental plots. The results showed that the seed treatment of bavistin gave maximum protection (88 per cent) without affecting much the microbial diversity.

Introduction

Brown leaf spot disease caused by *Drechsleria oryzae* (B.de Hann) Subrum and Jain is a major disease of rice (*Oryza sativa* L). The pathogen is seed and soil borne; however, the possible survival of the pathogen is the collateral hosts (Padmanabhan, 1974). Application of fungicides is the preferred control measures against this disease. Seed treatment and foliar sprays of Agrosun GN, tricyclazole and bavistin have been found to control this disease effectively (Viswanathan and Narayanaswamy, 1990). Frequent application of agrochemicals is hazardous to the environment as well as the various soil microbes beneficial to the crops. Effect of fungicides on the diversity of soil bacteria have been reported elsewhere (Foster and McQueen, 1977; Nicolson and Hirsch, 1998; Smith *et al.*, 2000). Hence, to counter these problems the concentration was made to investigate the possible methods to minimize the fungicide use.

Materials and Methods

The experiment was conducted in the field trial plots in Kulamangala, 20 km away from Madurai city, Tamil Nadu using the rice cv. IR 50. The optimum concentration, for seed treatment (6 g ai/Kg seed) and for foliar spray (2 per cent aqueous solution), of the fungicide bavistin was alone used in this study. The fungicide treatments were given as:

A: Control (crops without fungicide treatment).

B: Seed treatment alone.

C: Seed treatment with two foliar sprays at 30 and 50 days after sowing (DAS).

D: Foliar spray alone (four foliar sprays on the seedlings with the proportion of disease incidence ranging between 0.10–0.20 at 10 day intervals and the spray initiated at 20 DAS.

The seeds were coated with slurry of *D. oryzae* spores and dried. Then the seeds pretreated with fungicide and untreated seeds were sown directly in the field plot (1 m²) prepared in a randomized complete block design with triplicates. Ten hills per plot were randomly fixed and the disease incidence was assessed at 7-day intervals on the leaves. Area under disease progress curve (AUDPC) was then calculated for each treatment (Shaner and Finny, 1977). The percent disease protection (DP) was computed as given below

$$DP = \frac{DC - DT}{DC} \times 100$$

where,

DC = proportion of disease incidence in control plants and DT = proportion of disease incidence in fungicides treated plants. The statistical analyses of the data were done using statistical packages for social sciences (SPSS) v. 3.1 (Nie *et al.*, 1975). To enumerate the bacterial population, soil samples were suspended in sterile water and through serial dilution, reduced to a power of 10^{-5}. Nutrient agar was used in the isolation. Enumeration was done before sowing the seeds and after harvesting.

Results

Brown spots appeared on the leaves of 10 day old IR 50 rice seedlings. The results of disease incidence showed that the fungicide treatments (treatment B, C and D) suppressed the disease progress over the check (A). The seed treatment (treatment B) of bavistin gave the maximum protection till 48 DAS and the subsequent disease incidence was also very meager. The same trend was observed in the treatment C, where one spray was given in addition to seed treatment. Moreover, when the foliar sprays alone was used (treatment D), disease incidence progressed gradually and the maximum proportion attained was 0.32 (Table 30.1). The differences in the disease control was significant (P < 0.01) according to Duncan's Multiple Range Test.

Table 30.1: Effect of fungicide treatments on brown leaf spot disease in the rice cv. IR 50.

Treatment*	Proportion of Disease Incidence					AUDPC@
	Days after Sowing					
	20	27	41	48	55	
A	0.11a	0.24a	0.38a	0.49a	0.52a	48
B	0.00a	0.00b	0.00b	0.00b	0.16b	07 (85)
C	0.00a	0.00b	0.00b	0.00b	0.14b	06 (88)
D	0.10a	0.22c	0.26c	0.31c	0.32c	27 (42)

*A: Control (crops without fungicide treatment); B: Seed treatment alone; C: Seed treatment with two foliar sprays at 30 and 50 days after sowing (DAS); D: Foliar spray alone (four foliar sprays on the seedlings with the proportion of disease incidence ranging between 0.10–0.20 at 10 day intervals and the spray initiated at 20 DAS.

@ AUDPC: Area under disease progress curve.

Values in parenthesis indicate per cent disease protection.

Each value is the average of triplicates.

The AUDPC calculated for brown leaf spot against each treatment showed that the treatments B and C reduced the AUDPC with the protection of >85 per cent when compared to the foliar spray given alone (treatment D), where the maximum protection achieved was only 42 per cent (Table 30.1).

Microbial numbers were constantly high in the control plots (A) as well as in the plots sown with fungicide treated seeds (B) and the maximum number recorded was around 300×10^5. However, the plots that had received foliar sprays (treatment C and D) of agrochemical revealed the reduction in the microbial load. The treatment C showed 215×10^5t cfu and in the treatment D cfu decreased to $>100 \times 10^5$ (Table 30.2).

Discussion

Application of fungicides during the disease favorable conditions in the field is undoubtedly important for efficient control of crop disease (Madden *et al.,* 1978).

Hence, fungicide applications were timed to optimize fungicide use (Fry, 1977; Vincelli and Lorbeer, 1987) according to the conducive conditions for disease development and/or disease intensity.

Table 30.2: Changes in the soil bacterial population against fungicide treatments in the rice field.

Treatment*	Microbial Density (cfu X 10⁵)	
	Before Sowing	After Harvesting
A	282	302
B	280	288
C	278	215
D	280	98

*A: Control (crops without fungicide treatment); B: Seed treatment alone; C: Seed treatment with two foliar sprays at 30 and 50 days after sowing (DAS); D: Foliar spray alone (four foliar sprays on the seedlings with the proportion of disease incidence ranging between 0.10–0.20 at 10 day intervals and the spray initiated at 20 DAS.

Each value is the average of triplicates.

In the present study, the treatments B (seed treatment alone) and C (seed treatment with one foliar spray) reduced the incidence of brown leaf spot in the rice cv. IR 50 with the protection of >85 per cent when compared to the foliar spray given alone (treatment D), where the maximum protection achieved was only 42 per cent. Shoemaker and Lorbeer (1977) have also suggested fungicide-scheduling system based on critical disease level to control leaf blight of onions.

No significant differences in the diversity of soil microbial flora was noticed in the plots that had received no fungicides (treatment A and B). However, foliar spray of agrochemical reduced the microbial load. Same trend was also observed while studying the effect of fungicide in controlling the diseases of various crops (Smith and Hartnett, 2000; Viswanathan and Narayanaswamy, 1990).

Hence, the farmers are advised to adopt the modern technology to minimize the usage of agrochemicals so that our environment as well as the existing biodiversity would be safeguarded.

References

Foster, M.G. and McQueen, J., 1977. The effects of single and multiple applications of benomyl on non–target soil bacteria. *Bull. Environ. Contamin. Toxicology*, 17: 477.

Fry, W.E., 1977. Integrated control of potato late blight: Effects of polygenic resistance and techniques of timing of fungicide applications. *Phytopathology*. 67: 415–420.

Nicolson, P.S. and Hirsch, P.R., 1998. Effect of pesticide on the diversity of cultivable soil bacteria. *J. Appl. Microbiol.*, 84: 551.

Nie, N.H., Hull, C.H., Jenkinj, J.G., Steinbrenner, K. and Bent, D.H., 1975. *Statistical Package for the Social Sciences,* 2nd Edn. McGraw-Hill, NewYork, 675 pp.

Padmanabhan, Y., 1974. Control of rice diseases in India. *Indian Phytopathol.*, 27: 1.

Shanner, G.E. and Finney, R.E., 1977. The effect of nitrogen fertilizers on the expression of slow-mildewing resistance in knox wheat. *Phytopathology,* 67: 1051–1056.

Shoemaker, P.B. and Lorbeer, J.W., 1977. Timing of initial fungicide application to control botrytis leaf blight epidemics on onions. *Phytopathology,* 67: 409–414.

Smith, M.D., Hartnett, D.C. and Rice, C.W., 2000. Effect of long term fungicide applications on microbial properties in tall grass prairie soil. *Soil Biol. Biochem.,* 32: 935.

Vincelli, P.C and Lorbeer, J.W., 1987. Sequential sampling plan for timing initial fungicide application to control *Botrytis* leaf blight of onion. *Phytopathology,* 77: 1301–1303.

Viswanathan, S. and Narayanaswamy, S., 1990.Chemical control of brown leaf spot of rice. *Indian J. Mycol. Pl. Pathol.*, 20: 139.

2013, Perspectives in Plant Biodiversity   *Pages* **241–246**
Editor: **Dr. K. Muthuchelian,** *Vice Chancellor, Periyar University, Salem*
Published by: **Daya Publishing House, NEW DELHI**

Chapter 31

Improvement of Rooting and Plant Growth of Stem Cutting *Sauropus androgynous* L. with Inoculation of AM Fungus and Hormone (IBA) Treatment

M.A. Kadam and H.C. Lakshman
P.G. Department of Botany (Microbiology Lab),
Karnatak University, Dharwad – 580 003

Introduction

Propagation by stem cuttings is the most commonly used asexual method to regenerate many woody ornamental plants. There have been a number of attempts to propagate *Sauropus androgynous* L. by stem cuttings without much success. An interesting fact would be to examine the effect of rooting hormone and rooting media over rooting and subsequent over wintering. Higher concentration of auxins could lead to ethylene biosynthesis, increased defoliation, and poor bud break (Sun and Bassuk, 1993). Curtis *et al.* (1996) found that the optimum over wintering air temperature and storage duration for *Sauropus androgynous* L. ovata was 6°C for 10 weeks. Media plays an important role in the entire process of root growth and development once they have emerged. Rooting media could also affect the over wintering survival rate of the plant. Meldrum (1979), reported that adequate medium aeration was an important factor in successful rooting of camellia. in sand:perlite (1:1, by volume) and allowing the rooted cuttings to over winter in the same media. Media amendments like IBA plus AM fungus In this study, we focused on effect of rooting hormone (type and formulation), media characteristics, over on rooting and surviving of *Sauropus androgynous* L.

Materials and Methods

All new growth was included and the cutting length ranged from 6 to 15 cm. Cuttings were obtained from branches from the plants growing and placed in a plastic bag, which was immediately placed in coolers with ice and transported to the laboratory. Stem cuttings were prepared by cutting off half of the leaf area of individual leaves to reduce transpiration. Basal leaves were stripped off and the basal end of the cutting was slightly wounded by gently scoring with a hand clipper to facilitate the entry of rooting hormone and enhance root initiation. Rooting hormone IBA was prepared in the liquid formulation comprised of 8000 mg L^{-1} solution of IBA in water. The liquid treatment involved first dipping cuttings in a 2000 mg L^{-1} IBA solution followed by rolling in powder The chilling requirement of the cuttings was met by placing them in a cooler (5°C) for a period of 12 weeks.

Experimental pots were maintained in poly house at temperature 28-32° C. All the pots were randomized block design with in triplicate.

1. Control (Untreated)
2. Perlite+Sand.
3. *G. mosseae* (Mixed inoculum)
4. IBA (2000 mg/L).
5. IBA (4000 mg/L).
6. IBA (6000 mg/L).
7. IBA (8000 mg/L).
8. IBA (2000 mg/L) + *G. mosseae*.
9. IBA (6000 mg/L) + *G. mosseae*.
10. IBA (6000 mg/L) + *G. mosseae*.
11. IBA (8000 mg/L) + *G. mosseae*.

Results and Discussion

Sauropus androgynous L. is a multivitamin plant. All the parts of plant is used by man. Cultivation by seed germination considered to be problem and results in slow growth. Stem cuttings of this plant produce significant results with he treatment of arbuscular mycorrhizal fungi *Glomus mosseae* with IBA hormone treatment. The percentage of stem cuttings that produce roots is a limiting factor to the production with addition of perlite and sand, *Glomus mosseae* and different levels of hormone (IBA) treatment such as 2000mg/L, 4000mg/L, 6000mg/L and 8000mg/L was given to the plants of *Sauropus androgynous* L grown in experimental pots. The hormone IBA different levels as 2000mg/L to 8000mg/L with *Glomus mosseae* inoculation brought significant improvement in the production of rootlets with root hairs. This was clearly demonstrated at green house studies (Table 31.1). Plant height of *Sauropus androgynous* L. was not improved with treatment of perlite and sand, but there was

Table 31.1: Effect of *G. mosseae* and different level of IBA treatment in root initiation, plant growth, phosphorus and Nitrogen content of shoot of the *Sauropus androgynous* L. for 90 days.

Treatments	Plant Height	Root let No./plant	Per cent of Colonization	Shoot dry Weight	Root dry Weight	P-uptake	N-uptake
Control (Untreated)	9.07	4.5	–	1.81	0.79	0.6	12.4
Perlite+Sand.	11.14	6.2	2.7	2.2	0.85	12.8	14.1
G. mosseae (Mixed inoculam)	15.02	7.71	49.84	3.21	0.98	13.2	15.6
IBA (2000 mg/L)	19.19	10.31	36.71	4.11	1.01	14.1	16.6
IBA (4000 mg/L)	20.17	11.51	39.21	4.82	1.24	17.6	18.3
IBA (6000 mg/L)	48.52	13.81	46.00	8.32	3.15	14.9	32.5
IBA (8000 mg/L)	27.14	12.51	40.00	5.17	1.86	14.8	17.4
IBA (2000 mg/L) + *G. mosseae*	31.52	12.91	41.51	5.33	2.02	15.6	18.7
IBA (6000 mg/L) + *G. mosseae*	32.17	12.95	43.2	5.78	2.13	15.8	19.4
IBA (6000 mg/L) + *G. mosseae*	53.12	18.34	47.11	11.43	4.52	21.5	43.5
IBA (8000 mg/L) + *G. mosseae*	44.32	13.42	45.12	6.82	2.19	16.3	17.9

steadily increase was recorded in plants inoculated with *Glomus mosseae* but however, significant per cent of root colonization, root and shoot dry weight and number of rootlet initiation was recorded when the plants of *Sauropus androgynous* L. treated with 6000mg IBA per liter. The optimum plant growth, per cent of root colonization, number of rootlets, total root and shoot dry weight and phosphorous and nitrogen contents of shoot was demonstrated among the plants which receive 6000mg/L IBA with *Glomus mosseae* inoculation.

The inoculation of AM fungi alone may not most favorable in improving root initiation in the experimental plant *Sauropus androgynous* L. but however, AM fungi with 6000mg/L IBA with *Glomus mosseae* very significantly improved in biomass production, root initiation (Figure 31.1). *Sauropus androgynous* L. stem cuttings treated with perlite + sand, AM fungi *Glomus mosseae* alone and treated with plant hormone IBA in different levels and plants treated with IBA with AM fungi *Glomus mosseae,* mainly to improve rooting. The results shown that *Glomus mosseae* influenced root initiation of stem cuttings never the less increased root numbers and root hairs were not significant. The increased quality of rooting initiation with AM fungi *Glomus mosseae* inoculation brought significantly increased root number. This observation is strongly supporting the earlier workers (Powell and Bagyaraj, 1984; Verkade *et al.,* 1988 and Lakshman and Hosamani, 2005). During vegetative propagation the number of roots initiated influences the length of the production and quality of the roots at 90 days. This clearly seen in (Figure 31.1), compared to control and perlite and sand treated plants. In the present work AMF inoculation with out hormone treatment may not significantly increase the root number but 6000 mg/l IBA with *Glomus mossease* size of the root and functional root lets had influenced greatly compare to unionculated (Control) plants. Similar findings was documented by (Douds *et al.,* 1995; Parlada *et al.,* 1999;

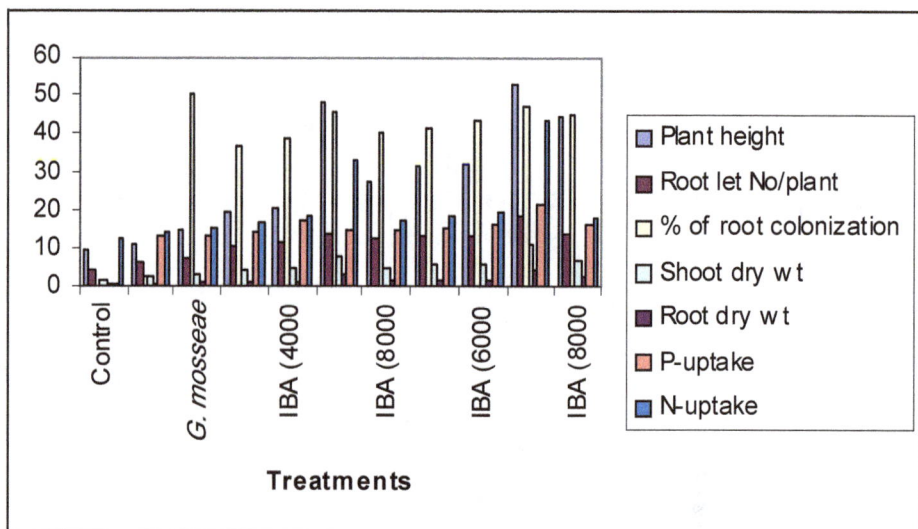

Figure 31.1: Effect of *G. mosseae* and different level of of IBA treatment in root initiation, plant growth, Phosphoros and Nitrogen content of shoot of the *Sauropus androgynous* L. for 90 days.

Scagel, 2001). Application of hormone to stem cuttings of *Sauropus androgynous* L. increased growth root and shoot biomass production, phosphorus uptake in shoot, this finding is parallel with (Chavez, 1999; Bhethalenfalvay and Linderman, 1992 and Wilson *et al.,* 1997). In the present investigation although AMF inoculum did not played role in increasing rooting but per cent of colonization significantly increased. This may help stem cutting that is batter able to with stand the stress of transplanting and increase growth during later stages of plant development (Hartman *et al.,* 1997; and Lakshman, 2009). The combination of IBA with AMF generally produced batter percentage of roots and phosphorus and nitrogen content in shoots in experimental plants. This will greatly reduce transplant morality and increase productivity of shoot and its organs (Linderman, 1994; Smith and Read, 1997).

Conclusion

Sauropus androgynous L. is a important multivitamin plant. Its rooting can be improved by treating with IBA 6000 mg/l for the initiation of early rooting. Significant colonization with increased biomass production, increased root number and root hairs improvement can be more important with *Glomus mosseae* inoculation and IBA 6000 mg/l treatment. Therefore hormone can increase root initiation with potentially increased number of roots with AM fungal inoculation of *Sauropus androgynous* L.

References

Bethlenfalvay, G.J. and Linderman, R.G., 1992. *Mycorrhizae in Sustainable Agriculture.* American Society of Agronomy Special Publication, 54. Madison, WI, 54: 124.

Chavez, M.C.G. and Cerrato, R.F., 1990. Effect of vesicular arbuscular mycorrhizae on tissue culture derived plants of strawberry. *Hort. Sci.,* 25: 903–905.

Curtis, D.L., Ranney, T.G., Blazich, F.A. and Whitman, E.P., 1996. Rooting and subsequent overwinter survival of stem cuttings of *Stewartia ovata. J. Environ. Hort.,* 14: 163–166.

Dirr, M.A. and Heuser, C.W. 1987. *The Reference Manual of Woody Plant Propagation.* Varsity Press, Athens, GA.

Douds, D.D., Bécard,, Pfeffer, P.E., Doner, L.W., Dymant, T.J. and Kayer, W.M., 1995. Effect of vesicular-arbuscular mycorrhizal fungi on rooting of *Sciadopitys verticillata* Sieb and Zucc. Cuttings. *HortSci.,* 30: 133–134.

Drew, J.J., Dirr, M.A. and Armitage, A.M., 1993. Effects of fertilizer and night interruption on overwinter survival of rooted cuttings of Quercus L. *J. Environ. Hort.,* 11:97–101.

Flemer, W., 1982. Propagating shade trees by cuttings and grafts. *Proc. Intern. Plant Prop. Soc.,* 32: 569–579.

Fordham, A.J., 1982. Stewartia: Propagation data for ten taxa. *Proc. Intern. Plant Prop. Soc.,* 32: 476–481.

Hartmann, H.T., Kester, D.E., Davies, F.T. and Geneve, R.L., 1997. *Plant Propagation: Principles and Practices,* 6th edn. Prentice Hall, N.J., 770 pp.

Hartmann, H.T., Kester, D.E., Davies, F.T. Jr. and Geneve, R.T., 2002. *Plant Propagation: Principles and Practices*, 6th Edn. Prentice Hall, Upper Saddle River, NJ.

Haynes, J.G., 1999. Improving vegetative propagation techniques and establishment practices for Stewartia koreana nakai rehder, Stewartia pseudocamellia Maximowicz, and Cornus Canadensis L, Univ. of Maine, Orono. *MS Thesis.*

Lakshman, H.C., 2009. Selection of suitable AM fungus to *Artocarpus heterophyllus* Lam. A fruit/timber for an ecofriendly nursery. M.D. Publisher, New Delhi, pp. 62–73.

Lakshman, H.C., and Hosamani, P.A., 2004. Effect of VAM and *Rhizobium* on *Erythrina indica* (Lam) Indian Coaral Tree under three salinity level. *J. of Current Science*, 5(1): 97–103.

Maynard, B.K., Sun, W.Q. and Bassuk, N., 1990 Encouraging bud break in newly-rooted soft wood cuttings. *Proc. Intl. Plant Prop. Soc.*, 40: 597–602.

Parladé, J.Pera, J., Alvarez, I.F., Bouchard, D., Genere, B. and Tacon, F., 1999. Effect of inoculation and substrate disinfection method on rooting and ectomycorrhiza formation of Douglas fir cuttings. *Ann. For. Sci.,* 56:35–40.

Perkins, A. and Bassuk, N., 1995. The effect of growth regulators and overwinter survival of rooted cuttings. *Proc. Intern. Plant Prop. Soc.,* 45: 450–458.

Powell, C.L. and Bagyaraj, D.J. 1984. Effect of mycorrhizal inoculation on the nursery production of blueberry cuttings: A note. *N.Z. J. Agric. Res.,* 27: 467–471.

Samartin, A., Vieitez, A.M. and Vietez, E., 1986. Rooting of tissue cultured camellias. *J. Hort. Sci.,* 61:113–120.

Smith, S.E. and Read, D.J., 1997. *Mycorrhizal Symbiosis*, 2nd edn. Academic Press, San Diego, CA, 606 pp.

Sun, W.Q. and Bassuk, N.L. 1993. Auxin induced ethylene synthesis during rooting and inhibition of budbreak of 'Royalty' rose cuttings. *J. Amer. Soc. Hort. Sci.,* 118: 638–643.

Verkade, S.D. and Hamilton, D.F., 1987. Effect of endomycorrhizal inoculum on root initiation and development of *Viburnum dentatum* L. cuttings. *J. Environ. Hort.,* 5: 80–81.

Verkade, S.D., Elson, L.C. and Hamilton, D.F., 1988. Effect of endomycorrhizal inoculation during propagation on growth following transplanting of *Cornus sericea* cuttings and seedlings. *Acta Hort.,* 27: 248–250.

Wilson, D., Nayar, N.K. and Sivaprasad, P., 1997. Hardening and *ex vitro* establishment of rose plantlets. *J. Trop. Agric.,* 37: 5–9.

2013, Perspectives in Plant Biodiversity   *Pages* **247–253**
Editor: **Dr. K. Muthuchelian,** *Vice Chancellor, Periyar University, Salem*
Published by: **Daya Publishing House, NEW DELHI**

Chapter 32

Diversity of AM Fungi on *Madhuca indica* G.Mel.: A Threatening Medicinal Plant

R.F. Inchal[1] and H.C. Lakshman[2]
[1]S.K. Arts College and H.S.K. Science Institute, Hubli
[2]P.G. Department of Botany (Microbiology Lab),
Karnatak University, Dharwad – 580 003

Introduction

Arbuscular mycorrhizal fungi are geographically ubiquitous and occur over a broad ecological range. About 85 per cent terrestrial plants colonized with arbuscular mycorrhizal fungi. The beneficial role of arbuscular mycorrhizal fungi in plant nutrition is well established. Knowledge and conservation of diversity of symbiotic arbuscular mycorrhizal fungi is of paramount importance for their efficient use in medicinal plants *i.e. Madhuca indica.* These fungi are the potential factor determining the diversity in ecosystems and are capable of modifying the structure and function of plant communities in a complex and unpredictable manner (Allen,1991; Read,1992; Lakshman and Geeta Patil,2004). A few earlier attempts demonstrated the association of AM fungi with medicinal plants (Taber and Trappe,1982; Lakshman and Raghavendra,1990). However, very less work has so for been directed on diversity of AM Fungi on *Madhuca indica.* Hence, the present was undertaken to evaluate the occurrence of AM fungi in threatened medicinal plant *Madhuca indica* from areas of Dharwad district.

Materials and Methods

Field survey has been undertaken in three different seasons (summer, winter and monsoon) by collecting rhizospheric soil samples, root samples of *Madhuca indica* in different localities of Dharwad district. Physioco chemical characteristics of soil was done according to (Jackson, 1973). Root samples were subjected to wet sieving and decanting technique following the method of (Gerdemann and Nicoloson, 1963). Per cent root colonization was carried out according to (Nicoloson, 1955). The AMF spores were identified by using manual (Schenck and Perez, 1990). Isolated indigenous AM fungi (*Glomus fasciculatum*) multiplied with host plant zea mays by using sterilized soil in separate earthen pot. This soil based inoculum was used for inoculation to the experimental plants.

Establishment of Test Plants and Green House Experiments

The studies were conducted under green house conditions with temperature ranges from 28–31° C. Arbuscular mycorrhizal fungal treatments were given on a layer below the germinated seedlings to the pot soil. Control pots were maintained without mycorrhiza inoculation. 15 g inoculum/10 kg of soil (*Glomus fasciculatum*) were given to each pot. The experiments were completely randomized with three replication per treatment of the plant. All the pots are maintained under green house condition. To maintain moisture, pots were watered on every alternate day 20 ml Hogland solution was given to all the pots once in 10 days.

Results

Surveys were conducted to collect roots and soil samples of *Madhuca indica* from various localities consists of agricultural land adjacent to forest regions. Physico chemical characteristics of soil in 10 places given in (Table 32.1). Most of the examined soil samples revealed varied amount of spore population and root colonization in natural conditions in different seasons. Highest per cent of colonization was observed in monsoon. On contrary lowest colonization was observed in winter but spore population was most significant during summer. In most of the soil samples organic matter does not alter the spore number. Similarly seasonal fluctuation of spore population was observed in three different seasons (Figure 32.1).

Table 32.1: Physico-chemical characteristics of selected places *Madhuca indica* plants.

Locality	Soil Type	pH	OM	EC	N	P	K
Kamalapur	Black cotton	6.5	2.52	2.89	5.90	0.67	0.72
Ramapur	Black cotton	6.7	2.49	3.12	5.88	0.68	0.75
Varur	Black cotton	6.5	2.39	2.87	6.02	0.82	0.81
Nigadi	Sandy loam	5.9	2.42	2.69	5.93	0.68	0.76
Adargunchi	Red laterite	5.8	2.18	3.40	6.07	0.93	0.84
Narendra	Red laterite	6.2	3.19	2.18	6.12	0.70	0.92
Mansur	Red loam	5.9	2.92	1.76	5.22	0.69	0.88
Mugud	Red gloomy	5.6	3.21	4.23	5.87	0.67	0.86
Mummigatti	Red laterite	5.7	3.64	2.83	5.79	0.65	0.95
Byahatti	Black cotton	5.6	3.44	3.12	5.47	0.69	0.98

Number of AM fungal spores were recovered in 10 places of Dharwad district. Amongest the recovered spores *Glomus* species were dominate in Mugud and Mummigatti were as *Scelerocystis* species recovered in Byahatti area. Lest number of *Aclospora* species were recovered in Kamlapur but moderate spores recovered in Varur and Mansur area.

Rhizosophere soil samples of *Madhuca indica* were screened for dominant native AM fungal spore isolation. *Glomus fasciculatum* spores were found to be the predominate AM fungus associated with *Madhuca indica*. They were therefore, used for pot experiments. The green house pot experiments results revealed that mycorrhiza inoculated plants showed significant increase in plant height, root and shoot biomass production per cent of root colonization, spore number and P uptake in shoots. The lateral feeder roots of potted plants inoculated with AM fungus (*Glomus fasciculatum)* were heavily colonized by mycelium, arbuscules and vesicles. The AM fungus (*Glomus fasciculatum)* treated plants appeared to be more vigorous than the non treated plants (Tables 32.2–32.4).

Discussion

Arbuscular mycorrhizal fungi hold greater potential for use as inoculants to improve plant production system. Most of the AM fungi resulted in significant increase in plant height plant biomass, stem diameter, number of leaves and P content of *Madhuca indica*. Host preference among AM fungi has been reported by earlier workers (Gracy, L. Sailo, Bagyaraj, D.J, 2005). Hence, the need for inoculating different mycotropic plants has been stressed (Jeffries, P. 1987; Bagyaraj. D.J. 2007). Improved plant height, because of AM fungal inoculation has been reported in other medicinal plants like coleus and Andrographis. The plant biomass is an important parameter for selecting fungus in its symbiotic efficiency. (Shoot + Root) was enhanced due to *Glomus fasciculatum* inoculation compared with that of uninoculated plants. Increased plant growth due to AM fungal inoculation is manly through improved uptake of diffusion limited nutrients such as P. AM fungi improving plant biomass were also good in increasing the P content of the host significantly highest being in plants inoculated

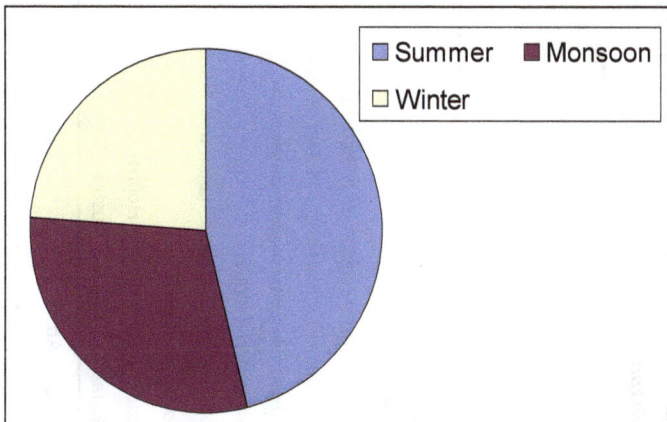

Figure 32.1: Seasonal changes of AMF spore number in rhizospheric soil *Madhuca indica*.

Table 32.2: AM fungi recovered from rhizosphere soil of different localities of Dharwad district in Karnataka.

AM Fungi Species	Rhizospheric Soils of Madhuca indica from Selected Areas									
	1	2	3	4	5	6	7	8	9	10
Acaulospora species										
Acaulospora bireticulat Rothwel and Trappe.	–	–	–	–	+	–	–	–	–	–
A. nicolsonii Walker, Reed and Sanders.	–	–	+	+	–	–	–	+	–	–
Gigaspora species										
G. margarita Becker and Hall.	–	–	+	–	+	–	–	+	+	+
G.gigantea (Nicolson and Gerdemann) Gerdemann and Trappe.	–	–	–	+	–	+	–	–	–	–
G. decipiens Hall and Abbott.	–	–	+	–	+	–	–	+	+	–
Glomus species										
Glomus mossae (Nicolson and Gerdemann) Gerdmann and Trappe.	+	+	–	+	–	+	–	–	–	+
G. fasciculatum Gerdemann and Trappe emend Walker and Koske.	–	–	+	+	–	+	–	–	+	+
G. ambisporum Schenck and Smith	–	–	–	+	–	–	+	+	+	+
G. constrictum Trappe.	–	–	–	–	–	+	–	–	–	–
Glomus reticulatum Bhattacharjee and Mukerji	–	–	+	–	–	–	–	+	–	–
Sclerocystis species										
S. dussii (Patouillard) Von Hohnel.	–	+	–	–	+	–	+	–	–	–
S. pakistanica Iqbal and Bushra.	–	–	+	–	–	–	+	–	+	–

+: Present; –: Absent.

1: Kamalapur; 2: Ramapur; 3: Varur; 4: Nigadi; 5: Adargunalli; 6: Naredra; 7: Mansur; 8: Mugada; 9: Mummigatti; 10: Byahatti.

with *Glomus fasciculatum*. Selected efficient fungi enhancing plant biomass and P uptake has been reported in other plants by several workers (Lakshman, 2009). Such higher P content in AM fungal inoculated plants attributed to higher influx of P in to the plant system which explores the soil volume beyond P depletion zone. The enhancement in growth and nutritional status was also related to mycorrhizal root colonization and spore numbers in the root zone soil. This upholds the observation made by earlier works in other plants (Gracy *et al.*, 2006). Conservation and efficient utilization of AM fungal biodiversity are of crucial importance for sustainable plant production system.

Table 32.3: *Glomus fasciculatum* on growth response, plant dry weight, per cent of colonization on *Maduca indica*.

Plant Age		Plant Height	Shoot Weight		Root Weight		Root Shoot Ratio (g)	Percent-age of coloni-zation	Spore per 50 gm Soil
			Fresh	Dry	Fresh	Dry			
60 days	NM	6.2	4.5	1.92	1.42	0.77	0.29	0.00	0.00
	M	10.3	9.25	2.53	2.76	1.05	0.78	52.4	70.2
100 days	NM	8.4	5.3	2.16	2.7	1.1	0.38	0.00	0.00
	M	15.7	18.4	3.84	8.23	2.1	1.12	78.5	92.4
180 days	NM	12.6	8.23	2.94	3.4	1.78	0.62	0.00	0.00
	M	28.7	27.93	4.18	12.78	3.12	2.14	82.7	90.7

Table 32.4: Mycorrhizal developments in *Madhuca indica*.

Plant Age (Days)	Per cent Root Colonization	Average Number of Vesicles cm^{-1}	Spores in 50 gm/soil		
			1	2	3
30	42.5 (4.9)	12.5 (2.8)	86 (3.1)	12 (2.1)	–
60	56.8 (6.2)	18.7 (2.1)	115 (4.2)	15 (4.1)	2 (0.5)
90	49.4 (9.3)	20.6 (3.1)	118 (2.2)	18 (2.3)	2 (1.6)
120	36.8 (7.2)	21.3 (2.3)	220 (4.3)	20 (4.6)	3 (0.6)

1: *Glomus* spp.; 2: *Gigaspora* spp.; 3: *Sclerocystis* spp.

Per cent root segments examined spore number in parameters denotes SD values (n=5)

Field study have shown a great deal that is know about vesicular arbuscular mycorrhizal fungi their population and communities. The *Glomus* species are more dominated compared to remaining four species such as *Aculospora, Gigaspora, Sclerocystis, Scutelsopora* similar observations made by earlier workers (Lakshman, 1996; Mukuraje *et al.*,2000, Inchal., R.F, 2002). To provide a complete picture field studies must be integrated with laboratory investigations which attempt to elucidate general principal from controlled environment (Inchal, R.F and Lakshman, 2006). To achieve this series of techniques is being applied including more advanced developments in microscopic techniques. Identification of AM fungi in conjunction with ecological information is of primary importance to interpret their distributions in natural and disturbed sites. This would require the steps such as accumulation collection of isolates of AM fungi from different geographic regions.

Conclusion

Madhuca indica is a most important threatened medicinal plant the present investigation clearly brought out that the mycorrhizal inoculation (*Glomus fasciculatum*) benefits *Madhuca indica* plants at nursery stage so that this fungus aids in improving water and nutrients uptake. These seedlings in the nursery can be transplanted in to denuded area in reforestation programme. Keeping in views that the production of threatened medicinal plant *Madhuca indica* seedlings at nursery level are most essential.

References

Allen, M.F., Smith, W.K., Moore, T.S. and Christensen, M., 1981 Comparative water relations and photosynthesis of mycorrhizal and nonmycorrhizal *Bouteloua gracilis* H. B. K. Lag ex stead. *New Phytol.,* 88: 683–693.

Gerdemann, J.W. and Nicolson, T.H., 1963. Spores of mycorrhizal endogene Species extracted from soil by wet sieving and decanting Transactions of the British Mycological Society, 46: 235–244.

Gracy, L. Sailo. and Bagyaraj, D.J., 2005. Influence of different AMF on growth, nutrition and Forskholin content of *Coleus forskolii. Mycol. Res.,* 109: 795–798.

Inchal, R.F. and Lakshman, H.C., 2006. The indigenous medicinal plants and their use. *Vignana,* 2: 97–101.

Jackson, M.L., 1973. *Soil Chemical Analysis.* Prentice Hall (India) Pvt, Ltd., New Delhi, 32: 34–36.

Lakshman, H.C., 2009. AM Fungi a promising bioinoculant for sustainable plant growth. In: *Proceeding of ICAR Conference,* April 14–16, NAL, Bangalore, pp. 117–121.

Lakshman, H.C. and Patil, G.B., 2004. Mycorrhizal (VAMF) dependency of sixty tree species. *Asian J. of Microbiology Biotechnology Env. Science,* 1(10): 147–153.

Lakshman, H.C., 1996. VA-Mycorrhizal studies in some Economicaly important Timber tree species. *Thesis,* pp. 247.

Lakshman, H.C. and Raghvendra, S., 1990. Occurrence of VA mycorrhizal fungi in medicinal plants. *Mycorrhizal Proceedings,* Bangalore, p. 21–23.

Mosse, B., 1973. Techniques for studying mycorrhiza in tropical countries. *Mycorrhiza News,* 9(2): 7–13.

Mukerji, K.G., Sabharwal, A., Kochar, B. and Ardey, J., 2000. Vesicular Arbuscular Mycorrhizae, concepts and advances. In: *Progress in Microbial Ecology*, (Eds.) K.G. Mukerji, V.P. Agnihotri and R.P. Singh. Printed House (India), Lucknow, pp. 489–525.

Phillips, J.M. and Hayman, D.S., 1970. Improved procedures for clearing roots and staining parasitic and vesicular arbuscular mycorrhizal fungi for rapid assessment of infection. *Transaction of the British Mycological Society,* 55: 158–161.

Schenck, N.C. and Perez, Y., 1990. *Manual for the Identification of VA Mycorrhizal Fungi,* 3rd Edn. Synergistic Publications, Gainesville, Florida, USA.

2013, Perspectives in Plant Biodiversity   *Pages* **254–260**
Editor: **Dr. K. Muthuchelian,** *Vice Chancellor, Periyar University, Salem*
Published by: **Daya Publishing House, NEW DELHI**

Chapter 33

Screening of Efficient AM Fungus to Improve the Growth and Nutrient Uptake of *Achyranthus aspera* L.: A Member of Amaranthaceae

C. Shwetha, Madgaonkar and H.C. Lakshman*
Post Graduate Studies in Botany, Microbiology Laboratory,
Karnatak University Dharwad – 580 003

Introduction

The family Amaranthaceae contains several economically important species used as vegetables or pot herbs in various parts of the world. The members of the family amaranthaceae often considered to be nonmycorrhizal (Peterson *et al.,* 1985; Mohankumar and Mahadevan, 1987) and a large number of species belonging to this family is not colonized by mycorrhiza (Meada, 1954; Gerdemann, 1968; Harley and Harley, 1987). However this may not always be true as there are infrequent reports of occurrence of mycorrhiza in amaranthaceae (Lakshman *et al.,* 2001). *Achyranthus aspera* L., a member of amaranthaceae is one of the most important medicinal plants with various medicinal properties. The whole plant, root and seeds of *Achyranthus aspera* L. act as a herbal agent used to control fever and inflammation, used in the treatment of diarrhoea, gastric disorders. The herbal agent contains chemicals such

---

* Corresponding Author: E-mail: shweta.madgaonkar@gmail.com as a saponin and achyranthine. Repots of *Achyranthus aspera* L. used in the treatment of snakebites. It also helps with sprains, constipation, malaria and bronchitis and used as a source of alkali for dyeing.

It has long been thought that plants of amaranthaceae are nonmycorrhizal but recent research suggests that, they may be capable of forming AM association (Neeraj *et al.,* 1991). Arbuscular mycorrhizal (AM) fungi are key players in soil structure. Soil structure and aggregation is tremendously important in the health of the vegetation because it facilitates water infiltration, aeration, root growth and movement of nutrients. AM fungi are known to improve the nutritional status, growth and development of plants. The contradictory in the statement of mycorrhiza in this family prompted us to evaluate the effective mycorrhizal status of *Achyranthus aspera* L. in selected places of Dharwad district, Karnataka. The objective of present study was to screen an efficient AM fungi species on *Achyranthus aspera* L. growth, biomass and nutrient uptake.

Materials and Methods

Root and rhizospheric soil sample of *Achyranthus aspera* L. were collected. The survey is carried out in Dharwad district, Karnataka. Geographically situated in between 14° 15' and 15° 50' North longitude and 74° 48' and 76° 20' East latitude. The roots were carefully washed under running water, cut into 1cm segment and autoclaved in 10 per cent KOH aqueous solution then cleared in distilled water, neutralized with 2 per cent HCl and stained in 0.05 per cent tryphan blue in lacto phenol (Phillips and Haymann, 1970). The percent of root colonization was determined based on the number of root segments colonized by AM fungi.

Spore population in the rhizospheric soil was counted following the wet sieving and decanting method (Gerdemann and Nicolson, 1963). Spores were mounted in polyvinyl alcohol lacto phenol. The species level identification of different AM fungi was done following the keys produced by Trappe (1982) and Schenck and Perez (1990). A green house experiment was conducted in the Department of Botany, Karnatak University at Dharwad. Six species of AM fungi *i.e. Gigaspora margarita, Glomus macrocarpum, Glomus bagyarajii, Glomus mosseae, Acaulospora laevis* and *Sclerocystis dussii* were dominant in studied area. Therefore, these species were selected for efficacy experiments. Isolated spores were multiplied using *Sorghum vulgare* as a host plants, maintained in earthen pots measuring 15 cm height: 30 cm diameter, pots were filled with autoclaved soil in the ratio of 3:1 (3 parts soil: 1 part sand). After 60 days *Sorghum vulgare* plant was cut at ground level, the roots were chopped into 1cm pieces and mixed with the rhizosperic soil of host plant. This soil based inoculums were used for inoculation. 15 gram of air dried of different AM fungal inoculums were given to each pot of *Achyranthus aspera* L. as thin layer 2cm below the soil surface. Plants were watered on alternate days. Six treatments with noninoculated (control plant) were carried out as following (1) Control, (2) *Gigaspora margarita,* (3) *Acaulospora laevis,* (4) *Sclerocystis dussii,* (5) *Glomus bagyarajii,* (6) *Glomus mossese* and (7) *Glomus macrocarpum* (Figure 33.1). Hogland plant nutrient solution was given to the seedling at the intervals of 15 days. The plants were harvested, after 30, 60 and 90 days intervals. The plant growth parameter such as

Figure 33.1: Effect of different AMF strains on growth of *Achyranthus aspera* L.

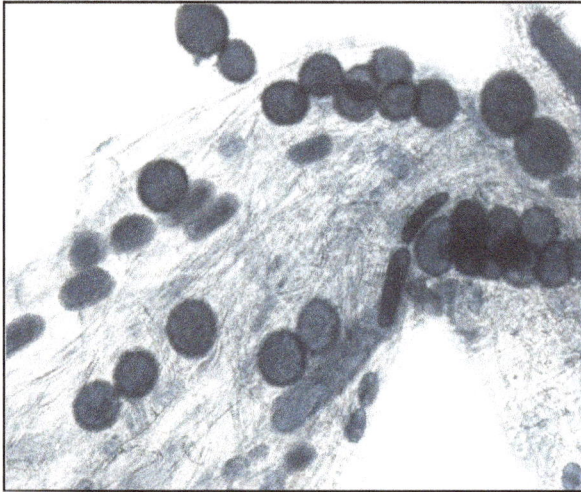

Figure 33.2: Macerated root of *Achyranthus aspera* L. showing numerous vesicles.

plant height, dry wet of shoot, per cent of colonization, spore number, P and N content of shoot were recorded.

The root and shoot sample were air dried in hot air oven at 60°C till it attained constant weight and the dry weight of shoot were recorded. The per cent of root colonization was determined by Girdline intersect method (Giovanetti and Mosse, 1980) and number of spore were counted as mentioned method. The shoot phosphorus and nitrogen content was determined by Vanadomolybdate yellow colour and micro kjeldahl method (Jackson, 1973).

Table 33.1: Showing the effect of different AMF inoculation on plant height, dry weight of shoot and root, per cent of root colonization, spore number, P and N content of shoot of *Achyranthus aspera* L. for different intervals.

Treatment	Plant height (cm)	Dry Weight of Shoot (g/plant)	Dry Weight of Root (g/plant)	Per cent of Root coloni-zation (per cent)	Spore Number	P Uptake mg/plant	N Uptake mg/plant
30 days							
Control	18.3	2.86	1.8	0.00	0.00	0.02	0.04
Gigaspora margarita	25.1	4.9	2.7	29.36	27	0.07	0.42
Glomus macrocarpum	41.7	10.66	6.07	51.2	55	0.19	1.13
Glomus bagyarajii	30.1	9.01	5.02	41.03	38	0.12	1.02
Glpmus mosseae	33	9.89	5.41	45.7	42	0.14	1.09
Acalospora laevis	27.68	5	3.21	35.41	30	0.09	0.57
Sclerocystis dussii	28.7	7.32	4.32	38.12.	32	0.10	0.87
60 days							
Control	28.5	3.1	2.8	0.00	0.00	0.02	0.04
Gigaspora margarita	36	5.6	4.2	58.23	40	0.09	0.47
Glomus macrocarpum	67	13	8.01	79.2	78	0.25	1.23
Glomus bagyarajii	55	09.6	6.4	69.5	58	0.19	1.07
Glpmus mosseae	58.10	10.11	7.15	71.67	62	0.21	1.16
Acalospora laevis	39	6.8	4.5	61.2	42	0.11	0.61
Sclerocystis dussii	41	7.5	4.8	63.17	47	0.15	0.86
90 days							
Control	31	5.02	3.4	0.00	0.00	0.06	0.05
Gigaspora margarita	47	7.42	4.8	59.41	60	0.12	0.57
Glomus macrocarpum	86	16.3	10.2	91.26	98	0.28	1.47
Glomus bagyarajii	71.15	11.26	7.1	68.19	74	0.20	1.05
Glpmus mosseae	78	13.71	8.3	76.27	85	.0.25	1.20
Acalospora laevis	53.2	8.9	6.1	60.08	62	0.14	0.73
Sclerocystis dussii	60	10.33	6.9	64.39	68	0.17	0.97

Results and Discussion

Most of the Amaranthaceae members are growing in waste lands as a weed and they are also cultivated as a green vegetable. Green house experiments on these members are very scanty. *Achyranthus aspera* L. plants of its rhizosphere and root samples recovered in the agricultural soil. The soil has low P content. There was significant increase in the plant height when inoculated with AM fungi compare to control plant (Figure 33.1). Screening of 6 different AM fungi showed (Table 33.1) varied effect on *Achyranthus aspera* L. Response of experimental plant to different

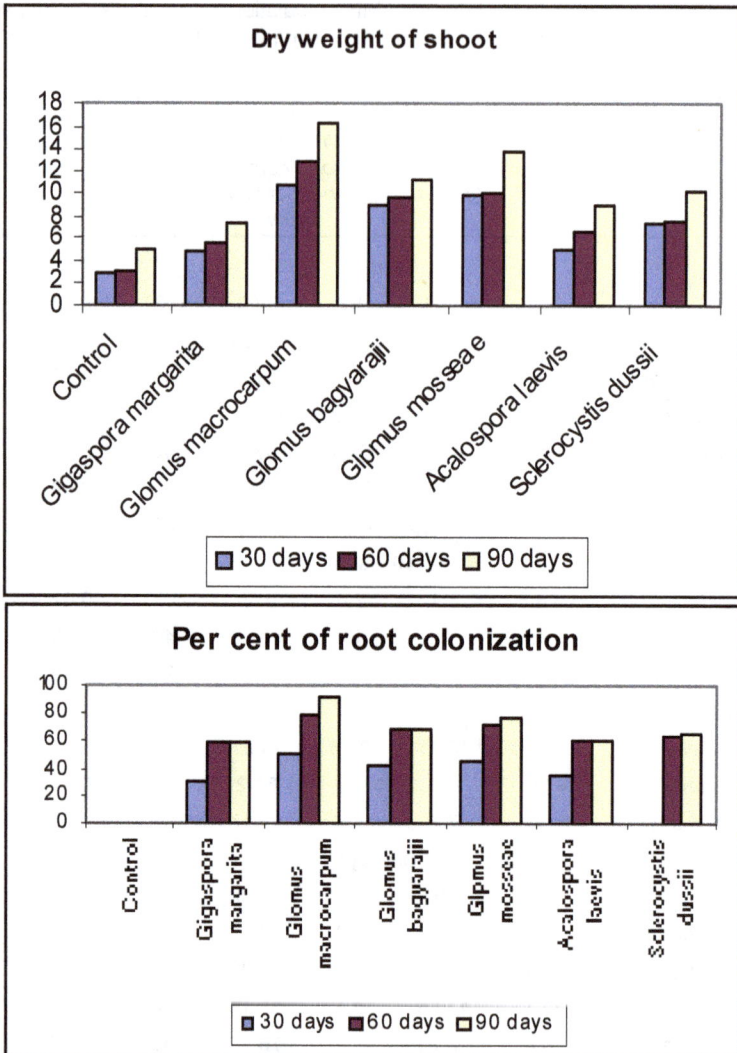

Figure 33.3: Effect of different strains of AMF on dry weight of shoot and per cent of root colonization of _Achyranthus aspera_ L.

AM fungal inoculation demonstrate varied results. Plants inoculated with _Glomus macrocarpum_ recorded significantly higher in plant height and dry weight of root and shoot compare to all other treatments. This was followed by _Glomus mossese, Glomus bagyarajii, Sclerocystis dussii, Acaulospora laevis_ and it is least in _Gigaspora margarita_.

The AMF structures in roots were illustrated in _Achyranthus aspera_ L., there was many type of structure such as intraradical hyphae, arbuscules, vesicle, and extraradical hyphae were found when plants inoculated with AMF strains. In contrast there was not any structure of AMF infected to plant root in control plant. Colonization of AMF on the root is the first indication of relationships between the two symbionts. More number of vesicles found in *Achyranthus aspera* L. (Figure 33.2) with the inoculation of *Glomus macrocarpum*. The plant shown highest root colonization, spore number, P and N content of root with *Glomus macrocarpum* followed by *Glomus mossese, Glomus bagyarajii, Sclerocystis dussii, Acaulospora laevis* and *Gigaspora margarita*. The increase in biomass could be due to enhanced P uptake. Increase in P uptake and biomass growth response because of AM fungi is well documented (Gracy Sailo and Bhagyaraj, 2005; Lakshman, 2009). The results indicate that the natural population of *Achyranthus aspera* L. is associated with AM fungi predominantly by species of *Glomus macrocarpum*.

Conclusion

Therefore, from this study giving weight age to biomass which is the economically important part of the plant, it can be concluded that an efficient indigenous AM fungi *i.e. Glomus macrocarpum* is the best mycorrhizal symbiont and more suitable biofertilizer for inoculating *Achyranthus aspera* L. in sterilized sandy loam soil.

Acknowledgements

First author is grateful to U.G.C. New Delhi for awarding meritorious scholarship for doing Ph.D. Second author is also indebted for sanctioning major research project on rare millets of North Karnataka.

References

Gerdemann, J. W., 1968 and Nicolson, T.H., 1963. Spores of mycorrhizal endogene species extracted from the soil by wet sieving and decanting. *Trans Br. Mycol. Soc.*, 46: 235–244.

Giovannetti, M. and Mosse, B., 1980. An evaluation of technologies for measuring vesicular arbuscular mycorrhizal infection in roots. *New Phytol.*, 84: 489–500.

Gracy, L. Sailo and Bagyaraj, D.J., 2005. Influence of different AMF on growth, nutrition and Forskholin content of *Coleus forskolii. Mycol. Res.*, 109: 795–798.

Harley, J.I. and Harley, F.I., 1987. A Checklist of mycorrhiza in the British flora. *The New Phytologist*, 105(2 supplements): 1–102.

Jackson, M.L., 1973. *Soil Chemical Analysis*. Prentice Hall (India) Pvt, Ltd., New Delhi, 32: 34–36.

Lakshman, H.C., 2009. AM Fungi a promising bioinoculant for sustainable plant growth. In: *Proceeding of ICAR Conference*, April 14–16, NAL, Bangalore, pp. 117–121.

Lakshman, H.C., Mulla, F.I., Inchal, R.F. and Srinivasalu, S., 2001. Prevalence of Arbuscular mycorrhizal fungal colonization in some disputed plants. *Mycorrhizal News*, 13(3): 17–21.

Meada, M.I., 1954. The meaning of mycorrhiza in regard to systematic botany. *Kamoto Journal Science,* Ser. B3: 57–84.

Mohankumar, V. and Mahadevan, A., 1987. Vesicular arbuscular mycorrhizal association in plants of Kalakad reserve forest, India. *Angew Bot.*, 61: 225–274.

Neeraj, A.S., Mathew, J. and Varma, A., 1991. Occurrence of vesicular arbuscular mycorrhizae with Amaranthaceae in soils of the Indian semi-arid region. *Biol. Fertile. Soils*, 11: 140–144.

Peterson, R.L., Ashford, A.E. and Allaway, W.G., 1985. Vesicular arbuscular mycorrhizal association of vasicular plants on Heron islands, a great barrier reef coral ray. *Aust J Bot.,* 33: 669–676.

Phillips, J.M., 1982. Mycorrhizae and productivity of arid and semi-arid range lands. In: *Advances in Food Productivity Systems for Arid and Semiarid Lands.* Academic Press, New York.

Schenck, N.C. and Perez, V., 1987. *Manual for the Identification of VA Mycorrhizal Fungi.* University of Florida, Florida.

Trappe, J.M., 1982. Mycorrhizae and productivity of arid and semiarid range lands. In: *Advances in Food Productivity Systems for Arid and Semiarid Lands.* Acadamic Press. New York.

2013, Perspectives in Plant Biodiversity   *Pages* **261–265**
Editor: **Dr. K. Muthuchelian,** *Vice Chancellor, Periyar University, Salem*
Published by: **Daya Publishing House, NEW DELHI**

Chapter 34

Antifungal Activities of Some Selected Medicinal Plants Against Red Rot Disease in Sugarcane (*Saccharum officinarum*)

N. Siva, R. Manimaran, C. Pitchairamu,
R. Muthuramalingam, K. Irulandi, K. Muralidharan,
M. Thiyagarajan, and M. Guru Moorthi
Department of Botany, PTMTM College, Kamuthi

Introduction

Red rot of sugarcane (*Saccharum officinarum*) is a fungal disease it is a serious disease for several varieties of sugarcane in the USA, India, Australia and Hawaii. It is caused by the fungus *Collectotrichum falcatum* and it is comes under the class Deueteromycetes. This disease is characterized by wringling of the canes, red-dening a of pith and red patches on leaves. Red rot attacks the standing sugarcane crop and wipes out the entire crop in the field. The pathogen converts the sucrose in the cane into glucose and alcohol by enzymatic reaction. So the amount at reconvertable sugar gets reduced upto 30 per cent. Red rot of sugarcane was finst discovered in JAVA in 1893. It was reported to be present in India by Butter in 1906, and named red rot disease.

In India it is prevalent in Tamil Nadu, Bihar, U.P, M.P, Haryana and Punjab. In order to prevent the plant disease and to protect the crop plants against pathogen chemical control methods were in practice, agriculturist adopts the use at synthetic chemical for controlling crop plant diseases, the chemical substances that kill all pests are commonly referred as, pesticides. Which include fungicide, bactericide, nematicide, insecticide etc., these pesticides kill the pathogen by inhibiting spore germination, growth and multiplication. The adverse use of chemical substances produce many side effects and several disadvantages, such as effecting soil micro organisms, contributing pollution, developing immunity in pests, further these chemical are not easily degradable.

In view of the high cost of chemical pesticides and their hazardous consequences, use of biodegradaple materials like fresh plant extract gained importance during last three decades for plant disease control (Fowcett and Spenser, 1970; Mitra *et al.*, 1984; Grainge and Ahamed, 1988; Jespers and Ward, 1993).

Objectives

The available literature is pertaining to the use of plant extracts in the control of various fungal pathogens but there is not much information regarding the control of *Collectorichum falcatum* causing red rot disease of sugarcane using plant extracts, keeping this view, the present study is conducted in order to fined out the effect select plant extracts on *Collectotrichum falcatum* isolated from infected leaves of sugarcane.

Materials and Methods

The present study include the following aspects

1. Isolation, identification and maintenance of fungal pathogen.
2. *In vitro* effect of plant extract on the pathogen
3. Effect of different concentration of selected plant extracts on the pathogen.

Isolation of Pathogen

Collection of Infected Plant Materials

The plant *Saccharum officinarum*.L (sugarcane) was selected for the present study the infected leaves of sugarcane showing red rot disease were collected from Pasumpon village at Ramanathapuram district (5 km away from east at Kamuthi).

Surface Sterilization and Method of Isolation

The symptom of the disease is the red rot of plant leaves. The red rot is characterized by red dening of pith and red patches on leaves and drying of leaves. The infected leaf materials were cut into small pieces (1cm x1cm) and washed in running tap water surface sterilization was done using 0.1 per cent. Mercuric chloride for one minute. the leaf pits were repeatedly washed in sterile distilled water under aseptic condition.

Culture Medium

Potato Dextrose Agar medium (PDA) was used to culture the pathogen from the infected leaf materials.

Composition

Potato	–	200 gms
Dextrose	–	20 gms
Agar	–	15 gms
Distilled water	–	1000 ml
pH	–	5.5

Preparation of Medium

Potato was cut into small pieces and boiled in 300ml distilled water for 20 min in hot micro oven and extract was taken the agar was melted in 500ml of water the potato extract was mixed with melted agar. The known amount of dextrose was added and the medium was made up to one litre. The medium (18ml) was distributed in boiling tubes for the sake of convenience. The petridishes and medium were sterilized at 15 psi are 120°C for 20 minutes.

Plating

Potato dextrose agar medium was distributed in petriplates of 9 cm diameter under aseptic condition in order to arrest the bacterial growth 1 per cent streptomycin (5 ml/liter Medium) was added to the sterilized and cooled medium (45°C) before pouring the medium in petri plates. After solidification surface sterilized infected leaf bits (4 plates) were placed on the agar medium. The plates were incubated for 5 days at room temperature.

Identification of Pathogen

After 5 days, the plates were examined for the growth at the fungus from infected leaf materials. The fungus was mounted on glass slide using cotton blue in lacto phenol and observed under the microscope.

Maintenance of Fungal Culture

The pathogen that appeared from the infected host leaf was sub cultured on potato dextrose agar medium in Petri plates in order to get the pure culture of the pathogen. Then the culture was maintained on potato dextrose agar slants and stored in refrigerator for future use.

Results and Discussion

In vitro Effect of Plant Extracts on the Pathogen

Selection of Plants

The antifungal activity was carried out on the isolated pathogen using extracts of different plants belonging to various families. Three different solvents (water, ethanol and acetone) were used for taking extracts. The common and locally available plants having medicinal properties were selected for the present study and those plants were collected in and around Kamuthi.

Totally 20 plants extract, were analyzed using *in vitro* study. Among 20 plants 19 were prepared from fresh leaves and one from succulent stem. The details regarding the name of the plants, family vernacular name and parts used are enumerated in Table 34.1.

Table 34.1: Effect of plant extract in different solvents at 50 per cent concentration on growth of *Collectotrichum.*

Sl.No.	Plant Species	Family	Percentage of Inhibition		
			Water Extract	Ethanol Extract	Acetone Extract
1.	*Acalypha indica* L	Euphorbiaceae	82	86	80
2.	*Adhatoda vasica* nees	Acanthaceae	100	100	100
3.	*Aloe vera* (L)Burm.f	Liliaceae	79	82	92
4.	*Antrograpis paniculata* nees	Acanthaceae	80	84	79
5.	*Azadirachta indica* Adr.juss	Meliaceae	96	96	98
6.	*Azima tetracantha* Lam	Salvadoraceae	97	84	84
7.	*Cadaba indica* Lam	Capparidaceae	84	97	81
8.	*Cissus quad rangularis* L	Vitaceae	62	82	72
9.	*Coleus aromaticus* Benth	Lamiaceae	81	94	84
10.	*Crateva religiosa* L	Capparidaceae	84	92	82
11.	*Datura metal* L	Solanaceae	60	74	73
12.	*Jatropha curcas* L	Euphorbiaceae	100	100	100
13.	*Leucas aspera* spreng	Lamiaceae	64	74	83
14.	*Mimusops elengi* L	Sapotaceae	78	82	78
15.	*Notonia grandiflora* DC	Asteraceae	84	82	78
16.	*Ocimum sanctum* L	Lamiceae	96	98	96
17.	*Ricinus communis* L	Euphorbiaceae	84	72	86
18.	*Santalum album* L	Santalaceae	92	94	89
19.	*Sapindus emarginatus* vahl	Sapindaceae	100	100	100
20.	*Vitex negundo* L.	Verbenaceae	97	96	94

In vitro Effect of Plant Extract in Different Solvents against *Collectotrichum* Species

Totally 20 plants extracts using different solvents (water, ethanol and acetone) were examined for anti fungal activity against *Collectotrichum* species. This *in vitro* study revealed that all the plant extract of 50 per cent concentration were effective in reducing the mycelial growth. Among the 20 plant extracts in different solvents 100 per cent inhibition was noticed only in 3 plant extract namely *Adhatoda vasica*, *Jatropha curcus* and *Sapindus emarginatus.*

Conclusion

Infected leaf samples of sugarcane showing red rot disease were plated on PDA medium and the pathogen *Collectotrichum falcatum* was isolated. The antifungal effect of crude plant extracts at 20 plants species was determined *in vitro* study using water, ethanol and acetone as a solvent by teod poisoned techniques. it was found that all the plant extracts at 50 per cent concentration were effective in reducing the mycelial growth of *Collectotricheum falcatum*.

Among the 20 plant extracts in different solvent higher inhibition was noticed in 4 plant extracts namely *Adhatoda vasica, Jatropha curcas, Sapindus emarginatus,* and *Vitex negundo*. These were selected for different concentrations of 10 per cent,20 per cent,30 per cent and 40 per cent. Among them adhatoda vasica of 40 per cent alone recorded 100 per cent inhibition and remaining there plants produced almost similar inhibitory effect.at the low concentration at 10 per cent vitex negundo had more inhibitory effect (82 per cent) while jatropha curcas extracts showed very low inhibition (25 per cent). There was not much differences in the inhibition between the extract of *Adhatoda vasica, Jatropha curcas* and *Sapindus emaginatus*.

2013, Perspectives in Plant Biodiversity   *Pages* **266–279**
Editor: **Dr. K. Muthuchelian,** *Vice Chancellor, Periyar University, Salem*
Published by: **Daya Publishing House, NEW DELHI**

Chapter 35

Short-Term Dynamics of Two Tropical Dry Evergreen Forests of South India

*R. Venkateswaran[1] *, N. Parthasarathy[2] and K. Muthuchelian[1]*

[1]Centre for Biodiversity and Forest Studies,
School of Energy Environment and Natural Resources,
Madurai Kamaraj University, Madurai – 625 021
[2]Department of Ecology and Environmental Sciences,
School of Life Sciences, Pondicherry University, Puducherry – 605 014

ABSTRACT

To understand short-term changes in woody species composition and patterns of species recruitment, mortality and growth rates, all trees and lianas ≥10 cm gbh (girth at breast height) were monitored in two 1-ha (100 m × 100 m) permanent plots of tropical dry evergreen forest located at Kuzhanthaikuppam (KK) and Thirumanikkuzhi (TM) on the Coromandel coast of south India over 3 years (1999 and 2002). The results indicated species richness, density and basal area of the total woody species increased at both sites. The study revealed that the smaller stems (10-30 cm girth at breast height) have greater mortality of trees and there was no reduction in species number in either forests. Trees experienced more reduction in basal area than lianas. The mean annual diameter growth rate was more or less equal for small trees and lianas. The mean annual diameter growth was high for the upperstory species. Among different modes of natural tree death, the 'standing dead' category was predominant. Overall, tree species have greater mortality than lianas. As

* Corresponding Author: E-mail: drvenkeyenbio@gmail.com

discernible changes were documented within a three-year period with respect to changes in diversity, tree growth and mortality rates and this three-year period between two census appears to be an ideal time interval to understand short-term dynamics in tropical dry evergreen forests.

Keywords: *Diversity changes, Mortality, Permanent plot, Trees and lianas, Short-term dynamics, Species recruitment, Tropical forest.*

Introduction

Forest communities are dynamic and changes occur continuously at individual, species, and population levels throughout time, even though the community as whole is expected to be stable, due to a balance between growth, recruitment and mortality (Felfili 1995). Interest in tree mortality and forest dynamics has increased recently, because forest dynamics is thought to be involved in the maintenance of tree species diversity (Phillips *et al.*, 1994) and also to be related to global climate change in particular (Phillips and Gentry 1994). Most importantly, long-term forest inventory can provide information about forest characteristics, but decades are needed to measure dynamics of forest stands (Carey *et al.*, 1994). The systematic measurement of permanent plots, both at short and long term intervals is needed to understand the processes through which the changes occur at species and community levels (Felfili 1995). Hence, an attempt has been made to understand short-term diversity changes, growth, recruitment and mortality of trees and lianas in two tropical dry evergreen forests on the Coromandel coast, south India. In future, this study will facilitate in re-monitoring the studied forest sites to understand the long-term changes/dynamics of the dry evergreen forests.

Although quantitative plant diversity inventories have been undertaken in these forests (Visalakshi 1995, Parthasarathy and Karthikeyan 1997, Parthasarathy and Sethi 1997, Ramanujam and Kadamban 2001, Reddy and Parthasarathy 2003, Venkateswaran and Parthasarathy 2003), but quantitative information on tree mortality, recruitment and growth is lacking (Venkateswaran 2004, Venkateswaran, and Parthasarathy 2005). In this study we present short-term (three-year) changes in diversity, recruitment, mortality and growth rates of woody species ≥10 cm girth at breast height (gbh) in two tropical dry evergreen forests and tried to understand the changes in woody species composition of two tropical dry evergreen forests over a three-year period, differences in mortality patterns between the two life-forms; and differences in mortality, recruitment and growth rates among trees of lowerstory, middlestory and upperstory species.

Methods

Study Area

This study was conducted in two tropical dry evergreen forests, namely Kuzhanthaikuppam (KK) and Thirumanikkuzhi (TM) located on the Coromandel coast of south India. These are sacred groves or temple forests, protected by religious belief. Site KK lies at 11° 45' N lat. and 79° 38' E long. and TM at 11° 43' N lat. and 79°

41' E long and are located respectively about 26 km and 28 km south of Pondicherry (Figure 35.1). The climate data of Pondicherry (12° 03' lat. and 79° 52' E long). available for 21 years (1980-2000) reveal a mean annual temperature of 28.5° C and a mean annual rainfall of 1311 mm. The mean annual rainy days are 55.5. The mean monthly temperature ranges from 24.1° C to 32.4° C for the same period. The climate is tropical dissymmetric, with bulk rainfall received during the northeast monsoon (October to December). The period from March to May marks the summer followed by southwest monsoon up to September (Parthasarathy and Karthikeyan 1997, Venkateswaran 2004, Venkateswaran, and Parthasarathy, 2005).

The tropical dry evergreen forests on the Coromandel coast of India, which occur as patches, are short-statured (mean tree height 7m), largely three-layered, tree-dominated evergreen forests with a sparse and patchy ground flora composed largely of the succulent herb *Sansevieria roxburghiana*. They occur in climatically dry areas that experience six dry months in a year with dew as a source of precipitation for six months (September–March). Trees are mostly evergreen (72 per cent) or brevi-deciduous (23 per cent) species in the lower–and middlestories, and a few species (5 per cent) are of deciduous type constituting the upperstorey (Venkateswaran and Parthasarathy 2003; Venkateswaran, and Parthasarathy, 2005). These forests also harbor a considerable diversity and density of lianas (For more details please refer Reddy and Parthasarathy 2003, Venkateswaran 2004).

Field Methods

Although a majority of forest studies consider10 cm diameter at breast height (dbh) as the lower limit for the inventories, in the present study a lower girth threshold of ≥10 cm gbh (girth at breast height) was chosen because of the small diameter of trees in tropical dry evergreen forests mature early.

In 1995, at each of the two forest sites a one-ha plot (100 m × 100 m) was established. Each hectare was divided into 100 sub-plots, each 10 m x10 m in size. All woody tree individuals (≥10 cm gbh) occurring in each quadrat were tagged and measured for girth (circumference) at 1.3 m from the ground level (Parthasarathy and Karthikeyan 1997). In 2002 all stems ≥10 cm gbh of both the plots were recensused. The mean annual recruitment, mortality and growth rates were calculated for all stems ≥10 cm gbh as well as for stems ≥31.4 cm gbh (10 cm diameter at breast height), to facilitate comparison with other tropical forests.

Recruitment was regarded as those individuals which entered the minimum girth threshold (10 cm gbh) during re-census. Mortality was considered as the number of dead stems including cut, standing dead, missing stems, stand broken stems and uprooted categories. Annual rates of mortality and recruitment were calculated from the equations (Alder 1995, Sheil *et al.,* 1995, Sheil and May 1996, Newbery *et al.,* 1999):

$$\text{Mortality rate } (m) = 1 - [1 - (n_d/n_{99})]^{1/t}$$

and $$\text{recruitment rate } (r) = [1 + (n_r/n_{99})]^{1/t} - 1$$

where, t = relevant time interval, n_d = number of death, n_r = number of recruits and n_{99} = population size in 1999.

Results

Changes in Woody Species Richness and Population Density

After a three-year of biomonitoring, an increase in species richness, population density and basal area of all woody species ≥10 cm gbh (as well as for ≥31.4 cm gbh 10 cm dbh) was evident in the two tropical dry evergreen forest sites (Table 35.1). For stems ≥10 cm gbh, the change in woody species richness in site KK was nearly twice (22 per cent increase) as those in site TM (14 per cent). While for stems ≥31.4 cm gbh the species addition was two thirds more in 2002 in site KK than TM (6 and 2 species respectively). The percent positive change in stem density (for stems ≥10 cm gbh) was more or less equal in both the forests (25.7 per cent in KK and 25.4 per cent in TM), whereas there was a marked difference for stems ≥31.4 cm gbh in the two sites (19.8 per cent increment in KK and 8.2 per cent in TM). The per cent increment in basal area for stems ≥10 cm gbh and 31.4 cm gbh was more in KK (11.8 per cent for each) than in TM (8.3 per cent and 4.9 per cent, respectively).

Table 35.1: Changes in species richness, density and basal area of all woody species ≥10 cm gbh and ≥31.4 cm gbh in the 1-ha plot of tropical dry evergreen forests at sites Kuzhanthaikuppam (KK) and Thirumanikkuzhi (TM) on the Coromandel coast of India.

Variable	KK		TM	
	1999	*2002*	*1999*	*2002*
Number of species				
≥10 cm gbh	46	56	37	42
≥31.4 cm gbh	25	31	21	23
Number of individuals				
≥10 cm gbh	1460	1835	1295	1624
≥31.4 cm gbh	379	454	267	289
Basal area (m²ha⁻¹)*				
≥10 cm gbh	16.9	18.9	25.3	27.4
≥31.4 cm gbh	11.9	13.3	22.3	23.4

*Excludes basal area of *Ficus benghalensis.*

Woody Species Recruitment, Mortality and Growth Rates

In the 2002 recensus, for stems ≥10 cm gbh, 30 per cent and 28 per cent of stems were recruited by attaining 10 cm minimum gbh in sites KK and TM, respectively and 4 per cent and 3 per cent of stems died in sites KK and TM (Table 35.2). The top five species with high recruitment were *Memecylon umbellatum* (22.8 per cent), *Tricalysia sphaerocarpa* (14 per cent), *Miliusa montana* (5.8 per cent), *Diospyros ebenum* (5.3 per cent) and *Combretum albidum* (4.8 per cent) in site KK and the top five recruits in site TM were *Tricalysia sphaerocarpa* (24.1 per cent), *Glycosmis pentaphylla* (10.4 per cent), *Miliusa montana* (9.9 per cent), *Lepisanthes tetraphylla* (9.9 per cent) and *Mallotus rhamnifolius* (7.8 per cent) (Table 35.3). Most of the dead stems were in the smaller girth class (10-30 cm gbh), representing 80 per cent and 89

Table 35.2. Short-term population changes and growth rate of woody species (stems ≥10 cm gbh and ≥31.4 cm gbh) at sites Kuzhanthaikuppam (KK) and Thirumanikkuzhi (TM).

Variable	KK	TM
Number of recruits		
≥10 cm gbh	434	365
≥31.4 cm gbh	86	25
Recruitment rate (per cent yr⁻¹)		
≥10 cm gbh	9.1	8.6
≥31.4 cm gbh	7.0	3.0
Number of deaths		
≥10 cm gbh	59	36
≥31.4 cm gbh	11	3
Mortality rate (per cent yr⁻¹)		
≥10 cm gbh	1.4	0.9
≥31.4 cm gbh	1.0	0.4
DBH growth rate (mm yr⁻¹)		
≥10 cm gbh	4.3	3.8
≥31.4 cm gbh	4.8	4.2

Table 35.3: Number of individuals of top five tree/liana species recorded in 2002 within the 1-ha permanent plot of the two study sites Kuzhanthaikuppam (KK) and Thirumanikkuzhi (TM) that were not enumerated in 1999, arranged in descending order of stem density.

Species Name	1999 Stem Density	No. of Recruits	No. of Deaths	2002 Stem Density	Net Change
		KK			
Memecylon umbellatum	562	112	21	653	+91
Tricalysia sphaerocarpa	252	61	11	302	+50
Miliusa montana	3	25	0	28	+25
Diospyros ebenum	100	23	2	121	+21
Combretum albidum	36	21	1	56	+20
		TM			
Tricalysia sphaerocarpa	351	88	8	431	+80
Glycosmis pentaphylla	9	38	0	47	+38
Miliusa montana	16	36	0	52	+36
Lepisanthes tetraphylla	273	36	5	304	+31
Mallotus rhamnifolius	76	34	4	106	+30

per cent of the total number of dead stems in sites KK and TM respectively (Figure 35.1). The mortality of smaller stems was considerably greater in site TM and KK. Recruitment and mortality rates marginally differed between the two sites (9.1 per

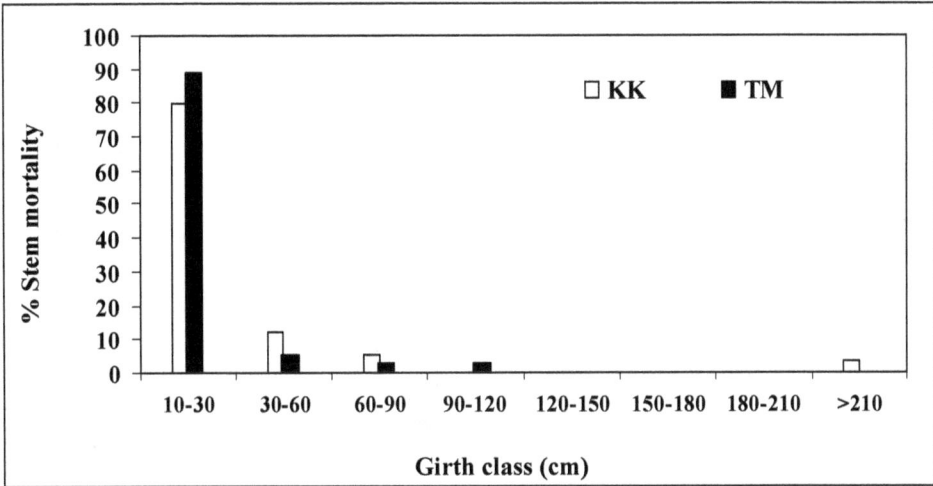

Figure 35.1: Per cent stem mortality in different girth class of woody species (≥10 cm gbh) in the two tropical dry evergreen forest sites Kuzhanthaikuppam (KK) and Thirumanikkuzhi (TM).

cent yr^{-1} and 1.4 per cent yr^{-1} in site KK and 8.6 per cent yr^{-1} and 0.9 per cent yr^{-1} in TM, respectively). The stems that attained 31.4 cm gbh were more in site KK (86 stems) than in TM (25), with an annual recruitment rate of 7 per cent and 3 per cent for sites KK and TM, respectively. In both the forests recruitment rate exceeded mortality rate for all stems (≥10 cm gbh). The annual turnover rate (mean mortality and recruitment rates) for stems ≥10 cm gbh in site KK was slightly greater than site TM (5.25 per cent and 4.75 per cent respectively), but for stems ≥31.4 cm gbh, site KK registered nearly three-fold more than that of TM (4.75 per cent in KK and 1.7 per cent in TM). Mean dbh growth rates were greater for stems ≥31.4 cm gbh than stems ≥10 cm gbh in both the forests, particularly in site KK (4.8 mm yr^{-1}). Figure 35.2 indicated that the maximum stems have 2-4 mm diameter increment per year in both the sites and it was greater for TM site compared to KK site (Figure 35.2).

Population Changes, Recruitment and Mortality in Life-Form

The species richness, density and basal area of trees and lianas in the two sites increased in recensus (Table 35.4). In site KK, species addition in the two life-forms was equal (5 species each in trees and lianas), and in TM more trees were added (4 species in trees and only one species in lianas). Among the newly recruited species in the two sites, *Ficus hispida*, *Ipomoea carnea* and *Dimorphocalyx glabellus* were the most abundant in site KK and *Benkara malabarica* in TM. There was no species lost in the two sites. In site KK, the changes in woody species density under a 3-year period included an addition of 72 per cent liana density as recruits and 3.7 per cent reduction as deaths, while the corresponding figures for trees were 26.4 per cent and 4.1 per cent, respectively. In TM, 30.4 per cent tree density were added and 2.9 per cent were lost, while lianas registered 17.1 per cent recruitment and 2.3 per cent mortality (Table 35.4). Changes in basal area revealed that trees experienced a higher reduction in basal area in the two forests, particularly in KK than lianas.

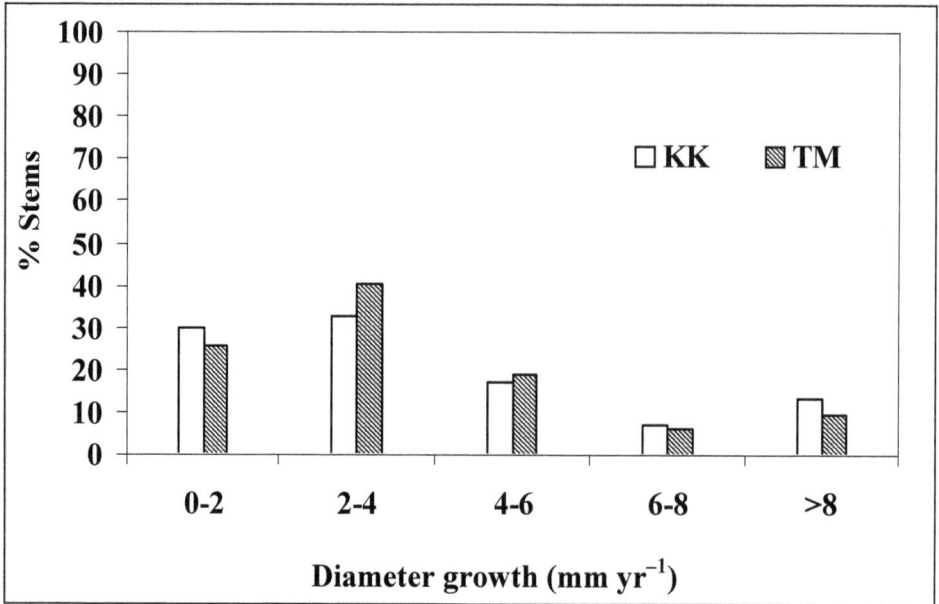

Figure 35.2: Per cent stem density in different diameter growth categories in sites Kuzhanthaikuppam (KK) and Thirumanikkuzhi (TM).

Table 35.4: Short-term changes in woody species categorized by life-forms, viz. trees and lianas (≥10 cm gbh) in two tropical dry evergreen forests Kuzhanthaikuppam (KK) and Thirumanikkuzhi (TM).

Life-form/Variable	KK				TM			
	1999	2002	↑	↓	1999	2002	↑	↓
Trees								
Species richness	29	34	5	0	23	27	4	0
Density	1353	1655	357	55	1079	1376	328	31
Basal area (m²ha⁻¹)	16.2	18.0	0.6	1.7	24.4	26.1	0.4	0.24
Lianas								
Species richness	17	22	5	0	14	15	1	0
Density	107	180	77	4	216	248	37	5
Basal area (m²ha⁻¹)	0.7	0.9	0.1	0.02	0.9	1.2	0.05	0.01

↑ Indicates gain due to recruitment; ↓ indicates loss due to mortality.

The annual recruitment and mortality rates as well as diameter growth rates of trees and lianas in the two study sites KK and TM revealed a strikingly imbalanced recruitment and mortality (Table 35.5). In site KK, recruitment rate recorded for lianas was more than two fold greater than that of trees (19.8 per cent yr^{-1} and 8.1 per cent yr^{-1} respectively), but for TM it was more for trees (9.2 per cent yr^{-1}). There was no significant difference in mean annual diameter growth rates for the two life-forms in both the study sites, with the exception of little increment in case of trees at KK site.

Table 35.5: Tree and liana species addition for all stems ≥10 cm gbh between 1999 to 2002, arranged in descending order of stem abundance (stems ha⁻¹) in the two study sites Kuzhanthaikuppam (KK) and Thirumanikkuzhi (TM).

KK	Abundance	TM	Abundance
Trees		Trees	
1. *Ficus hispida*	15	1. *Benkara malabarica*	5
2. *Ipomoea carnea*	9	2. *Ficus hispida*	1
3. *Dimorphocalyx glabellus*	7	3. *Flacourtia indica*	1
4. *Maytenus emarginata*	3	4. *Maytenus emarginata*	1
5. *Lantana camara*	1		
Lianas		Liana	
1. *Cissus quadrangularis*	2	1. *Premna corymbosa*	1
2. *Gymnema sylvestre*	2		
3. *Carissa spinarum*	1		
4. *Cissus vitiginea*	1		
5. *Coccinia grandis*	1		

Table 35.6: Per cent recruitment rate, mortality rate and mean annual diameter growth rate for trees and lianas ≥10 cm gbh in two tropical dry evergreen forest sites Kuzhanthaikuppam (KK) and Thirumanikkuzhi (TM) on the Coromandel coast.

Variable	KK		TM	
	Trees	Lianas	Trees	Lianas
Recruitment rate (per cent yr⁻¹)	8.1	19.8	9.2	5.4
Mortality rate (per cent yr⁻¹)	1.4	1.3	1	0.8
Growth rate (mm yr⁻¹)	4.3	4.1	3.9	3.9

Population Changes, Recruitment, Mortality and Growth Rates in Forest Stratum

The population changes of trees ≥10 cm gbh in the two tropical dry evergreen orests revealed that the lowerstory species were more dynamic except in growth rates (Tables 35.7 and 35.8). The two forest sites increased in their species richness (from 14 species to 19 species in KK and from10 to 14 species in TM) in lowerstory and there was no change in the middle and upperstories. Tree densities increased for lower–and middlestory species, but there was a slight reduction in upperstory species. A remarkable increment in basal area was recorded for lowerstory trees in both the forests (Table 35.6). Recruitment rate exceed mortality rate in all the stories in both the forests, especially for lowerstory and middlestory trees in both the sites. The mean annual growth rate was high in upperstory species in both the study sites

(6.7 mm yr^{-1} in KK and 5.6 mm yr^{-1} in TM, Table 35.7). Of the eight woody species common to both the study sites, the upperstory species *Pterospermum canescens* showed the maximum mean annual dbh growth in both the forests (6.1 mm yr^{-1} and 5.9 mm yr^{-1} in KK and TM respectively), and it was greater than the forest mean (Table 35.2). The growth rates of *Mallotus rhamnifolius* (5 mm yr^{-1}) and *Lepisanthes tetraphylla* (4.4 mm yr^{-1}) in site KK and *Tricalysia sphaerocarpa* (4.6 mm yr^{-1}) and *Atalantia monophylla* (4.4 mm yr^{-1}) in TM were intermediate (Figure 35.3).

Table 35.7. Analysis of short-term population changes for trees (≥10 cm gbh) by forest strata in two tropical dry evergreen forest sites Kuzhanthaikuppam (KK) and Thirumanikkuzhi (TM) for the period between 1999 and 2002.

Variable	Lower Story		Middle Story		Upper Story	
	1999	2002	1999	2002	1999	2002
KK						
Species richness	14	19	13	13	2	2
Density	930	1184	284	326	139	145
Basal area (m² ha^{-1})	5.9	7.2	4.5	5.1	5.1	5.7
TM						
Species richness	10	14	10	10	3	3
Density	487	688	547	645	45	43
Basal area (m² ha^{-1})	1.5	2.1	15.8	16	7.1	7

Table 35.8: Tree recruitment rate, mortality and growth rate at sites Kuzhanthaikuppam (KK) and Thirumanikkuzhi (TM) by forest stratum.

Variable	Lower Story		Middle Story		Upper Story	
	KK	TM	KK	TM	KK	TM
Number of recruits	295	213	51	115	11	0
Recruitment rate (per cent yr^{-1})	9.6	12.8	5.7	6.6	2.6	0
Number died	41	12	9	17	5	2
Mortality rate (per cent yr^{-1})	1.5	0.8	1.1	1	1.2	1.5
Mean growth rate (mm yr^{-1})	4.1	4.1	3.8	3.6	6.7	5.6

Modes of Death

Five different modes of death of trees and lianas were recorded in the two sites (Table 35.9). Of the 59 dead stems recorded during the three-year study period in site KK, 36 per cent of stems were standing dead, 22 per cent were missing stems, 10 per cent were stand broken stems and 2 per cent were 'uprooted'. In TM, the prominent mode of tree death was for 'missing stems' (44.4 per cent), followed by standing dead (30.5 per cent), cut (19 per cent) and standing broken stems (5.5) for trees and lianas in both the sites. As for the mortality of lianas in site TM, 'missing stems' category accounted for greater loss than by other modes (Table 35.8).

Table 35.9: Modes of death recorded for trees and lianas in two tropical dry evergreen forests Kuzhanthaikuppam (KK) and Thirumanikkuzhi (TM).

Mode of death	KK			TM		
	Trees	Lianas	Total	Trees	Lianas	Total
Cut	19	2	21	7	0	7
Standing dead	17	1	18	11	0	11
Missing stems	12	1	13	11	5	16
Stand broken stems	6	0	6	2	0	2
Uprooted	1	0	1	0	0	0
Total	55	4	59	31	5	36

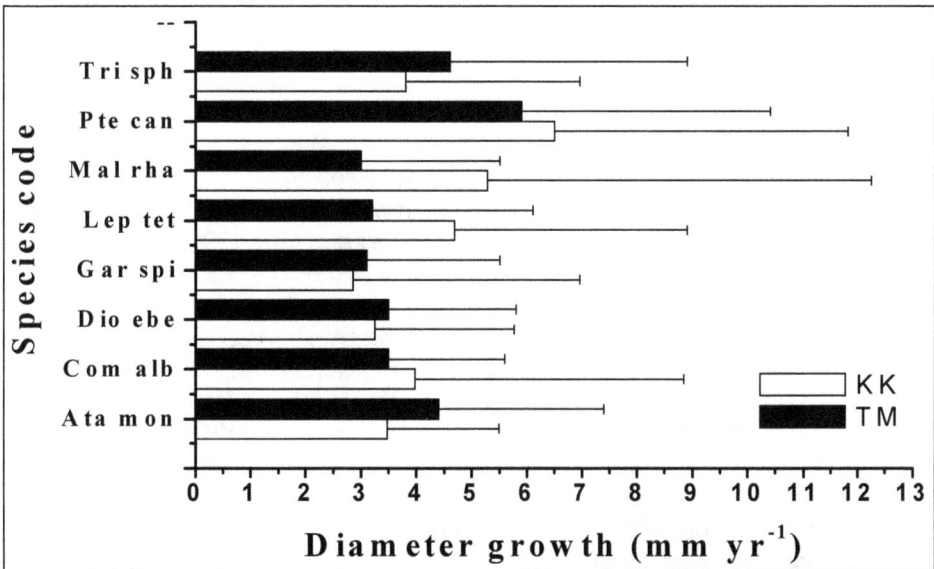

Figure 35.3: Mean annual growth of the common woody species (≥10 cm gbh) of the study two sites Kuzhanthaikuppam (KK) and Thirumanikkuzhi (TM) (Bars indicate standard deviation; species: *Ata mon–Atalantia monophylla*; *Com alb–Combretum albidum*; *Dio ebe–Diospyros ebenum*; *Gar spi–Garcinia spicata*; *Lep tet –Lepisanthes tetraphylla*; *Mal rha–Mallotus rhamnifolius*; *Pte can–Pterospermum canescens* and *Tri sph–Tricalysia sphaerocarpa*).

Discussion

Recruitment

Species recruitment, growth and mortality are continuous processes in community dynamics, but the rate if changes would vary with forest types. In re-census the number of species increases in our study sites that may be due to hurricane damage in November 2000, that could have created canopy gaps in which recruitment of species occurred, especially, in the lowerstory species, as also reported in Barro Colorado Island (BCI) and northeastern Thailand tropical forests (Condit *et al.*, 1992, Bunyavejchewin 1999). The other reason may be the high proportion of standing mode of death that creates small gaps, which are likely to favor recruitment of shade tolerant understory species (Carey *et al.*, 1994). An addition of 72 per cent liana density in site KK may be due to anthropogenic disturbances. Increase in human disturbances in recent decades was reported to accelerate development of lianas (Brown and Lugo 1990, Phillips and Gentry 1994, Laurance *et al.*, 2001).

Mortality

The short time span between the two measurements (3 years) might account for the imbalance between mortality and recruitment rates (Felfili 1995, Carey *et al.*, 1994). Longer-term measurements would reveal the actual dynamic process of these forests. The mortality rate recorded for stems ≥10 cm dbh in site KK (1 per cent yr^{-1}) was more or less equal to that in the *Hopea* plot in the tropical dry evergreen forest of northeastern Thailand (1.1 per cent yr^{-1}) and less than half that of the *Shorea* plot value (2.4 per cent yr^{-1}) of the same forest type (Bunyavejchewin 1999). The higher mortality recorded for small stems is in accordance with the results of Pelissier *et al.* (1998) at Uppangala site and Sukumar *et al.* (1998) at Mudhumalai site in the Indian Western Ghats, and Tayler *et al.* (1996) in Mpanga Research Forest Reserve, Uganda; but in opposition to many other tropical studies in which mortality did not significantly differ with tree size (see Swaine *et al.*, 1987, Sheil and May 1996, Sundaram and Parthasarathy 2002, Ayyappan and Parthasarathy 2003). The higher number of tree deaths in small trees suggests that these trees may be dying due to normal stand thinning (Carey *et al.*, 1994) or competition among smaller trees (Felfili 1995), and these are general features of steady-state forests (Swaine *et al.*, 1987). The reported annual mortality rates of trees in tropical rain forests ranged from 1 to 2 per cent (Swaine *et al.*, 1987), but in our study it was 1 per cent (in site KK) and 0.4 per cent (in site TM) (for stems ≥10 cm dbh). The present value is low compared to other tropical forests. This may be attributed to the low to moderate disturbance due to a hurricane in November 2000, which was the only notable natural disturbance that affected these tropical dry evergreen forests during the study period. In general, low to moderate hurricane damage leads to low mortality as it was documented in Central American and Caribbean forests (Frangi and Lugo 1991, Walker 1991).

Like in lower montane moist forests of Venezuela (Carey *et al.*1994) and in La Selva, Costa Rica (Lieberman and Lieberman 1987), the predominant type of natural tree death in our tropical dry evergreen forests was standing death, that creates tiny gaps (Whitmore 1978) or "gradual gaps" (Krasny and Whitmore 1992), thereby releasing resources such as light and soil nutrients gradually to the surrounding area (Carey *et al.*, 1994). These results suggest that to gain a better understanding of species dynamics in tropical dry evergreen forests, more focus on gap dynamics is needed.

Growth

To facilitate a valid comparison with other tropical forests, all the parameters were unified for the ≥31.4 cm gbh (10 dbh) girth threshold. The mean dbh growth increment of 4.8 and 4.2 mm yr⁻¹ in sites KK and TM falls within the range reported for other tropical forests of 1.0 to 7.0 mm yr⁻¹ (Silva *et al.,* 1989), but relatively high when compared to other tropical forests particularly the montane tropical rain forests (Herwitz and Young 1994). The possible explanation for the relatively high mean dbh growth increments recorded in our dry evergreen forests could be an accelerated growth, associated with stem recovery following low to moderate storm damage as also reported by Brokaw and Walker (1991) in Puerto Rico. The high growth rate of the upperstory species supports the findings of Felfili (1995) that light-demanding upperstory species tend to have high increment rates. The lower growth rate was recorded for lianas than trees, may be because, tree growth is mainly directed to lateral or circular growth, while for lianas it is directed to apical or elongated growth for getting host support and light.

As discernible changes were documented within a 3-year period with respect to diversity changes, tree growth and mortality rates, a 3-year period between two measurements appears to be an ideal time to understand short-term dynamics in tropical dry evergreen forests. Therefore, these forests deserve more attention with respect to research on forest dynamics than is currently received.

In conclusion, it can be revealed that, the site KK is more dynamic than the site TM, with higher rates of mortality, recruitment, turnover and mean annual diameter growth increments and tree species suffered more mortality than lianas in both the study sites.

Acknowledgements

We (RV and NP) thank the Department of Science and Technology, New Delhi for financial support (grant No. SP/SO/A-07/99, members of our research team Dr. N. Ayyappan, Dr. S. Muthuramkumar, Mr. J. Annaselvam and Mr. S. Kadiresan during the 1999 re-census of the two sites, and Mr. Manickam and family for local support during the field work. We thank Dr. N. Brokaw, Institute for Tropical Ecosystem studies, University of Puerto Rico, U.S.A. and an anonymous reviewer for offering valuable comments on this paper.

References

Alder, D., 1995. Growth modeling for mixed tropical forests. *Tropical forestry paper* No. 30. Oxford Forestry Institute.

Ayyappan, N. and Parthasarathy, N., 2000. Short-term changes in tree populations in a tropical evergreen forest at Varagalaiar, Western Ghats, India. *Biodiversity and Conservation,* 13: 1843–1851.

Brokaw, N.V.L. and Walker, L.R., 1991. Summary of the effects of Caribbean hurricanes on vegetation. *Biotropica,* 23: 386–392.

Brown, S. and Lugo, A.E., 1990. Tropical secondary forests. *Journal of Tropical Ecology,* 6: 1–32.

Bunyavejchewin, S., 1999. Structure and dynamics in seasonal dry evergreen forests in northeastern Thailand. *Journal of Vegetation Science,* 10: 787–792.

Condit, R., Hubbell, S.P. and Foster, R.B., 1992. Short-term dynamics of a neotropical forest. *BioScience*, 42: 822–828.

Carey, E.V., Brown, S., Gillespie, A.J.R. and Lugo, A.E., 1994. Tree mortality in mature lowland moist and tropical lower montane moist forests on Venezuela. *Biotropica*, 26: 255–265.

Connell J.H., Tracey, J.G. and Webb, L.J., 1984. Compensatory recruitment, growth and mortality as factors maintaining rain forest tree diversity. *Ecological Monographs,* 54: 141–164.

Felfili, J.M., 1995. Growth, recruitment and mortality in the Gama gallery forest in central Brazil over a six-year period (1985–1991). *Journal of Tropical Ecology,* 11: 67–83.

Frangi, J.L. and Lugo, A.E., 1991. Hurricane damage to a flood plain forest in the Luquillo Mountains of Puerto Rico. *Biotropica*, 23: 324–335.

Herwitz, S.R. and Young, S.S. 1994. Mortality, recruitment and growth rates of montane tropical rain forest canopy trees on Mount Bellenden-Ker, northeast Queensland, Australia. *Biotropica,* 26: 350–361.

Krasny, M.E. and Whitmore, M.C., 1992. Gradual and sudden forest ganopy gaps in Allegheny northern hardwood forests. *Can. J. For. Res.,* 22: 139–143.

Laurance, W.F., Perez-salicrup, D. and Delamonica, P., *et al.,* 2001. Rain forest fragmentation and the structure of Amazonian liana communities. *Ecology*, 82: 105–116.

Lieberman, D. and Lieberman, M., 1987. Forest tree growth and dynamics at La Selva, Costa Rica (1969–1982). *Journal of Tropical Ecology,* 3: 347–358.

Newbery, D.M., Kennedy, D.N., Petol, G.H., Madani, L. and Ridsdale, C.E., 1999. Primary forest dynamics in lowland dipterocarp forest at Danum valley, Sabah, Malaysia, and the role of the understory. *Phil. Trans. R. Soc. Lond. B* 354: 1763–1782.

Parthasarathy, N. and Karthikeyan, R., 1997. Plant biodiversity inventory and conservation of two tropical dry evergreen forests on the Coromandel coast, south India. *Biodiversity and Conservation,* 6: 1063–1083.

Parthasarathy, N. and Sethi, P., 1997. Tree and liana species diversity and population structure in a tropical dry evergreen forests in south India. *Tropical Ecology*, 38: 19–30.

Pelissier, R., Pascal, J., Houllier, F. and Laborde, H., 1998. Impact of selective logging on the dynamics of a low elevation dense moist evergreen forest in the Western Ghats (south India). *Forest Ecology and Management,* 105: 107–119.

Phillips, O.L. and Gentry, A.H., 1994. Increasing turnover through time in tropical forests. *Science*, 263: 954–958.

Phillips, O.L., Hall, P., Gentry, A.H., Sawyer, S.A. and Vasquez, R., 1994. Dynamics and species richness of tropical rain forests. *Proc. Natl. Acad. Sci., USA*, 91: 2805–2809.

Ramanujam, M.P. and Kadamban, D., 2001. Plant biodiversity of two tropical dry evergreen forests in the Pondicherry region of south India and the role of belief systems in their conservation. *Biodiversity and Conservation*, 10: 1203–1217.

Reddy, M.S. and Parthasarathy, N., 2003. Liana diversity and distribution in four tropical dry evergreen forests on the Coromandel coast of south India. *Biodiversity and Conservation*, 12: 1609–1627.

Sheil, D. and May, R.M., 1996. Mortality and recruitment rate evaluations in heterogeneous tropical forests. *Journal of Ecology*, 84: 91–100.

Sheil, D., Burslem, D.F.R.P. and Alder, D., 1995. The interpretation and misinterpretation of mortality rate measures. *Journal of Ecology*, 83: 331–333.

Silva, J.N.M., Carvalho, de J.O.P. and Lopes, J. do C.A., 1989. Growth of a logged-over tropical rain forest of the Brazilian Amazon. In: *Proceedings of the Seminar on Growth and Yield in Tropical Mixed/Moist Forests*, (Eds.) W.R. Was Mohd, H.T. Chan and S.Appanah. IUFRO, Kuala Lumpur, pp. 117–136.

Sukumar R., Suresh, H.S., Dattaraja, H.S. and Joshi, N.V., 1998. Dynamics of a tropical deciduous forest: population changes (1988 through 1993) in a 50-hectare plot at Mudumalai, southern India. In: *Forest Biodiversity Research, Monitoring and Modeling: Conceptual Background and Old World Case Studies*, (Eds.) F. Dallmeier and J.A. Comiskey. Parthenon Publishing, Paris, pp. 495–506.

Sundaram, B. and Parthasarathy, N., 2002. Tree growth, mortality and recruitment in four tropical wet evergreen forest sites of the Kolli hills, Eastern ghats, India. *Tropical Ecology*, 43: 275–286.

Swaine, M.D., Lieberman, D. and Putz, F.E., 1987. The dynamics of tree populations in tropical forest: a review. *Journal of Tropical Ecology*, 3: 359–366.

Taylor, D.M., Hamilton, A.C., Whyatt, J.D., Mucunguzi, P. and Bukenya-ziraba, R., 1996. Stand dynamics in Mpanga Research Forest Reserve, Uganda, 1968–1993. *Journal of Tropical Ecology*, 12: 583–597.

Venkateswaran, R. and Parthasarathy, N., 2003. Tropical dry evergreen forests on the Coromandel coast of India: structure, composition and human disturbance. *Ecotropica*, 9: 45–58.

Venkateswaran, R., 2004. Short-term tree population changes, growth and phenology of woody species in tropical dry evergreen forests on the Coromandel coast of India. *Ph.D. Thesis*, Pondicherry University, Pondicherry, India.

Venkateswaran, R. and Parthasarathy, N., 2005. Tree population changes in a tropical dry evergreen forest of south India over a decade (1992–2002). *Biodiversity and Conservation*, 14: 1335–1344.

Visalakshi, N., 1995. Vegetation analysis of two tropical dry evergreen forests in southern India. *Tropical Ecology*, 36: 117–127.

Walker, L.R., 1991. Tree damage and recovery from Hurricane Hugo in Luquillo Experimental Forest, Puerto Rico. *Biotropica*, 23: 379–385.

Whitmore, T.C., 1978. Gaps in the forest canopy. In: *Tropical Trees as Living Systems*, (Eds.) F.B. Golley and E. Medina. Springer-Verlag, New York, pp. 639–655.

2013, Perspectives in Plant Biodiversity   *Pages* 280–287
Editor: Dr. K. Muthuchelian, *Vice Chancellor, Periyar University, Salem*
Published by: Daya Publishing House, NEW DELHI

Chapter 36

Removal of Malachite Green on *Equcalptus glubulous* Bark Carbon: A Kinetic and Equilibrium Study

R. Pagutharivalan* and N. Kannan

Department of Chemistry,
Ayya Nadar Janaki Ammal College (Autonomous),
Sivakasi – 626 124, Tamil Nadu

ABSTRACT

Removal of Malachite Green in aqueous solution on *Equcalptus glubulous* Bark Carbon (EGBC) has been studied. The effect of various experimental parameters has been investigated using a Batch Adsorption Technique (BAT) to obtain information on treating effluents from the dye industry. The extent of dye removal increased with decrease in the initial concentration of the dye and increased with increase in contact time, amount of adsorbent used and the initial pH of the dye solution. Adsorption data were modeled using the Freundlich and Langmuir adsorption isotherms and first order kinetic equations. The kinetics of adsorption was found to be first order with regard to intra-particle diffusion as the rate determining step. The adsorption capacity of dye has been compared. The results indicate that EGBC is one of the best adsorbent that can be used in wastewater treatment for the removal of colors and dyes.

Keywords: *Adsorption of malachite green, Batch adsorption technique, Eucalyptus globules bark carbon, Adsorption isotherms, Kinetics of adsorption.*

_____

* Corresponding Author: E-mail: paguthuchem@yahoo.com

Introduction

Water pollution is a very persistent problem; the intensive disposal of different toxic substance without control constitutes a real danger. Wastewaters from the textile, cosmetics, printing, dyeing, food colouring, paper making, etc., are polluted by dyes. Most of the dyes are stable to biological degradation. Coloured waters are often objectionable on aesthetic grounds for drinking and other agricultural purposes. Color affects the nature of water by inhibiting sunlight penetration, thus reducing photosynthetic action. Some dyes are carcinogenic and mutagenic [1, 2]. Therefore, there is a considerable need to treat such element prior to discharge. Most of the used dyes are stable to photo degradation, Bio-degradation and Oxidizing agent [3, 4]. Currently, several physical or chemical processes are used to treat dye-laden wastewaters. However, these processes are costly and cannot effectively be used to treat the wide range of dye wastewater. The adsorption process is one of the efficient methods to remove dyes from effluent [5, 6]. The adsorption process has an advantage over the other methods due to the excellent adsorption efficiency of activated carbon. It over comes the problem of the water treatment techniques[7] by taking advantage of an adsorbent's surface having an affinity for a particular molecular or ionic species coming onto contact with it. A further benefit is that adsorption can be very simple and offers sludge free operation. The evaluation of activated carbon for color removal has been extensive [8] and effluent treatment systems using activated carbon have been successful. Some works of low cost, non-conventional adsorbents have been carried out which include, agricultural solid waste, Such as Coir pith [9], Banana pith [10], Coconut husk [11], Sawdust [12], Peat moss [13], Paddy straw [14] and industrial solid wastes such as fly ash and coal, Red mud and Fe (III)/Cr (III) hydroxide. The objective of this present study is to explore the feasibility of using carbonized Date Seed and Tamarind Shell as an adsorbent for the removal of Metanil Yellow which is most widely used in various textile-processing industries by varying parameters like Initial concentration, contact time, dose variation pH and particle size.

Materials

Raw materials for the preparation of carbon such as *Eucalyptus Globules* Bark Carbon (EGBC) were collected locally, washed, dried, cut into small pieces, carbonized (at 300 °C) and steam digested (at 900 °C) acid treated and washed. The materials were then sieved to discrete particle sizes and dried at 120 °C for 5 hr in an air oven. Malachite Green (MG) supplied by BDH (India) was used as an adsorbate. All the other chemicals used in this study were of reagent grade and obtained commercially. Double distilled water was employed for preparing all the solutions and reagents. Adsorption data of the replicates (with in ± 1 per cent) were reported.

Methods–Adsorption Experiments

Adsorption experiments were carried out at room temperature (30 ± 1°C) under batch mode. A stock solution of MG (1000 mgL^{-1}) was prepared and suitably diluted to the required initial concentrations. The initial concentrations (C$_i$) of MG were also checked by measuring absorbance at 614 nm (λ max) in uv-visible double beam spectrophotometer (model no: uv-1700 made by shimadzu in Japan) and then interpolating it into the standard curve. Exactly 50 mL of MG solution of known initial concentration was shaken with a required dose of carbon of a fixed particle size for a specific period of contact time in a thermostatic orbit incubator shaker (NEOLAB, India) at 200 rpm, after nothing down the initial pH of the solution (pH = 7.8). The initial pH of the dye solutions were adjusted to the required value by adding either 1M HCl or 1M NaOH solution. After equilibration, the final concentrations (C_e) of the dye solutions were measured by spectrophotometer model–SL 207, India). The various experimental conditions are given in Table 36.1. The values of percentage removal of dye and amount adsorbed (q in mg g^{-1}) were calculated using the following relationships:

Percentage removal = $100\,(C_i–C_e)/C_i$ (1)

Amount adsorbed (q) = $(C_i–C_e)/m$ (2)

where,

C_i and C_e are the initial and equilibrium (final) concentration of dye (in mgL^{-1}), respectively and m is the mass of adsorbent, in gL^{-1}.

Results and Discussion

The adsorption experiments were carried out at different experimental conditions (Table 36.1) and the results obtained are discussed below:

Table 36.1: Effect of various parameters for the removal of dyes by adsorption technique.

Parameters	Variation (in range)	Per cent Removal of Dye
Initial concentration (ppm)	250-520	99.7–96.3
Contact time (in min)	5-50	94.7–99.6
Dose (gL^{-1})	50-140	96.4 -99.3
pH	2-11	89.6–96.7
Particle size variation (Micron)	45-280	96.2–84.8

Effect of Initial Concentration of Mg

The effect of initial concentration of dye on the extent of removal of MG (In terms of percentage removal) on various adsorbent EGBC is studied at 30 ± 1°C and the relevant data are given in Table 36.2. The percentage removal decreased with the increase in initial concentration of MG. This indicates that there exists a reduction in immediate solute (dye) adsorption, owing to the lack of available active sites required for the high initial concentration of MG Similar results have been reported in literature on the extent of removal of dyes. Although, the adsorption capacities of IPACs (as revealed by Q_o values in Table 36.2) were less, but still they could be considered as alternatives to CAC for the removal of dyes (15).

Adsorption Isotherms

In order to determine the adsorption potential, the study of sorption isotherm is essential in selecting an adsorbent for the removal of dyes (Majeswka–Nowak, 1989). The adsorption data were analyzed with the help of Freundlich and Langmuir isotherms.

Freundlich isotherms: $\log q = \log k + (1/n) \log C_e$ (3)

Langmuir isotherms: $(C_e/q) = (1/Q_o b) + (C_e/Q_o)$ (4)

where,

k and 1/n are the measures of adsorption capacity and intensity of adsorption, respectively. q is the amount dye adsorbed per unit mass of adsorbent(in mgg^{-1}) and C_e is the equilibrium concentration of dye(in mgL^{-1} or ppm); Q_o and b are Langmuir constants, which are the measures of monolayer adsorption capacity(in mgg^{-1}) and surface energy (in $g\,L^{-1}$), respectively. The adsorption data were fitted to these isotherm equations by carrying out correlation analysis and the values of slope (1/n and $1/Q_o$) and intercept ($\log K$ and $1/Q_o b$) were obtained. The adsorption isotherm parameter along with the correlation coefficients is presented in Table 36.3. The observed linear relationships are statistically significant at 95 per cent confidence as evidenced by the r-values (very close to unity), which indicate the applicability of these two adsorption isotherms and the monolayer coverage of adsorbate on adsorbent surface (16).

Table 36.2: Adsorption isotherm for the removal of basic dye by adsorbents.

Parameters	Adsorbent (EGBC)
Freundlich isotherm	
Slope (1/n)	0.103
Intercept (log K)	1.423
Correlation Coefficient (r)	0.984
Langmuir isotherm	
Slope ($1/Q_0$)	0.027
Intercept ($1/Q_0 b$)	0.022
Correlation Coefficient (r)	0.994
Q_0 (mg g^{-1})	36.49
b (gL^{-1})	1.274
R_L	0.006

Further, the essential characteristics of Langmuir isotherm can be described by a separation factor R_L; which is defined by the following equation:

$R_L = 1/(1 + bC_i)$ (5)

where,

C_i is the initial concentration of dye (in mg L^{-1} or in ppm) and b is the Langmuir constant (in $g\,L^{-1}$). The separation factor R_L, indicates the shape of the isotherm and nature of the adsorption process as given below;

R_L Value	Nature of the Process
$R_L > 1$	Unfavorable
$R_L = 1$	Linear
$0 < R_L < 1$	Favourable
$R_L = 0$	Irreversible

In the Present study, the computed values of R_L (Table 36.3) were found to be in the range 0-1 indicating that the adsorption process was favorable for this low cost adsorbent.

Effect of Contact Time

The effect of contact time on the amount of dye adsorbed (q, in mg g^{-1}) was studied at the optimum initial concentration of dye (Table 36.1). The amount of MG adsorbed by these carbons increased and reached a constant value with the increase in contact time (17). The decrease in the removal of dye adsorbed after reaching a constant value (in some cases) may be due to the desorption process. The increase in extent of removal of dye after a particular contact time is less and hence it is fixed as the optimum contact time. Similar results have been reported in literature for the removal of dyes. The adsorbate species normally forms a surface layer, which is only one molecule thick, *i.e.*, a monolayer on the surface of the adsorbent.

Kinetics of Adsorption

The Kinetics of adsorption of MG by EGBC have been studied by applying various first order kinetic equations proposed by Natarajan and Khalaf as cited by (18), Lagergren as cited by (19) and Bhattacharya and Venkobachar (20).

Natarajan and Khalaf equation: $\text{Log} [C_i/C_t] = (k/2.303) t$ (6)

Lagergren's equation: $\text{Log} (q_e-q_t) = \text{Log} q_e-[k_{ad}/2.303] t$ (7)

Bhattacharya and Venkobachar equation: $\text{Log} (1-U(T)] =-[k_{ad}/2.303] t$ (8)

where,

C_i and C_t are the concentration of dye (in mg L^{-1} or ppm), at time Zero (initial concentration) and at time t respectively; q_e and q_t are the amount of dye adsorbed per unit mass of the adsorbent (in mg g^{-1}) and at time t respectively, $U(T) = (C_i-C_t)/(C_i-C_e)$, C_e is equilibrium dye concentration (in ppm) and k and k_{ad} are the first order adsorption rate constants (in min^{-1}). The values of Log (C_i/C_t), Log (q_e/q_t) and Log (1-U (T)) were correlated with time. The values of first order rate constants along with the correlation coefficients are given in Table 36.3.

Intraparticle Diffusion Model

The adsorbate (MG) species are most probably transported from the bulk of the solution to the solid phase through intra-particle diffusion/transport process, which is often rate limiting step in many adsorption processes, especially in a rapidly stirred batch reactor.

The possibility of intra–particle diffusion was explored by using the intra–particle diffusion model (Deo Mall, 2005).

$q_t = k_p t^{½} + C$ (9)

where,

q_t is the amount of dye adsorbed (in mg g^{-1}) at time t; C is the intercept and k_p is the intra–particle diffusion rate constant (in mg g^{-1} min$^{-½}$). The values of q_t were found to be linearly correlated with values of t$^{-½}$ the k_p values are calculated and given in

Table 36.3. The results indicate the presence of intra–particle diffusion process as rate determining step. The values of intercept (C) give an idea about the boundary layer thickness *i.e.,* the larger the intercept, the greater is the boundary layer effect (21).

Table 36.3: Kinetics and dynamics of adsorption of basic dye by adsorbents.

Parameters	*Adsorbent*
Natarajan and Khalaf equation	
Correlation coefficient (r)	0.961
10^2 K (min^{-1})	3.100
Lagergren equation	
Correlation coefficient (r)	0.977
10^2 K (min^{-1})	17.50
Bhattacharya and Venkobacher equation	
Correlation coefficient (r)	0.977
10^2 K (min^{-1})	17.50
Intra-particle diffusion study	
K_p	0.024
Correlation coefficient (r)	0.968
Intercept	1.823
Log (per cent of Removal) Vs Log (Time)	
Slope	0.006
Intercept	1.962
Correlation coefficient (r)	0..926

Effect of Dose of Adsorbent

The effect of dose of adsorbent on the amount of dye adsorbed was studied (Table 36.1). The relative extent of removal of MG (in terms of q) is found and 4gL^{-1} for EGBC, which fixed as the optimum dose of adsorbent. The amount of dye adsorbed was observed to vary exponentially in accordance with a fractional power term of the dose of adsorbent *i.e.* (dose)$^{-n}$ where n = fraction. The plots of log (dose) vs log (per cent removal) are found to be linear (rH≈1.0). This suggests that the adsorbed species/solute may either block the access to the internal pores or cause particles to aggregate and thereby resulting in decrease in the availability of active sites for adsorption (22).

Effect of Initial pH

The effect of initial pH of the dye solution on the amount of dye adsorbed was studied by varying initial pH of dye solution and keeping the other process parameters as constant. The increase in initial pH of the dye solution increased the amount of dye solution adsorbed. This result is in harmony with the literature reports (23), the final pH of the dye (MG) solution after adsorption was found be increases, due to adsorption of the basic form of dye molecule.

Effect of Particle Size

The effect of particle size of the adsorbent on the extent of removal of dyes by indigenously prepared activated of (EGBC) was studied under constant optimum experimental conditions by varying the particle size. Since its particle size is uniform and constant at 90 microns. The effect of particle size variation of an adsorbent on the extent of removal of dye, indicates that the rate of dye uptake increase with the decrease in particle size. This is due to the increase in the availability of surface area of the adsorbent with the decrease in particle size (24).

Conclusion

The conclusions derived from the present investigation are the percentage of removal of MG increased with decrease in initial concentration of dye, particle size of EGBC and increases in contact time, dose of adsorbent and initial pH of the dye solution. Adsorption data obeyed Frendlich and Langmuir adsorption isotherms and first order kinetic equations. The intra-particle diffusion is one of the rate determining steps, and prepared EGBC could be employed as adsorbent for the removal of dye/color in general and Malachite Green (MG) in particular.

References

Annadurai, G. and Krishnan, M.R.V., 1996. Adsorption of basic dye using chitin. *Indian J. Environ. Protect.*, 16: 44.

Annadurai, G. and Krishnan, M.R.V., 1997. Adsorption of acid dye from aqueous solution by chitin: Equilibrium studies. *Indian J. Chem. Technol.*, 4: 217–222.

Asfour, H.M., Nasser, N.M., Fadali, O.A. and El-Geundi, M.S., 1985. Color removal from textile effluents using hardwood sawdust as an adsorbent. *J. Chem. Technol. Biotechnol.*, 35: 23–34.

Bhattacharya, A.K. and Venkobachar, C., 1984. Removal of cadmium by low-cost adsorbent. *J. Am. Civ. Engg.*, 110: 110–116.

Deo, N. and Ali, M., 1993. Dye adsorption by a new low cost material congo red. *Indian J. Environ. Prot.*, 13: 496–508.

Dutta, P.K., 1994. An overview of textile pollution and its remedy. *Indian J. Environ. Pollut.*, 14: 443–446.

Garg, V.K., Amita, M., Kumar, R. and Gupta, R., 2004. Basic dye (Methylene Blue) removal from simulated wastewater by adsorption using Indian rosewood Saw dust: A timer industry waste. *Dyes and Pigments*, 63: 243–250.

Juang, R.S., Tseng, R.L., Wu, C. and Lee, S.H., 1997. Adsorption behaviour of reactive dyes from aqueous solution on chitosan. *J Chem. Technol. Biotechnol.*, 70: 391–399.

Kannan, N. and Meenakshisundaram, M., 2002. Adsorption of congo red on various activated carbons: A comparative study. *Water, Air, Soil Poll.*, 138: 289–305.

Kannan, N. and Murugael, S., 2008. Comparative study on the removal of Acid Violet by adsorption on various low cost adsorbents. *Global Nest Journal*, 10(3): 395–403.

Kannan, N. and Vanangamudi, A., 1991. A study on removal of chromium (IV) by adsorption on lignite coal. *Indian J Environ Protec.*, 11(4): 241–245.

Kavitha, D. and Namsivayam, C., 2007. Exprimental and kinetic studies on methylene blue adsorption by coir pith carbon. *Bioresource Technology*, 98: 14–21.

Low, K.S. and Lee, C.K., 1990. The removal of cationic dyes using coconut husk as an adsorbent. *Pertanika*, 13: 221–228.

Malik, P.K., 2003. Use of activated carbon prepared from sawdust and rice husk for adsorption of acid dyes: A case study of acid yellow 36. *Dyes and Pigments,* 56: 239–249.

Mckay, G., 1982. Adsorption of dyestuffs from aqueous solutions with activated carbon. I. Equilibrium and batch contact-time studies. *J. Chem. Technol.*, 32: 759–772.

Mckay, G., Jamal, M.S. and Aga, J.A., 1985. Fuller's earth and fired clay as adsorbents for dye stuffs. Equilibrium and rate constants. *Water, Air, Soil Pollution*, 24: 307–322.

Namasivayam, C and Kadirvelu, K., 1997. Agricultural solid water for removal of heavy metal: Adsorption of Cu(II) by coir pith carbon. *Chemosphere*, 34: 377–399.

Namasivayam, C., Prabha, D. and Kumutha, M., 1998. Removal of dyes by adsorption onto agricultural solid waste. *Biores. Technol.*, 66: 223–228.

Nawar, S.S and Doma, H.S., 1989. Removal of dyes from effluents using low cost agricultural by products. *Sci. Tot. Environ.*, 79: 271–279.

Nigam, P., Banat, L.M., Singh, D. and Muchant, R., 1996. Microbial process for decolourisation of textile effluents containing azo, diazo and reactive dyes. *Process Biochemistry*, 31: 435–442.

Pandy, K.K., Prasad, G. and Singh, V.N., 1986. Mixed adsorbents for Cu(II) removal from aqueous solution. *Environ. Technol. Letts.*, 7: 547–554.

Poots, V.J.P., Mckay, G. and Healy, J.J., 1976. The removal of acid dye from effluent using natural adsorbents. *J. Peat. Water Res.*, 10: 1061–1066.

Purkait, M.K., 2007. Removal of congo red using activated carbon and its regeneration. *Journal of Hazardous Materials*, 145: 287–295.

Ramkrishna, K.R and Viraraghavan, T., 1997. Use of slag for dye removal waste. *Management*, 17(8): 483–488.

Yeh, R. and Thomas, A., 1997. Color difference measurement and color removal from dye wastewaters using different adsorbents. *J. Chem. Biotechnol.*, 63: 55–59.

2013, Perspectives in Plant Biodiversity   *Pages* **288–294**

Editor: **Dr. K. Muthuchelian,** *Vice Chancellor, Periyar University, Salem*

Published by: **Daya Publishing House, NEW DELHI**

Chapter 37

Morphological Responses of *Salvinia molesta* (Mitchell) by the Accumulation of Lead (Pb)

N.M. Rolli [1] ****, S.S. Suvarnakhandi*** [1]***,***
M.G. Nadagouda [2] ***and T.C. Taranath*** [3]

[1]*BLDEA's Comm, BHS Arts and TGP Science College,*
Jamkhandi – 587 301, Karnataka
[2]*A.S.M. College for Women, Bellary, Karnataka*
[3]*Department of Botany (Environmental Biology Lab)*
Karnatak University Dharwad – 580 003, Karnataka

ABSTRACT

Water is a natural resource and forms the basis of life. Industrial development coupled with population growth has resulted in the overexploitation of natural resources. Life support systems viz, water, air and soil are, thus, getting expose to an array of pollutants, especially heavy metals released by anthropogenic activities. Tolerant species of aquatic plants are able to survive and withstand the pollution stress and serve as pollution indicators. Thus, aquatic plants are employed to remove toxic contaminants from the aquatic system. This is an emerging biogeotechnological application based on "Green Liver Concept" and operates on the principles of biogeochemical cycling. The present study focused on morphological responses of *Salvinia molesta* to the lead stresses and its accumulation. The laboratory experiments were conducted for the assessment of morphological responses and accumulation of lead in plants at its various concentrations (0.2 ppm, 0.6 ppm 1.0 ppm 1.4 ppm and 2.0

* Corresponding Author: E-mail: drnmrolli@rediffmail.com

ppm) at regular intervals for 12 days exposure. *Salvinia* showed visible symptoms like withering of roots and necrotic conditions at higher concentrations (2.0 ppm), however, the test plants showed normal growth at lower concentration (0.2 ppm). The toxic effect of lead (Pb) is directly proportional to its concentration and exposure duration. The accumulation of lead by the *Salvinia* was maximum at 3 days exposure and gradually decreases.

Keywords*: Lead, Heavy metal, Accumulation, Toxicity.*

Introduction

Heavy metal pollution is an important environmental problem. In contrast with most organic materials, metals cannot be transformed by micro-organisms and therefore, accumulate in water, soil bottom sediments and living organisms (Miretzky *et al.*, 2004). These pollutants are present are in the environment as a natural compounds or as a result of anthropogenic activities (agricultural and industrial activities). Industries such as smelters, metal refineries and mining operations have been indicated as major sources of metal release into the environment (Gardea–Torresdey *et al.*, 1997; Srivastava *et al.*, 2007). Most of the heavy metals are toxic or carcinogenic in nature and pose threat to human health and the environment (Shakibaie *et al.*, 2008; Vinodhini and Narayanan, 2009). The nonessential ions like Pb and Cd can inhibit variety of metabolic processes even in small quantities (Carventes *et al.*, 2001; Dinakar *et al.*, 2009; Choudhary and Sharma, 2009), however, the essential metals like Zn, Cu, Fe are toxic at high concentrations. The metals are responsible for many alterations of the plant cell (Photosynthesis, chlorophyll production, pigment synthesis and enzyme activity) (Teisseire and Vernet, 2000; Vaillant *et al.*, 2005; Zhou *et al.*, 2009). The waste water emanating from sources contains metals which could be toxic to flora and founa. There are several methods of removing heavy metals from water based on ion exchange or chemical and microbiological precipitation which have been used with some success. These technologies have different efficiencies for different metals and may be very costly (Dushenkov *et al.*, 1995).

Aquatic macrophytes accumulate considerable account of toxic metals and make the environment free from the pollutants. Many aquatic plants have been successfully utilized for removing toxic metals from aquatic environment (Satyakala and Kaiser Jamil, 1992). The metal tolerance of plants may be attributed to different enzymes, stress proteins and phytochelatins (Van-Asche and Clijsters, 1990). The accumulation of metals at high concentrations causes morphological changes, biochemical activities and also generation of –SH group containing enzymes (Weckx and Clijsters, 1996).

Hence, in the present investigation, *Salvinia molesta*, a common aquatic floating macrophyte, is used to study the effect of different concentrations of Pb on morphology of the test plant due to the accumulation of Pb from the experimental pond under laboratory conditions.

Materials and Methods

Laboratory experiments were conducted for the study of morphological responses and accumulation of 'Pb' by *Salvinia molesta* from the experimental pond. The test plants were sampled from Kelageri pond near Karnatak University Dharwad. The plant stock was maintained under laboratory conditions, experiments were carried out in triplicates. The young and healthy *Salvinia* species were selected and acclimatized for 15 days in 3 per cent Hoagland solution in the experimental ponds of 10 lts capacity. About 50 gm plant material was introduced into each experimental ponds containing 0.2, 0.6. 1.0 and 1.4 ppm concentrations of Lead and tap water as control. Morphological index parameters (MIP) viz root length, leaf length and breadth were observed for 12 days at the interval of 3 days. Photographs of the *Salvinia* treated with different concentrations of 'Pb' were taken using canons power shot G_2 Digital camera. Plants were harvested at the end of 3, 6, 9 and 12 days exposure and thoroughly washed with distilled water and plants were dried in oven at 48° C to get a constant weight for metal extraction.

The estimation of 'Pb' in the test plant was carried out by using standard method (Allen *et al*, 1974). The dried and powdered I gm plant material was digested by using mixed acid digestion method in Gerhardt digestion unit. The digested samples were diluted with double distilled water and filtered thorough Whatman filter paper No. 44 into 100 ml volumetric flask and volume was made to mark. The estimation of 'Pb' was done by Atomic Absorption Spectrophotometer (GBC 932 plus Austrelia) with air acetylene oxidizing flame and metal hallow cathode lamp at the wavelength of (217.0 nm). Working standards were prepared by serial dilution of standard stock solution and used for calibration of instrument.

Results and Discussion

Morphological Responses

Morphometric assay is one of the qualitative tool for the assessment of toxicants *Salvinia* was measured using morphological index parameters (MIP). The rate of inhibition of growth in the root and leaf is directly proportional to the concentration of 'Pb'.

The plants showed normal growth with greenish fronds and normal root length similar observation was made by Garg *et al.* (1994) in *Limnanthemum cristatum* at the concentration of 1 ppm of Pb, Zn and Cr.

The higher concentration of Pb (0.6 ppm and above) shows toxicity symptoms which includes dry and thin leaf margin, chlorosis (a progression of green to yellow colour on the pond) and frond detachment from colonies. These signs progressed to necrosis (Nagoor and Vyas, 1997). Root is the primary organ to have direct contact with toxic metals and major amount of heavy metal accumulate and thus, inhibit the growth.

The reduction in the size and number of fronds was observed in *Salvinia natanus* by the accumulation of heavy metals (Yonggpisanhop *et al.,* 2005). The decrease in the root due to partial root damage and due to some enzyme systems (Page *et al.,* 1972). At higher concentrations (Pb) the test plant shows the drastic reduction in root length and leaf size, which is directly proportional to the concentrations of the media and exposure duration. According Teisseire and Vernet (2000), $CuSO_4$ at 10 µM was inhibitory for *Lemna minor*, activity of glutathione S–transference and glutathione reductase were inhibited Axtell *et al.* (2003) reported were the absorption of lead by *Lemna minor* and effect of Pb on tissues at high concentrations similar observations were also made by Zayed *et al.* (1998) reported that the toxicity of 'Cd' (5 mg/lt) results in the decreased growth when compared to control.

Accumulation of Lead (Pb)

The accumulation data reveals that 'Pb' accumulation in *Salvinia* is directly proportional to its exposure concentration and duration. The *Salvinia* grown in experimental pond containing 0.5 ppm found to accumulate 196 µg/gm, 238 µg/gm, 264 µg/gm and 276 µg/gm during 3, 6, 9 and 12 days exposure respectively. Accumulation of lead at 1.4 ppm was 433 µg/gm, 440 µg/gm, 447 µg/gm and 452 µg/gm respectively at 3, 6, 9 and 12 days exposure duration. It is observed that the rate of accumulation is maximum at 3 days. The accumulation of metal in plants at 3 days exposure more pronounced irrespective of concentrations of metal and exposure duration. However, at the remaining duration of exposure and concentration it remains marginal (Figure 37.1). There was dramatic uptake of Zinc in *Cladophora glomerata* within first minutes as reported by Brenda *et al.* (1990). The metal uptake increased gradually with increase in concentrations. The rate of Zn accumulation in *Potamogeton pectinatus* increases with increase in duration. Maximum accumulation of 'Pb' is linear and significantly correlated with Pb concentration in the effluent. Similar observation was made by Tripati *et al.* (2003). The accumulation of Zn, cd using *Eichornea* (Sridevi *et al.*, 2003) and *Lemna minor* with different heavy metals (Cu, Ni, Cd and Zn) at different concentrations shows the uptake was rapid in all the concentrations and gradually reduced by *Lemna minor* with increase in the number of days (Khellaf and Zerdaoui, 2009). The initial increase in the accumulation may be due to the availability of increased number of binding sites for the complexation of

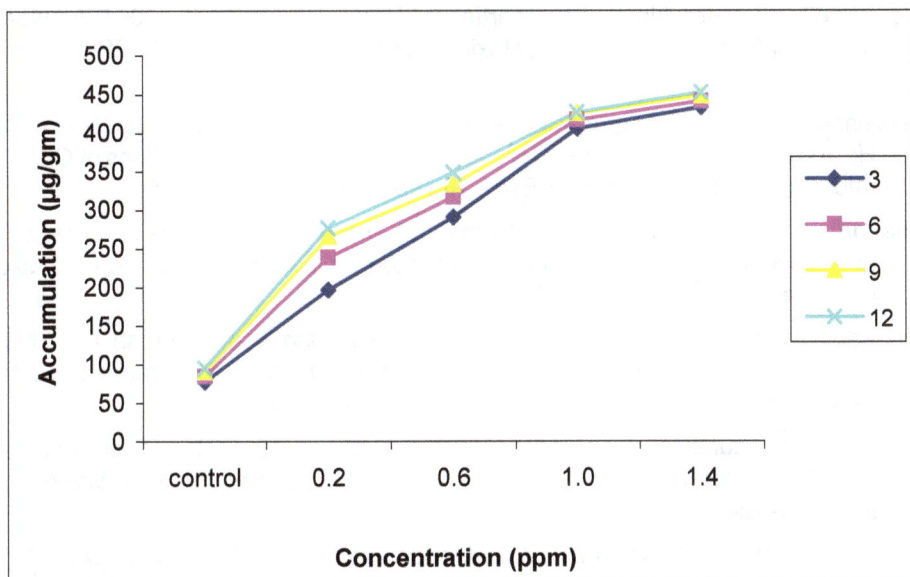

Figure 37.1: Accumulation profile of Lead by *Salvinia molesta*.

heavy metal ions leading to the increased absorption, however, slow accumulation may be attributed to the binding of almost all ions to the plants and establishment of equilibrium between adsorbent and adsorbate (Rai and Kumar, 1999). Khellaf and Zerdaoui (2009) and Osmolovskaya and Kurilenko (2005) reported that the accumulation of metal by aquatic macrophytes under controlled laboratory condition was dependent on metal concentration in the water.

It is concluded from the findings of the present investigation that morphological responses and accumulation of Pb by *Salvinia molesta* are proportional to concentration and exposure duration. Regular harvest of the plant at the interval of 3 days helps to cleanup aquatic environment.

Acknowledgements

The authors are thankful to the chairman P.G. Department of Botany Karnatak University Dharwad for proving necessary facilities to carry out the work and to the principal, BLDEA's Comm. BHS Arts and TGP Science college Jamkhandi and also to the principal G FG College Gokak (Karnataka).

References

Allen, S.E., Grimshaw, H.M., Parkinson, J.A. and Quarmby, C., 1974. *Chemical Analysis of Ecological Materials*. Blackwell Scientific Publications, Oxford.

Axtell, N.R., Stemberg, S.P.K. and Claussen, K., 2003. Lead and nickel removal using *Microspora* and *Lemna minor. Bioresour. Technol.*, 89: 41–48.

Bendra, M. McHardy and George, Jennifer J., 1990. Bioaccumulation and toxicity of zinc in the Green alga, *Clodophora glomerata. Environmental Pollution*, 66: 55–66.

Campanella, L., Cubadda, F., Sammartino, M.P. and Saoncella, A., 2001. An algal biosensor for monitoring of water toxicity in estuarine environments. *Water Res.*, 35: 69–76.

Cervanates, C., Campos-Garcia, J., Devars, S., Gutierrez-Corona, F., Loza-Tavera, H., Torres-Guzman, J.C. and Moreno-Sanchez, R., 2001. Interactions of Cr with microorganisms and plants. *FEMS. Microbial. Rev.,* 25: 335–347.

Chaudhary, S. and Sharma, Yogesh Kumar, 2009. Interactive studies of potassium and copper with cadmium on seed germination and early seeding growth in maize (Zea mays L). *J. Environ. Biol.*, 30: 427–432.

Dinakar, N., Nagajyothi, P.C., Suresh, S., Dhamodharam, T. and Suresh, C., 2009. Cadmium induced changes on proline, antioxidant enzymes, nitrate and nitrite reductases in *Arachis hypogaea* L. *J. Environ. Biol.*, 30: 289–294.

Dushenkov, V., Kumar P.B.A.N., Motto, H. and Raskin, I., 1995. Rhizofilteration: the use of plant to remove heavy metals from aqueous streams. *Environ. Sci. Tech.*, 29: 1239–1245.

Gardea-Torresdey, J.I., Tiemann, K.J., Gonzalez, J.H. and Rodriguez, O., 1997. Phytofilteration of hazardous metal ions by Alfalfa a study of calcium and magnesium interferences. *J. Hazardous Mater.*, 56: 169–179.

Garg, P., Chandra, P. and Devi, S., 1994. Cr (VI) induced morphological changes in *Limnanthemum cristatum* Griseb: A possible biondicator. *Phytomorphology*, 44(3 and 4): 201–206.

Khellaf, M. and Zerdaoui, M., 2009. Growth response of *Lemna gibba* I. (duckweed) to copper and nickel phytoaccumulation. *Water Sci. Technol.* Submitted paper.

Miretzky, P., Saralegm, A. and Cirelli, A.F., 2004. Aquatic macrophytes potential for the simultaneous removal of heavy metals (Buenos Aires Argentina). *Chemosphere*, 57: 997–1005.

Nagoor, S. and Vyas, A.V., 1997. Heavy metal induced changes in growth and carbohydrate metabolism in wheat seedlings. *Indian J. Environ. and Toxicol.*, 7(2): 98–103.

Osmolovskaya, N. and Kurilcnko, V., 2005. Macrophytes in phytoremediation of heavy metal contaminated water and sediments in urban in land ponds. *Geophysical Research Abstracts*, 7: 10510.

Page, A.L., Bingham, F.T. and Nelson, C., 1972. Cadmium absorption and growth of various plant species as influenced by solution cadmium concentration. *J. Environ. Qual.*, 1: 288–291.

Rai, A.K. and Kumar, S., 1999. Removal of Cr (VI) by low cost dust adsorbants. *Applied Microbiol. Biotechnol.*, 39: 661–667.

Satyakala, G. and Jamil, K., 1992. Cr-induced biochemical changes in *Eichhornea crassipes* (Mart) Solms and *Pistia stratiotes* L. *Bull. Environ. Contam. Toxicol.*, 48: 921–928.

Shakibaie, M.R., Khosravan, A., Frahmand, A. and Zare, S., 2008. Application of metal resistant bacteria by mutational enhancement technique for bioremediation of copper and zinc from industrial wastes Iran. *J. Environ. Health. Sci. Eng.*, 5(4). 251–256.

Sridevi, B.S., Dawood, S., Dawood, N., Noorjahan, C.M. and Prabhakar, K., 2003. Bioabsorption of Ni and Zn by water hyacinth-*Eichhornia sp. Eco. Env. and Cons.*, 9(3): 361–365.

Srivastava, R.K., Tiwari, R.P. and Bala Ramudu, P., 2007. Electrokinetic remediation study for cadmium contaminated soil Iran. *J. Environ. Health. Sci. Eng.*, 4(4): 207–214.

Teisseire, H. and Vernet, G., 2000. Copper induced changes in antioxidant enzymes activities in fronds of duckweed (Lemna minor). Plant. Sci., 153: 65–72.

Tripathi, R.D., Rai, U.N., Vajpayee, P., Ali, M.B., Khan, E., Gupta, D.K., Shukla, M.K., Mishra, S. and Singh, S.N., 2003. Biochemical responses of *Potamogeton pectinatus* L. exposed to higher concentration of Zn: *Bull. Environ. Contam. Toxicol.*, 71: 255–262.

Vaillant, N., Monnet, F., Hitmi, A., Sallamon, H. and Coudret, A., 2005. Comparative study of responses in four Datura species to Zinc stress. *Chemosphere*, 59: 1005–1013.

Van Assche, F. and Clijsters, H., 1990. Effects of metals on enzyme activity in plants, *Plant. Cell Environ.,* 13: 195–206.

Vinodhini, R. and Narayanan, M., 2009. The impact of toxic heavy metals on the haematological parameters in common carp (*Cyprinus carpio* L) Iran. *J. Environ. Health. Sci. Eng.*, 6(1): 23–28.

Weckx, J. and Clijsters, H., 1996. Oxidative damage and deference mechanisms in primary leaves of *Phaseolus vulgaris. Physiol. Plant.,* 96: 506–512.

Yongpisanphop, J., Chue, M.K. and Porethitiyook, P., 2005. Toxicity and accumulation of Lead and Chromium in *Hydrocotyle umbellate Journal of Environmental Biology,* 26(1): 79–89.

Zayed, A., Gowthaman, S. and Terry, N., 1998. Phytoaccumulation of trace elements by wetland plants. I. Duckweed. *J. Environ. Qual.*, 27: 715–721.

Zhou, X., Li, Q., Arita, A. Sun, H. and Costa, M., 2009. Effects of nickel chromatic and arsentic on histone 3 Tysme methylation. *Toxicol. Applied Pharmacology.* (In press).

2013, Perspectives in Plant Biodiversity Pages 295–304
Editor: Dr. K. Muthuchelian, *Vice Chancellor, Periyar University, Salem*
Published by: Daya Publishing House, NEW DELHI

Chapter 38

Elaeocarpus venustus Bedd. and their Plant Species Diversity in Mid-Elevation Wet Evergreen Forest of Agasthiyamalai Biosphere Reserve, Southern Western Ghats, India

S. Saravanan and K. Muthuchelian*
Centre for Biodiversity and Forest Studies,
School of Energy, Environment and Natural Resources,
Madurai Kamaraj University, Madurai – 625 021, Tamil Nadu, India

Introduction

Elaeocarpus a genus of about 360 species of family Elaeocarpaceae were found all around the world. Out of about 120 species of *Elaeocarpus* located in south East Asia, only 25 species found in India. The seeds of all *Elaeocarpus* members generally known as rudraksh and this word are precisely related with god Lord Shiva. Rudra means Shiva and Aksh means eyes the two words combine literally to mean "Eyes of Rudra". The mythological story in *Puranas* has plenty of information regarding the

* Corresponding Author: E-mail: drchelian1960@yahoo.co.in

importance of rudraksh. The fruits of Elaeocarpus species are endowed with a very hard and highly ornamental stony endocarp (Bhuyan *et al.,* 2002, ML Khan *et al.,* 2004; 2005).

Elaeocarpus venustus Bedd. locally named as Tamarai under the family Elaeocarpaceae is an endangered large tree species found in Agasthiyamalai Biosphere Reserve, as a part of Kalakad Mundanthurai Tiger Reserve (KMTR), southern Western Ghats. It is also categorized as vulnerable (IUCN, 2012) and endangered by Red Data Book (Nayar and Sastry, 1888). The tree was moist habitat nature and they were only found in the swampy region of the wet evergreen forest of Western Ghats. *E. venustus* were usually present in the second storey species and high altitude nature. However, tropical wet evergreen forests of India are increasingly affected and being degraded by physical alterations such as road building and dam constructions. Most of the species are become to endanger by both natural calamities and anthropogenic threats. In Western Ghats of India are heavily impacted by anthropogenic disturbances like expanding urbanization in forest area, extension of tea estates, tree felling by timbers, hunting the wild animal, tourist pressure, road and hydroelectric dam constructions. Monocultivation of tea plantation and Eucalyptus were most potent factors for changing the forest microenvironment and reducing the forest cover (Ramesh and Pascal, 1997). If the present situation is continue, the effective conservation measures not implemented and most of the existing forest species were destroyed. *E. venustus* have highly eroded last few decades by the anthropogenic pressure. Regeneration of the species were very poor and seeds were fell down the swampy soil thus will not allow to grow. Mostly the species were found in the road edges of the forests and because of the road clearing and other allied anthropogenic activities, the species being pushes to the endangered category. Hence, the species were having the high priority for conservation and restoration process. The study about the species, threat and conservation measures are crucially important for better management.

Species Description

E. venustus is an evergreen large tree and growing up to 10 m height, simple leaves, coriaceous with ovate-lanceolate and acuminate. The leaves have large glands in the axils of the nerves beneath. Leaf domatia are usually present on the lower surface of leaves, at the juncture of the midrib and the veins and this is the key factor for identification of *E. venustus.* Flowering is emerged during August to October with large white flowers, anthers mucronate, two celled ovary with drupe fruit. The fleshy fruits have bestowed with very hard and extremely ornamental stony endocarp.

Study Area

Agasthiyamalai Biosphere Reserve in southern Western Ghats as a part of KMTR is an important speciation centre for many plant species (Nayar, 1989). It is also an important biodiversity hot spot in India, known for many localized endemic plants (Henry *et al.,* 1984 and Anon, 1994). The global hot spot biodiversity of Western Ghats and extreme south are believed to harbor the highest levels of plant diversity and endemism. The swamp forests of the Western Ghats found in comprise the areas of about Upper kodayar, Muthukuzhlivoyal and Kakkachi.

Upper Kodayar (Site I)

The study was carried out in Upper Kodayar (N 08° 32 - 219' E 77° 21 - 43') of Kalakad Mundandurai Tiger Reserve (KMTR) at an altitude of 1250 m a.s.l. The study sites covered by contiguous tracts of mid-elevation wet evergreen forests, which are identified as *Cullenia exarillata - Aglaia bourdilloni - Palaquium elipticum* type on the basis of dominance. The average annual rainfall in the study sites during the study period was 2500 to 3500 mm. The mean maximum and minimum temperatures are 24°C and 14°C respectively with two dry seasons between southwest (June - August) and northeast monsoons (October - December) (Figure 38.1).

Muthukuzhivoyal (Site II)

The study site, Muthukuzhivoyal (N 08° 31' E 77° 23 ') is located around 10 km away from the Upper Kodayar of Agasthiyamalai Biosphere. The elevation of this area about 1350 m and is covered by dense mid elevation evergreen forest and shola forests surrounded by grasslands. Swamp amidst of wet evergreen forests is another distinct feature if this forest. The annual mean rainfall exceeds over 3000 mm, brought by South-West and North -East monsoons, with two peaks. February to May and August to September are dry periods (Figure 38.1)

Kakachi (Site III)

The study site, Kakachi (N 08° 42 - 314' E 77° 31') is about 1280 m. a.s.l. The annual rainfall over 3500 m from both Southwest and Northeast monsoons. The southwest monsoon is active during May to July and northeast monsoon during October to December. The study site is a part of Kalakad Mundanthurai Tiger Reserve. The vegetation is broadly classified as mid-elevation wet evergreen forest type (Figure 38.1)

Formation and Importance of the Swamps

In Upper Kodayar (Site I) and Muthukuzhlivoyal (Site II) and Kakkachi (Site III) at an altitude between 1250 to 1450 m, streams were naturally converted into swamps where the terrain was more or less flat. Wherever the streams were passed through relatively flat terrain, they formed swamps in few some areas. In some of the swamps were formed because of two or more streams joined in particular places and it will become a swamp. The sand, silt and litter material with water flow to get accumulated over some period and give rise to slushy bottom. During a year most of months, swamps remained water logged conditions and slushy bottom. The slush may be sometimes more than knee deep in some of the places even in summer also.

The swamps around the study area were three types. The first type has an open central part of swamp totally without vegetation. The boundary was occupied *E. venustus* and other shrub species. In second type a flat terrain was covered only with monospecific vegetation of *E. venustus* not other trees and shrubs. In third type free flowing streams with island of primary forests and swampy vegetation with *E. venustus* as only the tree species amidst thick shrubby undergrowths forms the swampy habitat. Each swamps has own characteristics features and plant species compositions. The second type swamp ie., monospecific vegetation swamp were found in both Upper

Kodayar (Site I) and Muthukuzhlivoyal (Site II) whereas in Kakkachi (Site III) primary forest vegetation swamp was occurred. The monospecific swamp was much wider and open canopy. No other tree species was found growing in the swampy region. Undergrowth was dominated by *Impatiens maculate, Xenoacanthus pulneyensis, Nilgirianthus perrtteanus* and *Didyplosandra lurida*. Also the grasses such as *Cyrotcoccum muricatum, C. deccanense, Oplismenus composites, Panicum brevifolium, Ischaemum hirtum* and *Eragrostis unioloides* and herbaceous plants which include the species *Eriocaulon, Utricularia caerulea* and *I. chinensis* were covered the ground vegetations. The third type swamps of primary forest vegetation and swampy vegetation were more than one stream entered into the swamp and was dotted with tiny islets of forests. This type of vegetation having the plant species such as *Cullenia exarilata, Myristica dactyloides, Garcinia echinocarpa, Epiprinus mallotiformis, Actinodaphane bourdillonii, Holigarna nigra, Litsea ligustrina, Persea sp. Elaeocarpus tuberculatus, E. serratus, E. munroii,* and *Syzygium ramavarmae* and other understory species of primary forests. Zingiberaceous species such as *Amoum* spp. were also found among the undergrowth of *E. venustus*.

These moist and water logged swampy habitats inhabited by *E. venustus* are much restricted to the forests in Upper Kodayar and Muthukuzhivoyal and Kakkachi. It was the only favourable habitat, which supports the endangered species *E. venustus* and many other types of vegetation in good abundance (Ganesan, 2002). These species only found in those study area of Agasthiyamalai Biosphere reserves. Probably this could be the only tree species, which proved to be successful through the selection process by nature to withstand the swampy habitat, or it might have preferred this habitat. It has stilt roots, developed from the trunk which grows down to balance the tree in the soft, slushy soil. Majority of the trees did not have a straight pole at least to a meter hight. From the base of the trunk, which is produced many new offshoots. From which few of them developed and become to multistemmed tress and few of them were destroyed and decomposed and some of them developed into a large trees.

Regeneration Potential

This species is mainly distributed only in the swampy region in all the three study sites. It was the only unique habitat, which supports this vulnerable tree species. Saplings of this species were absent in all the three study sites. It was noticed that flowering and fruiting were regular annual event in this species. August to October is the flowering season of the species and October to December is following the fruiting and dispersal events. Saplings of the species were totally absent in all the study area while the seedlings and trees were present in the study sites. Regeneration through seeds were rare phenomena of this species or very poor.

One hundred seedlings were labeled with aluminum tag to study the population flux of *E. venustus*. Height and leaf number of these seedlings were recorded monthly for a period of one year. The soil seed bank of the target species was estimated by using randomly placed 25 (50 X 50 cm) quadrates.

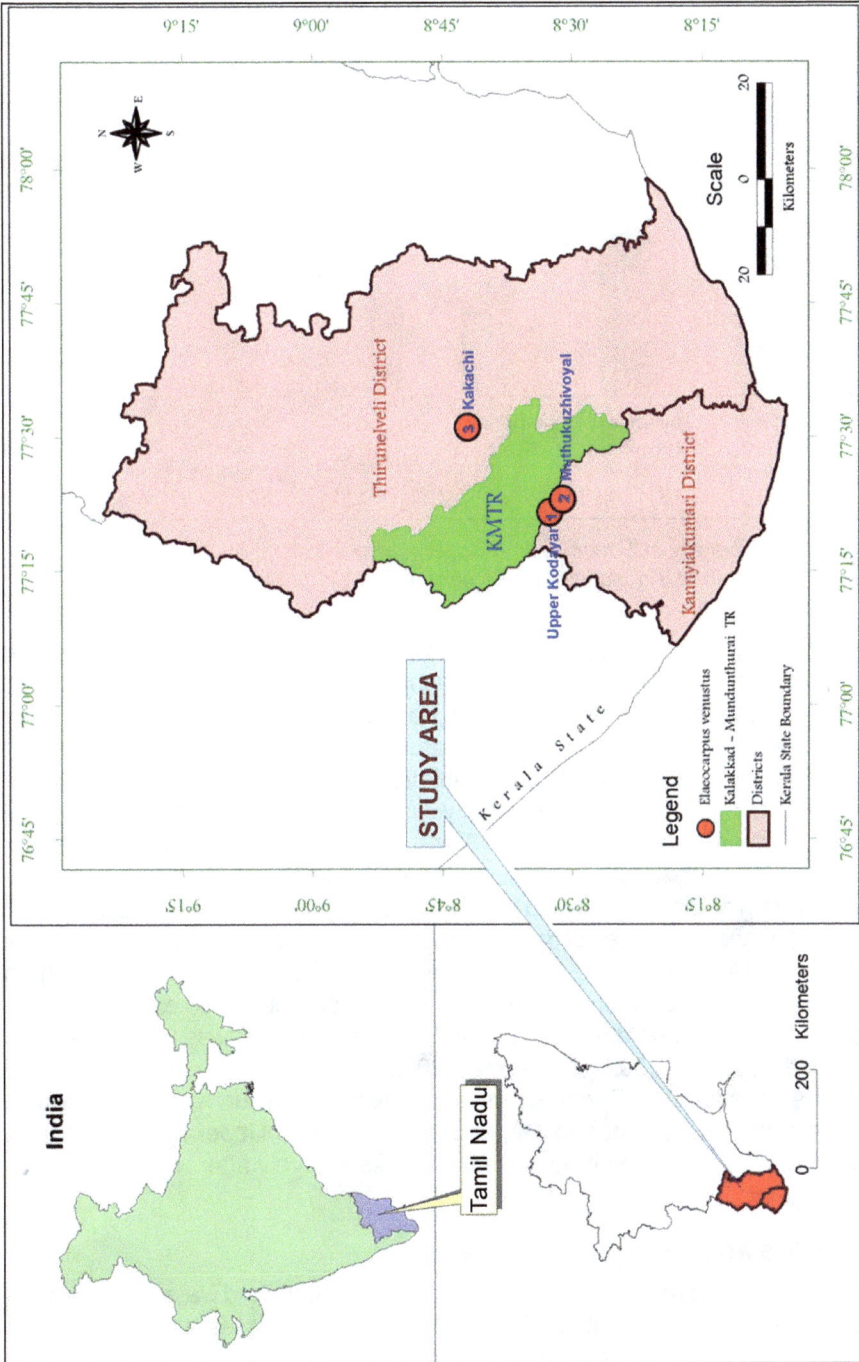

Figure 38.1: Map showing the distribution location and study sites of *Elaeocarpus venustus* in Agasthyamalai Biosphere Reserve, Southern Western Ghats, India

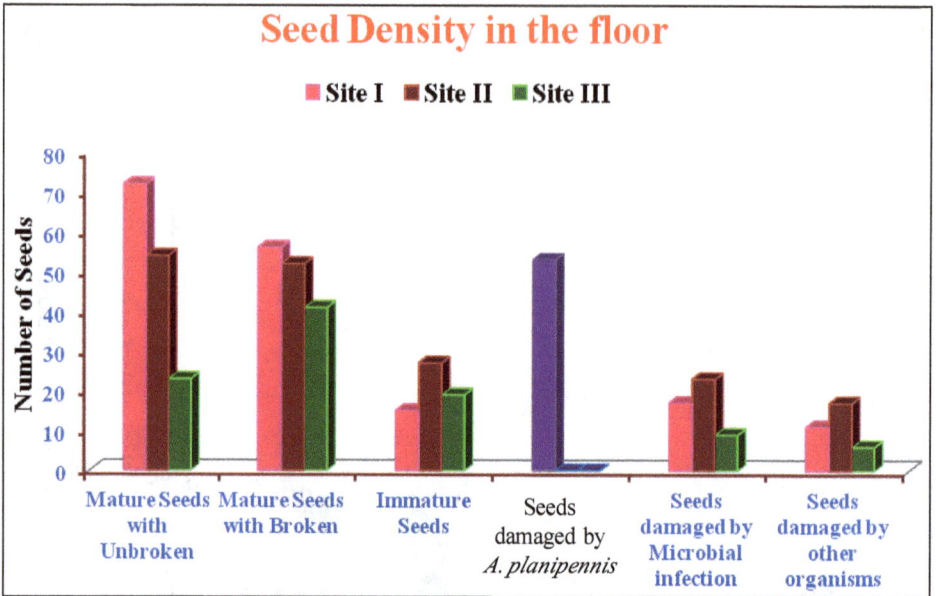

Figure 38.2: Seed density of *E. venustus* in Upper Kodayar (Site I), Muthukuzhivoyal (Site II) and Kakkachi (Site III) at Agasthyamalai Biosphere, southern Western Ghats.

Seed density of *E. venustus* was 10, 9 and 7 seeds per m² in Upper Kodayar (Site I), Muthukuzhivoyal (Site II) and Kakkachi (Site III) respectively. We recorded 72, 56 and 15 mature unbroken, mature broken and immature seeds are in Upper Kodayar (Site I), 54 mature unbroken, 52 mature broken and 27 immature seeds in Muthukuzhivoyal (Site II) and 23 unbroken, 41 mature broken and 19 immature seeds in Kakkachi (Site III) (Figure 38.2). Naturally some seeds are damaged by microbial infection and other organisms. Nearly 40 per cent of seeds were damaged by an exotic larva of *Agrillus planipennis* in site I. During the seedling establishment, most of the endosperm of the seeds would be damaged by invasive larvae *A. planipennis* (Plate 38.2 A). Generally these seeds have endowed with a hard and highly ornamental stony endocarp. However, these exotic larvae could damage and penetrate the hard endocarp and scavenge the endosperm for its development. These exotic beetle as commonly called as wood borer or Emerald Ash Borer (EAB) was first reported in Michigan and Ontario in 2002 (Haack *et al.*, 2002). In India, we are first report in this exotic beetle infestation in *E. venustus* seeds. The seedling survival of *E. venustus* was 38 per cent, 22 per cent and 14 per cent recorded in Upper Kodayar (site I), Muthukuzhivoyal (Site II) and Kakkachi (Site III) respectively after two year of study (Saravanan *et al.*, 2011).

E. venustus and its Epiphytic Plants

In order to getting the plentiful amount of light, moisture and water the tree trunk of *E. venustus* has more number of epiphytes includes mosses and orchids. The orchid species *viz.*, *Papilionanthe subulata, Eria pauciflora, Bulbophyllum elegantulum*,

Stand view of *E. venustus* in KMTR

Swampy habitat of *E. venustus* in KMTR

A twig with flowering of *E. venustus*

A twig with fruiting of *E. venustus*

Seedlings emerged from the seed

Tagged Seedling for monitoring the seedlings mortality

Plate 38.1: General view of *E. venustus* population at Agasthyamalai Biosphere Reserve.

B. fischeri, B. xylophyllum, Gastrochilus acaulis, Coelogyne nervosa and *Trichoglotis tenura* were recorded. The epiphytic species such as *Aeschenanthus perrottetti, Impatients auriculata, Medinella malabarica, Peperomia dindigulensis and P. wightiana* and all the *B. xylophyllum* was abundant on the trunks of *E. venustus* in the swampy forests.

During the flowering session the *E. venustus* has attracted the rock bees and hollow bees. Honey bees are the effective pollinator of the endangered species. The monocarpic plants of Acanthaceae member present along with the *E. venustus* were good abundance attracts the bees during the flowering (Devy, 1996). The fruits are eaten by bats and monkey, seeds were predated by giant squirrel, rat and the dormouse.

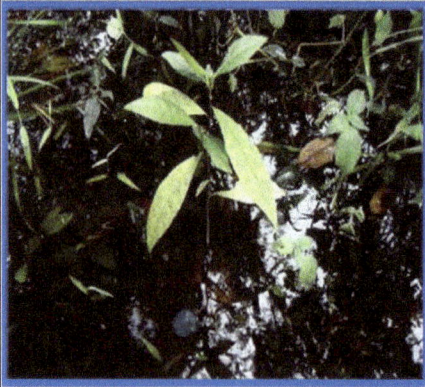

Tagged seedling in swampy area

Brokened seeds of *E. venustus*

Seed damaged by exotic *Agrillus planipennis*

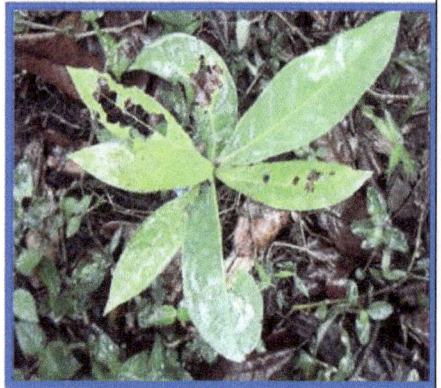

Leaf damaged by folivory or caterpillar

Wild animal disturbance in study area

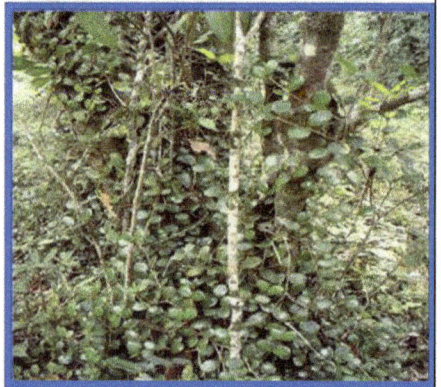

Epiphytic load on the *E. venustus*

Plate 38.2: Threats to the *E. venustus* in Upper Kodayar and Muthukuzhivoyal at Agasthyamalai Biosphere Reserve, Southern Western Ghats, India.

The swamps are islands in the middle of the evergreen forest and they can't extend their boundary. Though the streams in the hill forests are the quit common feature, swamps are the rare and study area they are highly specialized with unique plant species compositions. In all the study three sites swaps are very few in numbers and cover a small proportion of the area.

Threats

Formation of swamp and subsequent natural successional processes might take a long time. Mainly the species are found only in the moist and water logged area can't promote the species further. Most of the seeds were submerged the water and can't become to seedlings. The slushy soil could not allow to seedlings becomes to saplings. The trunks are heavily decomposed through this swampy nature.

The surrounding primary forest supports many other members of Elaeocarpaceae such as *Elaeocarpus munroii, E. serratus, E. tuberculatus* and *E. recurvatters*. But *E. venustus* and its adaptation for the swampy habitats makes it a highly vulnerable as it can't thrive anywhere else other than in swamps and streams of hill forests with high rainfall and long wet session for not less than 10 months.

Most of the *E. venustus* trees found along the forest edges of the tropical nature. These trees are mainly cutted by clearing the road buildings. Damming and deforestation in the past might have destroyed many swamps in Upper Kodayar forests. These streams before damming might have supported many swamps with *E. venustus* and other plant species mentioned here. In recent times the major threat to these habitats comes from damming the streams to make water holes for animals during summer.

The movement wild animals like elephants to make destroy of seedlings while they enter into the swamps. The epiphytic load on the tree is also heavily impacted to this species.

Generally these seeds have endowed with a hard and highly ornamental stony endocarp. However, these exotic larvae could damage and penetrate the hard endocarp and scavenge the endosperm for its development. These exotic beetle as commonly called as wood borer or Emerald Ash Borer (EAB) was first reported in Michigan and Ontario in 2002 (Haack *et al.,* 2002). In India, we are first report in this exotic beetle infestation in *E. venustus* seeds.

References

Anon. 1992. Global biodiversity – Status of the earth living resources, *World Conservation Monitoring Centre. Chapman and Hall,* London.

Bhuyan, P., Khan, M. L and Tripathi, R. S. 2002. Regeneration status and population structure of Rudraksh (*Elaeocarpus ganitrus* Roxb.) in relation to cultural disturbances in tropical wet evergreen forest of Arunachal Pradesh. *Current Science.* **83:** 1391-1394.

Ganesan, R. 2002. Evergreen forest swamps and their plant species diversity in Kalakad-Mundanthurai Tiger Reserve, South Western Ghats, India. *Indian Forester.* **128:**1351-1359.

Haack, R. A., Jendek, E., Houping Liu., Marchant, K.R., Petrice, T.R., Poland, T.M and Hui Ye. 2002. The emerald ash borer: A new exotic pest in North America. *Newsletter of the Michigan Entomological Society.* **47 (3-4):** 1-5.

Henry, A. N., Chandrabose, M., Swaminathan, M. S and Nair, N. C. 1984. Agasthiyamalai and its environs: A potential area for a Biosphere Reserve. *Journal for Bombay Natural History Society.* **81:**289-292.

Khan, M.L., Bhuyan, P and Tripathi, R.S. 2004. Survival and growth of seedlings of Rudraksh (*Elaeocarpus ganitrus*) under varied canopy conditions after transplant. *Tropical Ecology.* **45(2):** 233-239.

Khan, M. L., Bhuyan, P and Tripathi, R. S. 2005. Effects of forest disturbance on fruit set, seed dispersal and predation of Rudraksh (*Elaeocarpus ganitrus* Roxb.) in northeast India. *Current Science.* **88:** 133-142.

Nayar, M. P. and Sastry, A. R. K. 1988. Red data book of Indian plants Vol. II. *Botanical Survey of India*, Kolkata.

Nayar, M. P. 1989. In-situ conservation of wild flora resources. *Bulletin of Botanical Survey India.* **29:**319-333.

Ramesh, B.R. and Pascal, J. P. 1997. Atlas of endemics of the Western Ghats (India). Distribution of tree species in the evergreen and semi-evergreen forests. *Institute Francis de* Pondicherry.

Saravanan, S., Indra, M., Kamalraj, P., Venkatesh, D.R. and Muthuchelian, K. 2011. In -situ vegetative propagation of *Elaeocarpus venustus* Bedd. a threatened endemic tree of Agasthiamalai Biosphere Reserve, Western Ghats, India. *Journal for Bioscience Research.* **2(2):**46-49.

World Conservation Monitoring Centre 1998. *Elaeocarpus venustus.* In: IUCN 2012. IUCN Red List of Threatened Species. Version 2012.1. <www.iucnredlist.org>. Downloaded on 03 October 2012.

2013, Perspectives in Plant Biodiversity   *Pages* 305–311
Editor: **Dr. K. Muthuchelian,** *Vice Chancellor, Periyar University, Salem*
Published by: **Daya Publishing House, NEW DELHI**

Chapter 39

A Survey of Ethnomedicinal Plants Used by the *Palliyars* for the Treatment of Various Skin Diseases in Sirumalai Hills

A. Maruthupandian[1], V.R. Mohan[2] and
K. Muthuchelian[3]*

[1]Department of Botany, Periyar University, Salem – 636 011, Tamil Nadu
*[2]Ethnopharmacology Unit, Research Department of Botany,
V.O. Chidambaram College, Tuticorin – 628 008, Tamil Nadu*
*[3]Vice Chancellor, Periyar University, Periyar Palkalai Nagar,
Salem – 636 011, Tamil Nadu*

Introduction

India has one of the world's most complicated indigenous medical cultures, with an unbroken tradition coming down across more than four millennia. Though this medical heritage is several centuries old, even today people in the rural and remote areas depend upon it for their health care needs. According to the World Health Organization (WHO), as many as 80 per cent of the world's people depends on traditional medicine for their primary healthcare needs. There are considerable economic benefits in the development of indigenous medicines and in the use of medicinal plants for the treatment of various diseases. Even today, plant materials continue to play a major role in primary health care as therapeutic remedies in many developing

----

* Corresponding Author: E-mail: pandianmdu82@gmail.com

countries (Czygan, 1993). In India, medicinal plants are widely used by all sections of the population with an estimated 7500 species of plants used by several ethnic communities and it is known that, India has the second largest tribal population in the world after Africa (Kala, 2005; Jagtap *et al.*, 2006).

Skin diseases are of common occurrence among the rural masses due to poor hygienic conditions, poor sanitation facility and contaminated water. Traditional herbal medicines used by different communities play an important role in alleviating such skin diseases. They are safe, effective and inexpensive and in many cases, the only method of medication. The ethnomedicinal investigation of literature reveals that, remarkably inadequate attempts were made on *Palliyar* tribe and Sirumalai hills in Tamil Nadu. The main objective of this ethnomedicinal survey was undertaking particular importance on against various skin diseases of ethnomedicinal plants utilized by *Palliyars* in Sirumalai hills, Tamil Nadu.

Materials and Methods

The *Palliyars* are distributed along the Sirumalai hills of the Western Ghats, Madurai and part of Dindigul district, Tamil Nadu. The study area Sirumalai hills, lies lat: 77°.55′ long 78°.12′ E in Dindigul district of Tamil Nadu. In the present investigation, ethnobotanical studies were carried out in the following settlements of *Palliyar* tribals like, Ooradi, Madagamalai, Ponnuruvi, Kannadiparai and Talaikadu in Sirumalai hills, Dindigul District. Ethnobotanical information's were collected from the aged elders, Village vaidyas and experienced local *Palliyar* tribals through interview. The information was considered only after confirmation through two or more informants from different settlement of *Palliyars* in Sirumalai hills, Tamil Nadu. The medicinal plants were identified (Fischer and Gamble, 1957; Gamble, 1957a,b; Matthew, 1983a, b, c; Nair and Henry, 1983; Henry *et al.*, 1987; Henry *et al.*, 1989; Sharma *et al.*, 1993; Sharma and Balakrishnan, 1993; Sharma and Sanjappa, 1993; Sasidharan and Sivarajan, 1996; Hajra *et al.*, 1997; Matthew, 1999a, b, c; Pallithanam, 2001), photographed and sample specimens were collected for the preparation of herbarium. The identified plant specimens were confirmed and deposited in the herbarium of Ethnopharmacology Unit, Research Department of Botany, V.O.Chidambaram College, Tuticorin, Tamil Nadu.

Results and Discussion

The present ethnobotanical survey revealed that, the *Palliyar* tribal were using the medicinal plants to treat various skin diseases. In Table 39.1 data obtained from the medicinal survey are presented. The ethnomedicnal plants informations are arranged alphabetically with their botanical name, vernacular name, family, voucher specimen number (VOCB – V.O.Chidambaram College, Botany) and mode of administration. In this survey 33 plant species belonging to 31 genera and 24 families have been recorded for their herbal remedies against various skin diseases. Among the families like, Fabaceae (5), Mimosaceae (3), Acanthaceae, Convolvulaceae and Menispermaceae (2), Scrophulariaceae, Anacardiaceae, Ceasalpiniaceae, Moraceae, Agavaceae, Rutaceae, Verbenaceae, Thymeliaceae, Rubiaceae, Balsaminaceae, Oleaceae, Euphorbiaceae, Martyniaceae, Meliaceae, Ranunculaceae, Poaceae, Santalaceae,

Table 39.1: Description of the medicinal plants used by *Palliyar* in Sirumalai hills for the treatment of various skin diseases.

Sl.No.	Herbarium No.	Plant Name	Family	Vernacular Name	Mode of Administration
1.	VOCB 5403	*Acacia chundra* (Roxb. ex Pottl) Willd	Mimosaceae	Karungali	The stem bark paste mixed with goat's milk is used as an ointment to treat skin diseases
2.	VOCB 5404	*Acacia ferruginea* DC	Mimosaceae	Parampai	The stem bark paste is used as an ointment to treat skin diseases
3.	VOCB 5427	*Anamirta cocculus* (L.) Wight and Arn	Menispermaceae	Kakkakolivirai	The paste prepared from twenty grams of the fresh leaves and ten grams of the seed powder of *Cocculus hirsutus* (L.) Diels (sirukatunkodi) is used as an ointment to treat various skin infections.
4.	VOCB 5444	*Bacopa monnieri* (L.) Pennell	Scrophulariaceae	Nirbrami	The leaf extract is used as a lotion to heal cuts and wounds.
5.	VOCB 5445	*Barleria buxifolia* L.	Acanthaceae	Rosemulli	The leaf paste is used as an ointment to heal blisters.
6.	VOCB 5456	*Buchanania axillaris* (Ders.) Ramam	Anacardiaceae	Kattuma	The paste made from the seed kernel is used as an ointment to control intense itching.
7.	VOCB 5471	*Cassia tora* L.	Caesalpinaceae	Tagarai	The paste made from the seed kernel is used as an ointment to control intense itching.
8.	VOCB 5481	*Cissampelos pareira* L.	Menispermaceae	Appatta	The leaf paste is used as an ointment to treat scabies, skin inflammations
9.	VOCB 5494	*Crotalaria juncea* L.	Fabaceae	Sanapoo	The seed paste is used as an ointment to treat psoriasis.
10.	VOCB 5495	*Crotalaria verrucosa* L.	Fabaceae	Salangaisedi	An ounce of the leaf juice is taken with cow's milk twice a day for a period of one week to treat dyspepsia and scabies
11.	VOCB 5510	*Desmodium triflorum* (L.) DC	Fabaceae	Sirupulladi	The leaf paste is used as an ointment to heal abscesses and wounds.
12.	VOCB 5529	*Ficus microcarpa* L.f.	Moraceae	Kalitchi	The salted leaf paste is used as an ointment to treat skin diseases.
13.	VOCB 5532	*Furcraea foetida* (L.) Haw.	Agavaceae	Seemai katralai	The paste made by blending the extract of the aerial part of the plant with the seed powder of *Eleusine coracana* L. (khezhvaragu) is used as an ointment to heal burns and wounds.

Contd...

Table 39.1–Contd...

Sl.No.	Herbarium No.	Plant Name	Family	Vernacular Name	Mode of Administration
14.	VOCB 5537	*Glycosmis pentaphylla* (Retz.) DC	Rutaceae	Kulaparai	The stem bark paste is used as a cream to treat acne.
15.	VOCB 5538	*Gmelina arborea* Roxb.	Verbenaceae	Malai kumula	The leaf paste is used as an ointment to treat scabies.
16.	VOCB 5540	*Gnida glauca* (Fresen.) Gilg.	Thymeliaceae	Nachinar	The leaf paste is used as an ointment to treat skin inflammations
17.	VOCB 5542	*Hedyotis puberula* (G.Don) Arn.	Rubiaceae	Chiruver	The leaf extract is used as a lotion to control intense itching.
18.	VOCB 5550	*Impatiens balsamina* L.	Balsaminaceae	Kasithumbai	The flower paste is used as an ointment to heal burns and wounds.
19.	VOCB 5553	*Ipomoea obscura* (L.) Ker - Gawl	Convolvulaceae	Siruthaali	The fresh leaves are cooked with ghee and taken to treat skin diseases.
20.	VOCB 5555	*Jasminum auriculatum* Vahl	Oleaceae	Vsi malligai	The root paste is used as an ointment to heal burns.
21.	VOCB 5557	*Jatropha curcas* L.	Euphorbiaceae	Kattu amanakku	A handful of dried leaves are soaked in one hundred ml of coconut oil (oil obtained from dried copra of *Cocos nucifera* L.) for few days. The oil infusion is used as a lotion to treat eczema and scabies.
22.	VOCB 5569	*Martynia annua* L.	Martyniaceae	Thel kodukku	The seed oil is used as a lotion to treat urticaria.
23.	VOCB 5570	*Melia dubia* Cav.	Meliaceae	Malai vempu	An ounce of the boiled water extract of the stem bark is taken twice a day to treat leprosy.
24.	VOCB 5571	*Merremia tridentata* (L.) Hallier f.	Convolvulaceae	Muthiar kunthal	It is used as a poultice after blending with small amount of turmeric powder (powder made from the dried rhizome of *Curcuma longa* L. (manjal), few dried fruits of *Cuminum cyminum* L. (shiragam) and an ounce of cow's milk to take care of leprosy.
25.	VOCB 5586	*Naravelia zeylanica* (L.) DC.	Rannunculaceae	Vathom kolli	The leaf paste is applied as an ointment to treat rhinitis and boils.

Contd...

Table 39.1–*Contd...*

Sl.No.	Herbarium No.	Plant Name	Family	Vernacular Name	Mode of Administration
26.	VOCB 5591	*Panicum miliaceum* L.	Poaceae	Mani varagusamai	The paste made from the seeds with the tender leaves of *Azadirachta indica* A.Juss (neem) is used as an ointment to heal burns and wounds.
27.	VOCB 5605	*Pithecellobium dulce* (Roxb.) Benth.	Mimosaceae	Kodukka puli	The seed paste or stem bark paste is used as an ointment to treat eczema and scabies.
28.	VOCB 5616	*Pterocarpus marsupium* Roxb.	Fabaceae	Vengai	The leaf paste is used as an ointment to treat skin diseases.
29.	VOCB 5619	*Rhinocanthus nasutus* (L.) Kurz.	Acanthaceae	Nagamalli	The leaf or root juice blended with the fruit juice of *Citrus aurantifolia* (Christm) Swingle.) (elumitchai) is used as a lotion to treat skin diseases.
30.	VOCB 5623	*Santalum album* L.	Santalaceae	Santhanam	The paste made from the wood powder with the fruit juice of *Citrus aurantifolia* (Christm) Swingle (elumitchai) is used as antiseptic cream.
31.	VOCB 5626	*Sesbania sesban* (L.) Merr.	Fabaceae	Sithakathi	The leaf paste is used as an ointment to heal wounds.
32.	VOCB 5634	*Strychnos nux-vomica* L.	Loganiaceae	Etti	The stem bark paste blended with little amount of gingely oil (oil obtained from the dried seeds of *Sesamum indicum* L.) is used as an ointment to treat skin diseases.
33.	VOCB 5650	*Vernonia cinerea* (L.) Less.	Asteraceae	Siru sengalaneer	The leaf extract is used as a lotion to treat skin diseases.

Loganiaceae, and Asteraceae one species each were used by the *Palliyar*tribals in study area. The preparations were made from various plant parts like, leaves, root, root bark, stem bark, wood, seed, seed kernel, flower and whole plant; for the treatment of various skin diseases like blister, itching, scabies, inflammations, psoriasis, dyspepsia, abscess, burns, acne, eczema, urticaria, leprosy, rhinitis, cuts and wounds and boils. The leaves were the frequently used for the same. The above said plant parts were used in the form of extract or decoction, powder, paste, and juice. The tribal's preferred using herbal medicines either single or combining with other plant parts (seed powder of *Cocculus hirsutus* (L) Diels (sirukatunkodi), *Eleusine coracana* L. (khezhvaragu) dried copra of C*ocos nucifera* L. (Thennai), leaves of *Azadirachta indica* A. Juss (Vempu), and fruit juice of *Citrus aurantifolia* (Christm) Swingle.) and most of the remedies were preferred as external applications. The preparing medicines by combining several plants since the combination rapidly cure the diseases and also enhance the immunity power of the patients. This is constant with the other general observation which has been reported earlier in relation to medicinal plant studies by the Indian Traditional System of Medicine like Siddha and Ayurvedha (Gogate, 2000; Kirtikar and Basu, 2001). Some of workers have also reported medicinal plants which are commonly used by the Irular and tribal people in the Marudhamalai hills, Coimbatore, Tamil Nadu for the treatment of skin diseases (Senthilkumar *et al.,* 2006). Similar ethnobotanical studies have been supported in some parts of Tamil Nadu to protect the traditional knowledge from disappearing (Ayyanar and Ignacimuthu, 2005; Sivarajan and Ramakrishnan, 2012). This preliminary survey exhibits that medicinal plants continue to play an important role in the health care system of *Palliyar*communityin Sirumalai hills, Tamil Nadu. Many tribal communities in Tamil Nadu gather their healthcare needs by means of plant products and preparations based on traditional knowledge that has been gained indigenously over a period of time and/or by practice.

Conclusion

The documentation of the efficacy of traditional knowledge is necessary for the future generation and to prevent the cultural changes, socioeconomic problems and urbanization. Documenting the indigenous knowledge through ethnobotanical studies is important for the conservation of biological resources as well as their sustainable utilization.

References

Ayyanar M. and Ignacimuthu S. (2005b). Ethnomedicinal plants used by the tribals of Tirunelveli hills to treat poisonous bites and skin diseases. *Ind. J. Trad. Knowled.* 4: 229-236.

Czygan F.C. (1933). Kulturgeschichte and Mystie des Johanniskrautes. *Zeitschrift fur Phytotherapie.* 5: 276-282.

Fischer C.E.C. and Gamble J.S. (1957). The Flora of the Presidency of Madras III: (Rep. Ed.): 1347-2017.

Gamble, J.S. (1957a). The Flora of the Presidency of Madras I: (Rep. Ed.): 1-577.

Gamble, J.S. (1957b). The Flora of the Presidency of Madras II: (Rep. Ed.) : 578-1346.

Gogate, V.M. (2000). Ayurvedic Pharmacology and Therapeutic uses of Medicinal plants (Dravyagunavigyan), Frist ed. Bhartiyar Vidya Bhavan (SPARC), Mumbai Publications, pp.421-422.

Hajra P.K., Nair V.J. and Daniel P. (1997). Flora of India. IV: 1 – 561.

Henry A.N., Chithra V. and Balakrishnan N.P. (1989). Flora of Tamil Nadu, India,Series I: Analysis III: 1- 171.

Henry A.N., Kumari G.R. and Chithra V. (1987). Flora of Tamil Nadu, India, Series I: Analysis II: 1- 258.

Jagtap S.D., Deokule S.S. and Bhosle S.V. (2006). Some unique ethnomedicinal uses of plants used by the Korku tribe of Amaravati District of Maharashtra, India. *J. Ethnopharmacol.* 107: 463-469.

Kala C.P. (2005). Current status of medicinal plants used by traditional vaidyas in Uttaranchal state of India. *Ethnobot. Res. Appl.*: 3: 267-278.

Kirtikar K.R. and Basu, B.D. (2001). Indian Medicinal Plants. Vol.1. Lalit Mohan Basu, Allahabad, India, pp. 34-35.

Matthew K.M. (1983a). The flora of the Tamil Nadu Carnatic I: 1- 688.

Matthew K.M. (1983b). The flora of the Tamil Nadu Carnatic II: 689- 1540.

Matthew K.M. (1983c). The flora of the Tamil Nadu Carnatic III: 1541- 2155.

Matthew K.M. (1999a). The flora of the Palni Hills I: 1- 575.

Matthew K.M. (1999b). The flora of the Palni Hills II: 576- 1196.

Matthew K.M. (1999c). The flora of the Palni Hills III: 1197- 1635.

Nair N.C. and Henry A.N. (1983). Flora of Tamil Nadu, India, Series I: Analysis I: 1-184.

Pallithanam, J.M. (2001). A Pocket flora of the Sirumalai hills, South India: i-360.

Sasidharan N. and Sivarajan V.V. (1996). Flowering Plants of Thrissur Forests: 1-579. Series I: Analysis III: 1- 171.

Sharma B.D. and Balakrishnan N.P. (1993). Flora of India II: 1-625.

Sharma B.D. and Sanjappa M. (1993). Flora of India III: 1-639.

Sharma B.D., Balakrishnan N.P., Rao R.R. and Hajra P.K. (1993). Flora of India I: 1-467.

Sivarajan R and Ramakrishnan K. (2012). Traditional uses of medicinal plants in treating skin diseases in Nagapattinam district of Tamil Nadu, India. *Int. Res. J. Pharmacy.* 3(5): Online.

2013, Perspectives in Plant Biodiversity   *Pages* **312–317**
Editor: **Dr. K. Muthuchelian,** *Vice Chancellor, Periyar University, Salem*
Published by: **Daya Publishing House, NEW DELHI**

Chapter 40

Population Structure and Phenology of Critically Endangered Species (*Nothopegia aureo-fulva* Bedd. ex Hook.f.). Moist Deciduous Evergreen Forests of Western Ghats, India

S. Saravanan and K. Muthuchelian*

*Centre for Biodiversity and Forest Studies,
School of Energy, Environment and Natural Resources,
Madurai Kamaraj University, Madurai – 625 021, Tamil Nadu, India*

Introduction

Conservation of natural resources and potential of utilize them in a sustained manner are essential for the well being and continued survival of man. But, during the past hundred years or so the rate of decline of forest cover, the source of natural resources, has been very high throughout the world and various estimates available point out that at least 10 per cent of living species are extinct/vulnerable (Heywood, 1988, Puspangadan, 1991; Taylor, 2000). Biodiversity conservation is the millennium development goals agreed by virtually all countries over the past decade. Because

* Corresponding Author: E-mail: drchelian1960@yahoo.co.in

biodiversity loss is now recognized as a global-scale phenomenon and many conservation decisions are taken at an international level, conservation planning is increasingly done on a global scale (Barrett *et al.*, 2011).

Nothopegia aureo-fulva Beed. ex Hook.f. is a small tree belonging to the family Anacardiaceae. Anacardiaceae is widely distributed in temperate, subtropical, and tropical regions in East Asia including India and there is about 77 genera and 600 species. This species distributed in moist deciduous forest of Aryankavu at southern Western Ghats. It is critically endangered tree species and very narrow endemic of Agaasthyamalai Biosphere Reserve (IUCN, 2012). It is also categorized as critically endangered by Red Data Book (Nair and Sastry, 1988).

N. aureo-fulva is locally named as "Peccharu and Mogaraveenki" under the family Anacardiaceae. It is a critically endangered and endemic small tree or shrub of moist deciduous forests of Western Ghats, Kerala. It is also under the severe threat and categorized as critically endangered (IUCN, 2012) and extinct from the wild by Red Data Book (Nayar and Sastry, 1988). The tree was found only in Aryankavu Reserve forest which is lower elevation (~ 300 m) of Western Ghats, Kerala. The tropical forests were heavily affected by encroachment of rubber estate, extend the urbanization and house hold usage of forest. The degradation of existing over a period of years, affected the life support systems considerably in different ways such as loss of potential species for their valuable drugs, food and industrial commodities, soil erosion, loss of nutrients, destruction of ecosystems and habitats, over grazing and ultimately to reduction of biotic and abiotic resources in the region.

N. aureo-fulva are critically endangered and vulnerable species on the basis of population reduction and the population estimate of less than 200 mature trees in the wild (IUCN, 2012). Since they are endemic and critically endangered loss of these trees from this habitat would mean their extinction. Owing to its critical state in wild, there is an urgent need to conserve these taxa. The natural populations of *N. aureo-fulva* are becoming decreased year to year due to seed recalcitrance, poor seed germination and difficulties in the other vegetative propagation potentials. The conservation and recovery of threatened and endangered species is a tremendous and ever-increasing challenge.

Species Description

N. aureo-fulva is small tree of lower elevation stand and grows up to 2 to 3 m height. Leaves subopposite, linear-oblong or oblong lanceolate and gradually acuminate. The nerves are parallel to the midrib, 25 to 30 pairs, petiole stout, very shaggy and 0.3 cm long. The stout petiole has plenty of hairs which is key identification of *N. aureo-fulva*. Racemes very short, petals very glabrous and filaments slightly villous. The flowering period of the species was May to July and fruiting was June to September and flowering and fruiting was occurred annually (Gamble, 1957).

Study Area

Aryankavu

The study site, Aryankavu (N 09° 00 - 925' E 77° 06 - 803') is about 300 to 500 m a.s.l. in tropical moist deciduous evergreen forest. Mean annual ambient temperature is about 26.6°C. The maximum and minimum temperatures are 29°C to 22°C

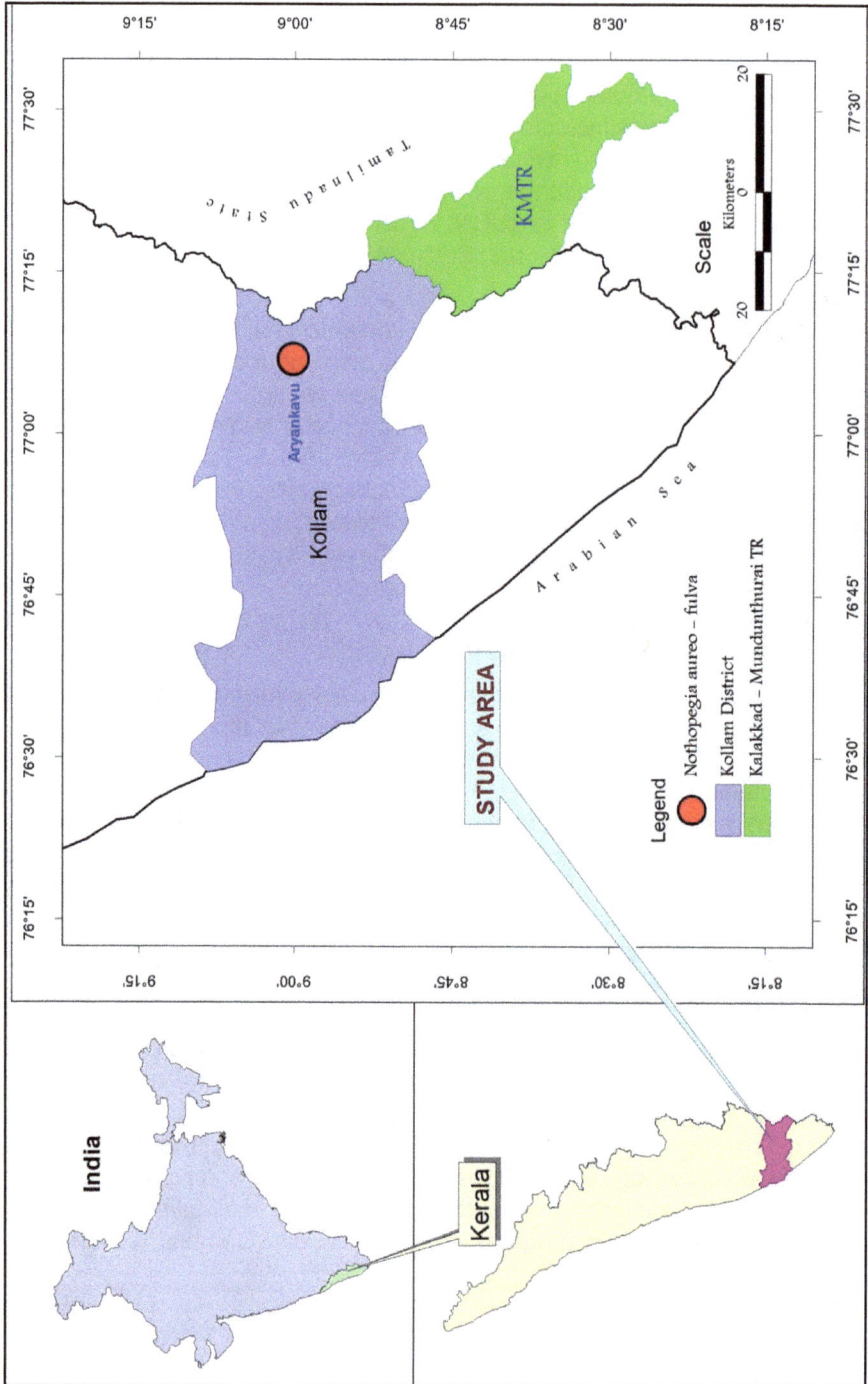

Figure 40.1: Map showing the distribution location and study sites of *Nothopegia aureo-fulva* in Agasthiyamalai Biosphere Reserve, Southern Western Ghats, India

respectively (Figure 40.1). This species distributed in moist deciduous forest of Aryankavu at southern Western Ghats. It is critically endangered tree species and very narrow endemic of Agaasthyamalai Biosphere Reserve (IUCN, 2012). It is also categorized as critically endangered by Red Data Book (Nair and Sastry, 1988).

Materials and Methods

Spatial Distribution

Ten quadrates were placed continuously at each stand for trees, saplings and seedlings of target species. The individuals of target species in each of the quadrates were listed and their circumference was measured. These individuals were grouped into three categories, *i.e.* (a) seedlings (< 10 cm collar circumference at the base), (b) saplings (10 to 30 cm collar circumference at the base) and (c) trees (> 30 cm (CBH) circumference at breast height, *i.e.* 1.37 m above ground level). Density (trees ha–1) and basal area values were calculated.

Phenology

Twenty five individuals of two selected target tree species (*N. aureo-fulva*) were marked for the phenological study. *N. aureo-fulva* with more than 5 cm DBH was marked with metal tags for phenological observations. Each individual was observed once in a month for two years period (2009-2011). The following phenological events were recorded for the target species, (i). Production of Young Leaves (YL), (ii) Maturation of Leaves (ML), (iii) Abscission of Leaves (AL), (iv) Production of Young Flowers (YF), (v) Maturation of Flowers (MF), (vi) Production of Young Fruits (YFR), (vii) Maturation of Fruits (MFR) and (viii) Ripening of Fruits (RFR). During the visits, all marked individuals were characterized and the phenostage of a species was determined by considering their status.

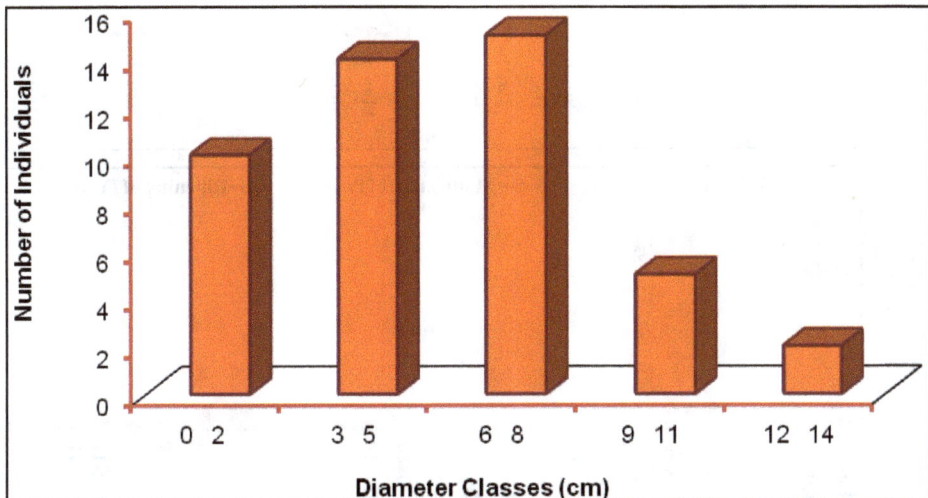

Figure 40.2: Population structure of *Nothopegia aureo-fulva* in Aryankavu at Agasthyamalai Biosphere Reserve, southern Western Ghats of Kerala.

Results

    N. aureo-fulva was recorded only in Kulirkadu forests of Aryankavu at Kollam districts of Western Ghats, Kerala, (Figure 40.1) with 46 individuals (>2 GBH) at Kulirkadu forest.

    The three life stages (seedlings/saplings/trees) for a particular species, suggested their possible future status of the species. The diameter distribution of trees has often been used to represent the population structure of the forests as well as the species. The status of the regeneration of species was determined based on population size of seedlings and saplings.

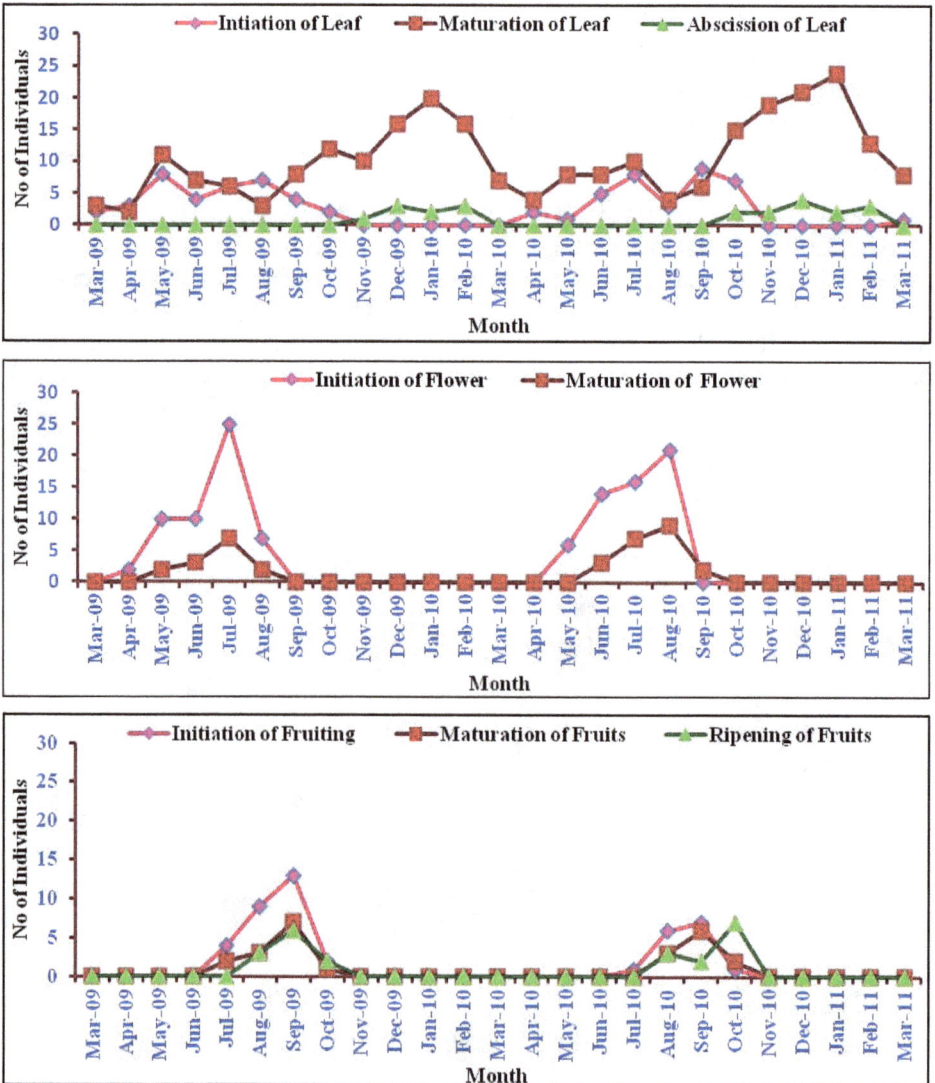

Figure 40.3: Phenology of *N. aureo-fulva* at Aryankavu, Kerala, Agasthiyamalai Biosphere Reserve.

Population Structure

The population density of *N. aureo-filva* was recorded 46 individuals at Kulirkadu forest of Aryankavu in southern Western Ghats. The age structure of *N. aureo-filva* indicates that the population has a healthy regeneration potential. All the life stages (seedlings, saplings and trees) of *N. aureo-fulva* were recorded. This is very small height class age structure tree. Most of the mature (15) individuals in the girth class of *N. aureo-filva* was 6–8 cm. One or two individuals were found in the diameter range upto >12 (Figure 40.2).

The activity of leaf production (YL) was recorded in May to July and it coincided with dry season in *N. aureo-fulva*. The leaf emergence of the species was seasonal and it occurred once in a year. The peak activity of leaf maturation (ML) was recorded in November to February. Abscission of leaf (AL) was recorded in January to February. The peak activity of flowering (YF) in April to July and flowering (MF) was recorded in June to August. Fruit initiation (YFR) peak was recorded in June to September. Fruit maturation (MFR) and fruit ripening activity (RFR) of the species was recorded in July to September and sometimes it extended to October (Figure 40.3).

Conclusion

N. aureo-fulva was recorded only in Kulirkadu forests of Aryankavu at southern Western Ghats of Kerala. Only 46 individuals were recorded in all the life forms (seedlings, saplings and trees). Phonological observations of the study show that the flowering and fruiting occurs annually. The activity of leaf production (YL) was recorded in May to July and it coincided with dry season in *N. aureo-fulva*. The peak activity of leaf maturation (ML) was recorded in November to February. Abscission of leaf (AL) was recorded in January to February. The peak activity of flowering (YF) in April to July and flowering (MF) was recorded in June to August.

References

Barrett, L. F., Mesquita, B., and Gendron, M. 2011. Emotion perception in context. *Current Directions in Psychological Science*. **20:** 286–290.

Gamble, J.S. 1957. *Flora of the Presidency of Madras*. Dehra Dun: Bishen Singh and Mahendra Pal Singh.

Heywood, V. H. 1988. Rarity: a privilege and threat. In: W. Greuter and Zimmer B. (eds.), *Proceeding of the Fourteen International Botanical Congress Koeltz. Konigstein/Tanus*. pp. 277-290.

Nayar, M. P., and Sastry, A. R. K. 1988. Red data book of Indian plants Vol. II. Botanical *Survey of India*, Calcutta.

Puspangadan, P. 1991. Aromatic Plants. Importance and need for their genetic resources conservation. In: K. L. Dhar, R. K. Thappa, and S. G. Agarwal (eds.). Newer trents in essential oils and flavours. *Tata Mc Grow Publishing Co.* New Delhi. pp. 308-321.

Taylor, H. C. 2000. 2000 IUCN Red list of threatened species. *International Union for Conservation of Nature and Natural Resources. Gland.* Switzerland.

World Conservation Monitoring Centre 1998. *Nothopegia aureo-fulva*. In: IUCN 2012. IUCN Red List of Threatened Species. Version 2012.1. <www.iucnredlist.org>. Downloaded on 03 October 2012.

Index